THE HISPANIC REPUBLICAN

☆ ☆ ☆

ALSO BY GERALDO CADAVA

Standing on Common Ground:
The Making of a Sunbelt Borderland

THE HISPANIC REPUBLICAN

THE SHAPING OF AN
AMERICAN POLITICAL IDENTITY,
FROM NIXON TO TRUMP

GERALDO CADAVA

An Imprint of HarperCollinsPublishers

HarperCollins books may be purchased for educational, business, or sales promotional use. For information, please email the Special Markets Department at SPsales@harpercollins.com.

Ecco® and HarperCollins® are trademarks of HarperCollins Publishers.

FIRST EDITION

Designed by Renata De Oliveira

Library of Congress Cataloging-in-Publication Data has been applied for.

ISBN 978-0-06-294634-8

20 21 22 23 24   LSC   10 9 8 7 6 5 4 3 2 1

To K & O, the best of companions

CONTENTS

INTRODUCTION

If you held American currency in the seventies and eighties, odds are you carried with you the autograph of one or more Mexican American treasurers of the United States. There were three of them—all women, all Republicans. Romana Acosta Bañuelos was from California and served under President Richard Nixon. Katherine Ortega was from New Mexico and served under Presidents Ronald Reagan and George H. W. Bush. And Catalina Vásquez Villalpando was from Texas and succeeded Ortega in the Bush administration. By the late twentieth century, it had become a tradition for Republican presidents to tap Hispanic women for the post. Their appointments were supposed to symbolize the greater inclusion of Hispanics—the ethnic label preferred by most Hispanic Republicans—within the Republican Party.

Romana Acosta Bañuelos. Katherine Ortega. Catalina Vásquez Villalpando. These aren't household names, even though their printed names were in every household in America for years. They remain largely unknown in the annals of American and Latino history, even though they were prominent political figures who helped elect Republican presidents, and helped build a Hispanic Republican base that plays an important role in politics today. These Hispanic Republicans are even alien to some fellow Latinos, who continue to insist that they're sellouts, race traitors, *pochos* (fruits that have lost their color), or *tio tacos* (Hispanic Uncle Toms). Students in my Latino history classes have questioned whether Hispanic Republicans should be considered "Latino" at all, as though the label itself implies a particular political identity.

Hispanic Republicans too often get secreted away in the archive of our family histories. Some Latinos roll their eyes as they explain how their conservative kin aren't pleasant to be around because they express their views too freely, complaining about illegal immigration, high taxes, and welfare, even if they were immigrants who at some point benefited from government services. The Hispanic Republicans in our own families demonstrate our closeness, our intimacy with them, and how we—as Latinos, as Americans, as Latino Americans—wrestle with the diversity of viewpoints in our communities.

Indeed, millions of Hispanic Republicans have been a part of the Hispanic community for a long time. Before the 1930s, most Hispanics, like most African Americans, were Republicans. Only afterward did the Hispanic Republican come to seem like a paradox. Between the 1930s and the 1960s, most Hispanics became Democrats, primarily, they said, because Franklin Roosevelt and the New Deal helped lift them out of the Depression. But then something interesting happened. Beginning in the late 1960s, and especially in the early 1970s, as African Americans continued their flight from the Republican Party, the percentage of Hispanics who voted for Republicans shot up. A graph comparing Hispanic and African American support for the Republican Party over the past half century would look like an X, with Hispanic support rising and African American support falling, and for intertwined reasons. Recognizing that they were losing African American votes, Republicans had to find support among new voters, especially Hispanics. Their numbers were small at first, too small to make a real difference. But as the Hispanic population grew, Hispanics began to influence elections, and politicians in both parties took notice.

Ever since Nixon's reelection in 1972, Hispanic Republicans have helped Republican presidential candidates win about a third of the

Hispanic vote. The exact number has been a little less or a little more, depending on a range of factors including the brand of the Democratic or Republican Party at a particular moment, the excitement or distaste Hispanic voters have for a particular candidate, the Hispanic rate of participation in a given election, and local, national, and world events. This consistency—the usual third, give or take—is especially remarkable given the variance in how Latinos and Latino voters have been counted. Only two Republican candidates dipped below 20 percent: Gerald Ford in 1976 and Bob Dole in 1996. Both lost, but even they did better than all Republicans before 1972. George W. Bush won some 40 percent of the Hispanic vote in 2000, and 44 percent in 2004. When he was president, Hispanic Republicans believed that a majority of Hispanics might vote for a Republican someday.

Hispanics even contributed to Donald Trump's surprise victory. In their 2014 book, two political scientists offered this prediction: "Both of us have a very difficult time seeing a Democrat lose the race for the White House in 2016," in part because of the impact they expected Latino voters to have. (Like them and others, I would have made the same prediction at that time.) But of course the Democrat did lose, and the Republican candidate won with just shy of 30 percent of the Hispanic vote, even though he launched his campaign with a tirade against Mexican immigrants. Trump's remarks were widely rebuked, making it seem unlikely that any Hispanic would support him. Yet they did, and by a slightly wider margin than the percentage of Hispanics who voted for establishment candidates such as John McCain in 2008 and Mitt Romney in 2012.

Since Trump's election, some evidence suggests that the Republican Party, and even the president himself, has gained—not lost—support among Hispanics, who, like many Americans, have been politically energized. The National Association of Latino Elected and Appointed Officials (NALEO) found that Hispanic voter turnout in the 2018 midterm elections was 50 percent higher than in 2014. Political analysts generally agreed that the Hispanic voter surge favored

Democrats, but in congressional races nationwide, the usual third of Hispanics voted for Republicans. In important races in Texas and Florida—two bellwether states we often look to in order to predict whether liberalism or conservatism are on the rise nationally— Republican candidates did better than that. Ted Cruz did a little better, winning 35 percent of the Hispanic vote in his race against Beto O'Rourke. Rick Scott did a lot better, winning 45 percent of the Hispanic vote in his race against Bill Nelson.

Perhaps it's easier for Republican candidates to find success locally, where they can have more direct and sustained contact with voters and sometimes distance themselves from national leaders. But Trump himself has also found support among Hispanics. One poll taken in early 2019, during the government shutdown, showed that more than half of Hispanics would definitely vote against him in 2020. More surprising was the poll's finding that he had a 50 percent approval rating among Hispanics, a full 19 percent higher than when he took office. "Thank you," Trump tweeted, "working hard!" To be sure, several polls taken since have shown Trump's approval among Latinos to be far lower, and even this one poll was disputed at the time. But it was still a marker of how Latino opinions of the president are diverse and have varied.

Trump's Hispanic supporters note that the unemployment rate is low, tax cuts have benefited them, and financial regulations have been slashed. They claim that socialism from Venezuela to the US Congress threatens freedom. These themes were repeated over and over again at a gathering of small-business owners in Washington, DC, where Cruz, Scott, and Mike Pence were featured speakers. Meanwhile, the border wall, family separations, and the Russia investigation didn't come up—the impeachment proceedings against Trump hadn't yet begun. Even after the devastating El Paso shooting in August 2019, which left twenty-two people dead and was widely regarded as an attack against Mexican immigrants and Hispanics that was brought on by Trump's rhetoric, Hispanic Republicans blamed only the shooter, not the president.

Why have Hispanics continued to support the Republican Party, even Trump's Republican Party? How has the Republican Party built a Hispanic base to withstand attacks by leaders who devalue them? This book seeks to answer these questions. It has long been presumed that Hispanics will play a growing role in the future of American politics, and in the United States more broadly. We need to understand their political behavior. The majority of Hispanics vote for Democrats, but Republicans keep winning a significant minority of their votes, even as many Americans believe there shouldn't be even a single Hispanic Republican. Hispanic support for the Republican Party isn't new, and, seen in historical perspective, it's not surprising. For the past half century, Hispanic Republicans and the Republican Party have been deliberate and methodical in their mutual, sometimes hesitant, embrace.

MANY AMERICANS CONTINUE TO ASSUME, THOUGH, THAT HISPANICS WILL HELP TO END the Republican Party. They have argued that the recent past for Hispanics should be like the sixties was for African Americans—the time when the vast majority of them finally, and permanently, rejected the Republican Party. Because of rising discrimination against them, Hispanics should be fleeing the Republican Party in droves, just as African Americans did. Because of their demographic growth, Hispanic Democrats are supposed to help usher in a permanent liberal majority. Those who believe this to be true have slept better imagining that, with Hispanics as the largest minority in the United States—who may yet become the majority—they will forever, at last, defeat the racist, backward-looking, individualistic policies favored by Republicans.

In the long run, even if growing numbers of Hispanics never vote like African Americans, the "semi-soft floor" of about 30 percent of their votes that Hispanics give Republicans won't be enough to save the GOP. So far, the Republican Party has been able to rely on its base of non-Hispanic white voters (and voter suppression) to eke out electoral victories, but as the Hispanic population continues to grow, consistently winning a third of the Hispanic vote won't be a blueprint for victory. It may well be that Hispanics will help Democrats recapture the majority in the near future and hold it for a long time. But it hasn't happened yet, and may not happen anytime soon, especially if Republicans are able to increase their support among Hispanics.

Meanwhile, Democrats look expectantly to states such as Arizona and Texas for signs of a liberal turn attributable to Hispanics. News reports after the 2018 midterms suggested that the surge of Hispanic voters did indeed help Democrats win in Arizona and Nevada. It appeared to be the fulfillment of predictions made over the past several years. Before Mitt Romney's loss to Barack Obama in 2012, Nate Silver's *FiveThirtyEight* blog cited the Republican Party's decreasing margin of victory in Arizona since the 1970s, and said that, even if Arizona wasn't yet a battleground state, "it may be soon." Right after that election, Ted Cruz, in an interview with the *New Yorker*'s Ryan Lizza, said, "In not too many years, Texas could switch from being all Republican to all Democrat," and, if that happened, "no Republican will ever again win the White House." Democrats have been encouraged by shifts in New Mexico, Virginia, and Colorado that have already helped them win. But the long-term liberalization of the United States because of Hispanics won't happen simply because Democrats say it will.

Election analysts and political scientists aren't guided by faith in the same way that partisans sometimes are, but the way they've discussed Hispanic support for the Republican Party hasn't always helped us understand it. Even though Hillary Clinton lost the 2016 election, she won the Latino vote in a "landslide" (65 to 29 percent), according to the *New York Times*. In his successful reelection campaign, Obama "crushed" Romney among Latino voters (71 to 27 percent), CNN reported. Four years earlier, Latino voters punched the Obama-Biden ticket over the McCain-Palin ticket "by a margin of more than two-to-one" (67 to 31 percent), stated a Pew Hispanic Center postelection analysis.

Their talk of overwhelming support for Democrats and the fundamental liberalism of Latinos dismisses a third of the Latino population, a significant slice of a rapidly growing electorate that amounts to millions of voters. In their efforts to capture how the majority of Hispanics voted, they've acknowledged but then skated past this sizable minority. They've also downplayed complicated dynamics between Latino liberalism and conservatism, which need to be seen together if we want to understand Latino politics—and ultimately American politics—writ large. Finally, they've missed the point. Hispanic Republicans and the Republican Party haven't sought support from a majority of Hispanic voters; their goals have been to erode the idea that Democrats have a monopoly on their votes, and to win enough Hispanic votes to win elections. Sometimes they've sought to neutralize Hispanic support for Democrats only by encouraging Hispanics to stay home.

To the extent that Democrats, election analysts, or political scientists dismiss Hispanic Republicans, or assume Hispanic support for liberals, it only helps Republicans on election day.

Focusing closely on numbers, ratios, and views of particular issues as explanations for what motivates Hispanics to vote for Democrats or Republicans has also prevented us from seeing a bigger picture. Latinos are "progressive or liberal," one political scientist has written, because more than 80 percent of them believe that more government is better than less government, less than 20 percent believe that the free market can solve all economic problems, and a greater percentage of Hispanics compared with non-Hispanic whites believes that the government should spend more on education, reduce inequality, and solve climate change. The only metric supporting the conclusion that Hispanics are conservative, he claimed, is the greater percentage of Hispanics compared with non-Hispanic whites who believe that "racial and ethnic minorities" should be "self-reliant," which is a proxy for their negative views of welfare. Similar statistics are presented on a range of issues including marriage, abortion, school vouchers, military service, and the construction of a border wall, the sum of which is supposed to explain why Hispanics are liberal or conservative. But survey findings about the percentages of Hispanics who embrace such labels are only snapshots of a particular moment that don't explain how Hispanic partisan identification and voting patterns have changed over time.

I would argue that the percentages of Hispanics who do or do not support certain issues, or who identify as liberal or conservative at a given moment, are insufficient as an explanation of Hispanic loyalty to the Republican Party, which has developed slowly, in fits and starts, and over several decades. More than any single issue, and more than the issues that have gotten the most attention, Hispanics have become loyal Republicans because of a nexus of beliefs about US–Latin American relations; the United States as the protector of freedom in the world; market-driven capitalism as the best path toward upward mobility; and a contrarian identity politics that is every bit as pronounced as the liberal identity politics they've spent decades criticizing. These themes suggest that explaining the persistence of Hispanic support for the Republican Party isn't only a question of what percentage of Hispanics do and do not support the issues that election analysts and political scientists have studied.

A RELATED IMPEDIMENT TO UNDERSTANDING HISPANIC PARTISAN IDENTITY IS THE FACT that, since the middle of the twentieth century, members of both major parties have argued that Hispanics are either naturally liberal or naturally conservative. Arguments about their natural leanings only blind us to seeing who Hispanics have been, who they are today, and who they may yet become. Such arguments are meant to persuade Americans, especially Hispanics, by force rather than reason, that their political inclinations should be obvious. In 2010, the former Democratic Senate majority leader Harry Reid told a mostly Hispanic crowd in his home state of Nevada, "I don't know how anyone of Hispanic heritage could be a Republican." He was on the campaign trail, running for reelection, and Republicans immediately jumped on his words as an example of ethnic and racial essentialism, just another example of how Democrats took Hispanic voters for granted because Hispanics have been presumed to be natural liberals.

Republicans, though, have also argued that Hispanics were, and are, natural conservatives. "The Hispanic-American of the Southwest, and the Mexican American in particular, is by nature a Republican" is how Manuel Ruiz, an important Hispanic conservative who worked on Barry Goldwater's campaign in 1964, put it. Fifteen years later, the first Hispanic presidential candidate, Benjamin Fernandez, a Republican who ran against Reagan in the 1980 primary, told the conservative columnist George Will that "Hispanics are natural Republicans because they are incorrigibly individualistic, have a centuries-old suspicion of oppressive central governments, and learned fiscal conservatism at the knees of their mothers, who assured them that if they didn't watch their nickels they wouldn't eat."

Republican presidents have echoed these sentiments. "The dedication to principles of loyalty, patriotism, strong religious faith, and devotion to family displayed by Hispanic Americans is basic to the American way of life," Reagan believed. He famously told the Hispanic advertising maven Lionel Sosa of San Antonio, Texas, that Hispanics are Republicans who "just don't know it yet." In 1989, President Bush said the "values" of the Hispanic community "are the very founding values of this nation and of the Republican Party." He continued, "Faith and family, hard work and individual responsibility, respect for others, and above all, an abiding love of freedom" characterized Hispanics and indeed all Republicans.

These ideas have been repeated so often, by Hispanic Republicans and non-Hispanic Republicans alike, that they've come to assume a static timelessness. Sometimes they've accurately characterized the beliefs of Hispanic Republicans, but sometimes they're only banal public pronouncements that feel so vague as to lack any specific meaning. They're ideas that Republicans—again, Hispanic and non-Hispanic alike—fall back on, as if to discourage questioning, to explain why Hispanics have supported, and will support, them.

Whichever way Hispanics "naturally" incline in politics, it clearly matters. For the past half century, both parties have courted Hispanics. They are a so-called sleeping giant; once awakened, they will reshape American politics. They're important swing voters who've already tipped elections, giving candidates narrow margins of victory in states with large Hispanic populations; the Republicans among them have accounted for 2, 3, or 4 percent of voters in elections decided by slim margins. Republican analysts credited Hispanic voters with Jimmy Carter's victory over Gerald Ford in 1976, George W. Bush's over Al Gore in 2000, and Obama's over Romney in 2012. After each of these losses, the Republican Party resolved to pursue Hispanic voters more aggressively. Hispanics are already the largest group of minority voters, and by 2050 they will represent a third to a half of the total US population, so many have argued that the future success of both parties depends on cultivating their loyalty.

My own perspective as a historian is a bit different. Latinos aren't naturally liberal or conservative. They aren't *naturally* anything. Their complex histories have given them good reasons to be Democrats, but also good reasons to be Republicans. The growth of the Latino population won't inevitably lead to a period of unending progressivism, as many would hope. Debates over the issues and policies that affect Latinos still need to be waged and won. Their long-term political allegiances continue to evolve. If Latinos hold the key to success for either party, it's not because they're naturally liberal or conservative. It's because they've fought, and will continue to fight, for political inclusion as both Democrats and Republicans. That is, their politics are a product of history and human action, not nature.

AS I WROTE THIS BOOK ABOUT HISPANIC REPUBLICANS, I DISCUSSED IT WITH MANY people who challenged its basic premise. Some were surprised that there was such a thing as a Hispanic Republican. Others challenged my early use of the term "conservative," because the individuals and groups I'm writing about can't be placed in such a box, especially since I'm talking about the United States and Latin America, where what it means to be conservative, or left wing and right wing, doesn't easily translate.

I agree. Conservatism isn't always the best label; there are differences between being conservative and being a Republican. One student told me that her grandmother was conservative but could never bring herself to vote for a Republican because of the party's positions on immigration. Her story drove home the point that "conservative" and "Republican" don't necessarily mean the same thing. These terms are like concentric circles with a large degree of overlap, but not total overlap; like a partial rather than a total eclipse.

I chose to focus on Hispanics and the Republican Party, and Hispanic partisan identification as Republicans instead of their conservatism—Hispanic Republicans instead of Hispanic conservatives—in part because that was more easily, but not perfectly, quantifiable and researched, and in part because I was interested in explaining electoral outcomes and why Hispanics have voted for, and continue to vote for, Republicans.

Others objected to my use of the term "Hispanic." They had various reasons, but above all was their argument that there's no such thing as a Hispanic. It's an invented category, and we can't lump together Mexicans and Mexican Americans; Cuban Americans and Cuban exiles; and Puerto Ricans. Some of the actors in this history have felt the same way. As Benjamin Fernandez testified before the Senate Watergate Committee, he told a joke about Hispanic disunity: "The Mexican

American does not talk to the Puerto Rican, the Puerto Rican does not talk to the Cubano, the Cubano talks to no one." If there's no such thing as a Hispanic, could there be such a thing as the Hispanic vote? Still, disunity didn't stop Fernandez and many other Hispanic Republicans from trying to bring Hispanics together. They argued that their similarities were greater than their differences to forge a Hispanic Republican identity. They generally supported one another's positions or interests, so long as these didn't directly oppose their own.

"Hispanic" is also the label that many Republicans have chosen for themselves. It is a term that liberal Latinos have associated with conservatism, since it was adopted by members of the Nixon administration to merge different national heritage groups, and because it implies a kinship with Spain, the Spanish Empire, Spanish colonizers, racial purity, and whiteness, instead of the mixed Spanish, African, and indigenous identities of most people of Latin American descent. Hispanic Republicans don't necessarily have a problem with these critiques. Their families did come from Spain, they say, and Spaniards brought to the Americas much that was good. Some Hispanic Republicans accept the term "Latino," but they draw the line at the relatively new "Latinx," which, dozens of Hispanic Republicans on Twitter argue, "isn't a word," even though *Merriam-Webster* added it in 2018, and the *Oxford English Dictionary* added it in 2019.

Other challenges were raised as well, including that politics is local, not national, so how can I talk about Hispanic Republicans nationally? Almost always, Republican candidates have found greater success among Hispanics locally and at the state level, depending on whether a candidate spoke Spanish on the campaign trail; whether being Hispanic helped or hurt their chances for election; and what their positions were on local issues that affect schools, housing, or businesses. Cubans in Miami relied on the federal government to help them settle in the United States and create opportunities for them, to a surprising extent considering their anticommunism. The situation in New Mexico was similar, where Hispanic Republicans also favored limited government but relied on federal aid. Even within states there were differences—between Cubans and Puerto Ricans in Florida, or service workers and ranchers in Nevada. Such particularities can get lost when talking about national politics.

These challenges are correct, but only partially. The goals and achievements of Hispanic Republicans have been in the arena of national politics as much as local or regional politics. They formed the Republican National Hispanic Assembly in the 1970s, for example, to gain influence within the Republican Party nationally as much as to encourage Hispanic participation in local Republican Party politics.

But most important, if I were discouraged from writing this book because of these challenges, where would that leave us? Pretty much where we are today, knowing little about the history of how the Republican Party developed a Hispanic base, and why Hispanics have become loyal Republicans even when the party has seemed, at least to outside observers, to reject them.

The fact remains that, while there may not be such a thing as *the* Hispanic or Latino vote, and while "conservative" may not be the best term for describing the politics of Hispanics who vote for Republicans, there are millions of Hispanics in every election who vote for Republicans. Although the majority of Hispanics or Latinos vote for Democrats, it is also important to understand the significant share who vote for Republicans, because if we do not, either because we continue to insist that they're part of the liberal majority, or because Hispanic identity itself is too diverse to describe categorically, then we'll miss something fundamental about Hispanics in the United States and therefore about the United States itself.

THE HISTORY OF HISPANIC REPUBLICANS HAS BEEN MISUNDERSTOOD, IF IT HAS BEEN recognized at all. Most people assume that Hispanics vote for Republicans either because they're Catholic or Cuban, or Catholic and Cuban. That's part of the story, but not the whole story.

Some of the characters in this book did indeed explain their support for the Republican Party as the natural result of their Catholic faith. But Hispanics have also argued that Catholicism was the root of their progressivism, especially with respect to labor rights and social justice. Consider the United Farm Workers (UFW), founded by César Chávez and Dolores Huerta, which drew inspiration from *Rerum Novarum*, the 1891 papal encyclical that called for the alleviation of the "misery and wretchedness pressing so unjustly on the majority of the working class." The union also marched under the banner of the Virgin of Guadalupe, the most important religious icon for Latin American Catholics. One Hispanic Republican told me his father quit the UFW because he was evangelical and didn't like Chávez's use of such religious symbols. He was also a representative of a broad group—evangelical Christians—that today constitutes the fastest-growing religious group among Hispanics and has been more reliably Republican than Catholics. Moreover, political scientists have argued that even if Hispanics are more religious than other Americans—as measured by church attendance—their religiosity isn't the determining factor in their political behavior.

What about Cubans? To be sure, Cuban American politicians today, such as Cruz and Florida senator Marco Rubio, fulminated against the Castro regime and socialism in general as important sources of the evolution of their partisan identities. A higher percentage of Cuban Americans compared with other Hispanic groups has identified as Republican, but there are plenty of Mexican American and Puerto Rican Republicans as well. In fact, Mexican Americans in the Southwest were the founders of a national Hispanic Republican movement, and Puerto Ricans in New York supported the Republican governor Nelson A. Rockefeller years before the Cuban Revolution of 1959. Cubans gained clout only in the late 1970s or early 1980s, when their rates of naturalization increased, and they supported Reagan because his anticommunism matched their own. The

Cuban exiles who opposed Fidel Castro also haven't had a monopoly on Hispanic anticommunism; they've shared it with Mexican Americans, Puerto Ricans, Nicaraguans, Venezuelans, and other groups of Hispanics.

To view Hispanic Republican identity only through the lens of Catholicism and anti-Castro Cubans, then, hides more than it reveals. It hides the Republican Party's efforts over a long time to develop a Hispanic base through patronage politics and grassroots organizing among Hispanics across the United States and its territories, including Puerto Rico. Hispanics don't vote for Republicans because they're Catholic or Cuban; they vote for Republicans because they've developed considerable loyalty to the Republican Party. To describe Hispanic Republicans as tradition-bound, freedom-loving capitalists gives weight to the idea that Hispanics are natural conservatives and focuses our attention on a particular institution, the Catholic Church, and a particular state, Florida. That is a limited view. It doesn't explain the political behavior of Hispanic Republicans as a whole in the past or the present.

PART OF WHAT SENT ME DOWN THE ROAD OF WRITING A BOOK ABOUT HISPANIC REPUBLI-cans was knowing, based on my own family history, that there was more to the story than Catholics and Cubans. It's no exaggeration to say that I've thought about Latino politics and political identity for decades, because of arguments I've had with my Republican grandfather and namesake, Geraldo Cadava Jr. He told me about his political conversion, from Democrat to Republican.

He was born in 1926 in Panama. His father was Filipino, his mother was Colombian. His father, named Gerardo Cadava, was born in the Philippines in a military town south of Manila called Cavite, then later joined the US Navy, which took him to Panama. His mother, whose maiden name was Lucila Gomez, "entertained soldiers" above a bar in Colón, Panama, my grandpa said. That's where his parents met, though they eventually made their way north, to San Diego, where they raised my grandfather.

Joining the US military became my grandfather's path to becoming an American citizen, after which he adopted the nickname "Jerry." He was stationed at bases in Texas; Nebraska; and finally Tucson, Arizona, where he retired and raised a family, and where I was born. He tells stories of having serviced General Curtis LeMay's airplane. When he saw LeMay smoking a cigarette too close to the plane, my grandfather, following protocol, warned him that it was dangerous to have a lit cigarette so close to a source of fuel. The plane might explode. "It wouldn't dare," said LeMay, who later served as Alabama governor George Wallace's running mate in the 1968 presidential election.

After my grandfather retired from the military, he washed dishes at the Forty Niner Country Club, a golf resort in Tucson, and worked in silver mines near the city. He voted for a Republican for the first time in 1980, becoming one among millions of Hispanics—including former Democrats, like him—who cast their ballots for Reagan. Explaining his decision, he said Reagan promised to put money back into his monthly paycheck by cutting taxes. Although he pledged himself to Reagan for this reason alone, he would, for decades, adopt Republican positions wholesale. Democrats were corrupt, wasteful spenders. Bill O'Reilly was "looking out" for him. The Fox News Channel was "fair and balanced." Immigrants should be welcomed into the United States, but they should come legally, as he did.

These are of course caricatures of my grandfather that don't fully capture the man; he looms much larger in my mind as an athlete with fingers made crooked by the impact of speeding baseballs, and a former serviceman who tells colorful stories about the people he met and the places he traveled while in the military. He is my grandpa. Yet I have vivid memories of our arguments about why he became a Republican, and his conservative views more generally. I wanted to understand, but didn't. I'm certain that this book is my way of carrying on our conversation, even though the days when we argued about politics are gone.

More to the point, he was Catholic in name only, and he wasn't Cuban. He wasn't even Mexican American in a region whose Hispanic population is dominated by Mexican Americans. Looking back on them now, all of our conversations were lessons that the history of Hispanic Republicans is not what we've assumed it to be.

I DIDN'T SET OUT TO WRITE A BOOK ABOUT HOW THE REPUBLICAN PARTY HAS BUILT A sizable base of Hispanic voters who even today continue to help elect Republicans, but that is indeed what this book has become. Hispanics of different backgrounds narrated their partisan evolution as Republicans as the result of their family's migration from Mexico, Puerto Rico, Cuba, or elsewhere. They came to see the United States as a symbol and protector of freedom, democracy, and opportunity, the antithesis of the authoritarian regimes they fled. Over time they became involved in US politics. They believed the Democratic Party took their votes for granted, so they voted for local and statewide Republican candidates before organizing for national candidates including Eisenhower, Nixon, and Goldwater. They started "Latinos con Eisenhower," "Viva Nixon," and "Latinos con Goldwater" groups that organized independently of their campaigns.

Hispanic Republicans felt like they gained real influence in the Republican Party, though, only during Nixon's first administration, when he relied on their advice and asked them to play an important role in his reelection campaign. Hispanic Republicans repaid him with their loyalty. During Ford's brief tenure as president, Hispanic Republicans consolidated their influence through the Republican National Hispanic Assembly, which promised to organize Hispanics nationally and locally. They became an effective lobbying group for Hispanic interests during the Ford presidency and beyond.

Ben Fernandez tested the limits of Hispanic influence when he became the first Hispanic to run for president of the United States. He received less than 1 percent of the primary votes in a handful of states where he made it onto the ballot, casting doubt on his belief that the Hispanic moment had arrived. Hispanic Republicans instead cast their lot with Reagan. He supported statehood for Puerto Rico, railed against Castro, granted amnesty to undocumented Mexicans, and insisted that the United States didn't need a wall to separate the United States from Mexico. During his presidency, though, the Republican Party's anti-immigrant fringe moved into the mainstream, and Hispanic Republicans were upset by the nativism and xenophobia of some in their own party.

They again doubted whether the Republican Party was the home for them that the founders of the Republican National Hispanic Assembly had said it would be. But by the time that immigration and border wars divided the Republican Party—not to mention Democrats and Republicans—Hispanic Republicans had already made up their minds that they were loyal Republicans who wouldn't easily be dissuaded from supporting the party, no matter who its leaders, or what their policies, were.

The far right of the Republican Party was represented by ideologues such as Pat Buchanan, who had worked in the Nixon, Ford, and Reagan administrations before he became the leader of an insurgency that challenged the Republican establishment. Hispanic Republicans found his views on immigration, the border, and ethnic diversity generally, to be detestable. They much preferred moderation and inclusion, so they supported the Bushes, who had a long history of relationships with Hispanics in Texas, Florida, and several countries in Latin America and the Caribbean.

But even as Hispanic Republicans supported George H. W. Bush and George W. Bush—and would have supported Jeb Bush if he had won the Republican nomination in 2016—the Republican Party moved farther to the right. The success of the Bushes among Hispanics was more the result of their family history than a sign of the Republican Party's momentum among Hispanics. All Hispanic Republicans may have left is their loyalty to a party that stands for some of the things they care about, but that in other respects seems to be pushing them away.

Over the past half century, therefore, Hispanic Republicans have shared many beliefs with Republicans in general, but they also have been different. Republicans and Democrats alike have seen the United States as a land of opportunity, dedicated to freedom and equality for all Americans. Such beliefs have compelled Puerto Rican Republicans to support statehood, motivated Cubans to oppose Castro, and spurred Mexican Americans to fight for immigration laws that are sympathetic to undocumented immigrants. Yet Hispanics have also exposed some of the deepest rifts within the Republican Party, especially when it comes to issues of race, immigration, and the construction and maintenance of exclusionary boundaries that divide the United States from Latin America. Only a small minority of Hispanics has supported the nativist and xenophobic views of Republican hard-liners.

There may come a time when Hispanics will reject the Republican Party, as almost nine in ten African Americans have done, but that time has not arrived, and nothing on the horizon suggests that it is imminent. During our conversation just before the 2016 election, one Hispanic Republican explained that he would vote for Trump even if he didn't like him, because he wouldn't let one man ruin what he and others before him had spent decades building: a national movement of Hispanics who were loyal Republicans.

PART I

AWAKENING

CHAPTER 1

BECOMING REPUBLICAN

Desi Arnaz's uninspiringly titled autobiography, *A Book*, is the story of his childhood in Cuba; his move to the United States; and his career as a musician, actor, and entertainer. His father had been the former mayor of Santiago de Cuba, a city in the Oriente province where Fidel Castro began his long march to Havana in the 1950s, by which point Arnaz was married to Lucille Ball and *I Love Lucy* was a smash hit. Desi's mother had been the daughter of a Bacardí rum executive. Desi remembered of his time in Cuba that his father and mother surrounded him with luxury. They owned three ranches, an island in Santiago Bay, and "several" race cars and speedboats.

They lost it all in 1933, he wrote, when a rebellion led by students, labor activists, and military officers ousted Cuban president Gerardo Machado. Desi's father backed Machado, who was often referred to as a dictator because of his reliance on police violence and the suppression of free speech to enforce his authority. The year before the coup, Desi's father was elected to the Cuban House of Representatives as a Machado supporter, so when the president was removed, Desi's father was imprisoned and the family's property was confiscated.

Desi and his mother fled to Miami to live in exile, and his father joined them when he was released from prison a few years later, when Desi was still a boy. According to Desi, the family started from scratch in the United States. In *A Book*, he claimed that the imprison-ment of his father and the theft of their property sparked his lifelong hatred of socialism and communism.

The day after Machado was overthrown, another Cuban boy in Santiago, Fidel Castro, turned seven years old. He was born in 1926 in Birán, where his father owned a twenty-three-thousand-acre plan-tation, but at the time of the coup he was a student in Desi's home-town. They weren't classmates because Arnaz was ten years Castro's senior, yet they walked the same streets in Santiago de Cuba some twenty-five years before the Cuban Revolution of 1959. That revolu-tion sent a much larger wave of Cuban exiles to Miami, who today are often thought of as the original Hispanic Republicans. Desi identified with their hatred of Castro.

One prominent exile said of Arnaz, "Anything against Fidel Cas-tro, he supported." More than anything else, his opposition to Castro led him to become a Republican. He gave money to the survivors of the Bay of Pigs invasion, which took place in 1961. He was the cochair of Richard Nixon's "Viva Nixon" campaign in 1968, and in 1970 Nixon appointed him to his Advisory Council for Minority Enterprises, which didn't do much but nevertheless gestured toward Nixon's efforts to include Hispanics and other minorities in his ad-ministration. Later, Desi gave "seed money" to several exile organiza-tions, including one started by Jorge Mas Canosa, the founder of the Cuban American National Foundation and leader of the Cuban exile community in the late twentieth century.

It is a coincidence of history that Desi's father and mother, Ge-rardo Machado, and Jorge Mas Canosa are all buried in Caballero Rivero Woodlawn Cemetery, near the heart of Miami's Little Havana neighborhood. The body of Nicaraguan president Anastasio Somoza Debayle, another prominent leader of the Latin American right whose family ruled over the country for much of the twentieth century, is interred there as well, just steps from the gravesites of these others. Desi and Fidel were laid to rest elsewhere. Arnaz's family scattered his

ashes in California, while Castro was cremated and returned to Santiago de Cuba. The ghosts of the Latin American right and left haunt
the Americas, from the Caribbean Sea and the Gulf of Mexico to the
Atlantic and Pacific Oceans.

The story of Desi Arnaz paints in broad brushstrokes the history
of Hispanic Republicans across the twentieth century. It shows how
Hispanics developed partisan identities over time, sometimes as part
of the process of immigrating to and settling in the United States,
and often in response to the politics of their home countries. Latin
American immigrants who became Hispanic Republicans forged political identities in the crucible of conflicts between right and left,
authoritarianism and democracy, and conservatism and liberalism in
Latin America and the United States.

Many of these conflicts predated the Cuban Revolution, but when
Hispanic Republican Cold Warriors looked back on earlier struggles
between right and left in the Americas—from their perspective in
the mid- to late twentieth century, at the height of the global Cold
War—their memories were refracted through the lens of these later
conflicts and often seemed indistinguishable from them. Above all,
Desi Arnaz's story shows how conflicts between right and left in the
Americas, which by the mid-twentieth century had become part of
the Cold War between the United States and Soviet Union, are central to the story of how Hispanics became Republicans.

HISPANICS FIRST RALLIED AROUND A REPUBLICAN PRESIDENTIAL CANDIDATE IN 1952,
when they supported Dwight D. Eisenhower. They pinned buttons
on their shirts that said "Me Gusta Ike" and "Viva Eisenhower." They
formed groups called "Latinos con Eisenhower" and "Latin American
Veterans and Volunteers for Eisenhower-Nixon." Hispanic veterans
were drawn to Eisenhower, in particular, because they recognized
him as their commander during World War II, and believed that he
would end the war in Korea. But other Hispanics who didn't serve in

these wars were drawn to Eisenhower as well. Lionel Sosa, a Mexican American from San Antonio who later worked on the campaigns of several prominent Republicans, was born in 1939 and therefore was too young to serve in either war. But in his 1998 memoir *The Americano Dream: How Latinos Can Achieve Success in Business and in Life*, he said he first identified as a Republican when he heard Eisenhower's acceptance speech at the Republican national convention in Chicago, in July 1952.

Sosa's parents were Democrats, like most other Hispanics at the time, but, he emphasized, they taught him the value of hard work. Sosa's mother shelled pecans in San Antonio, where twelve thousand shellers walked off their jobs in 1938 to protest poor pay and working conditions. She was one of the strikers. His father ran a laundry and dry-cleaning business, "working from 7 a.m. to 7 p.m. six days a week." His mother told him when he was six years old that even though he was "a Mexican," he was "going to succeed."

Eisenhower's acceptance speech echoed everything Sosa's parents had taught him. He said hard work was a core Republican value, along with freedom, prosperity, and spirituality. That was all the motivation Sosa needed to become a Republican. He first voted in the 1960 presidential election, when he cast a ballot for Richard Nixon instead of John F. Kennedy. Then Sosa voted for Arizona senator Barry Goldwater in 1964, Nixon again in 1968 and 1972, and every other Republican presidential candidate until 2016, when he left the GOP because he couldn't bring himself to vote for Donald Trump.

The young boy from San Antonio may not have been able to vote in 1952, but other Hispanics could. John Flores, a Mexican American from California, formed "Latinos con Eisenhower" in April 1952 and ran it out of a room in the Shoreham Hotel in Washington, DC. His business card claimed he ran a public relations firm with offices in Mexico City; Washington, DC; and Phoenix, Arizona. While he said Latinos con Eisenhower would seek to organize Hispanic voters in Florida, Illinois, and New York, its limited activities were concentrated in the Southwest, and even there didn't find much purchase. Still, Flores appears to have been the first Hispanic to articulate a national vision for Hispanic Republican mobilization. As a Republican himself, he observed how the Democrat Harry Truman defeated his Republican opponent, Thomas Dewey, narrowly, so he believed that Hispanics, according to one historian, "could impact a close election by increasing voter turnout and by getting normally Democratic voters to back the Republican nominee."

When Flores established Latinos con Eisenhower, Eisenhower hadn't declared himself a candidate yet and wasn't even in the United States. He had been called back into service the year before, in April 1951, so he left his post as the president of Columbia University and returned to Europe as the supreme commander of the North Atlantic Treaty Organization (NATO), a new military alliance among nations in Western Europe and North America designed to combat the spread of Soviet influence around the world. Flores reminded Eisenhower that he formed Latinos con Eisenhower before he even quit NATO, to let him know that his group had been with him from the beginning. Eisenhower returned to the United States in June 1952 and jumped right into his campaign for president.

Before the 1950s, Hispanic partisan identity swung like a pendulum. From the Civil War until the Great Depression, most Hispanics, like most African Americans, were Republicans, including eleven of the fifteen Hispanics who served in the US Congress. The pendulum swung in the other direction in the 1930s—again, as it also did for African Americans—when more Hispanics began to support Democrats because of Franklin Roosevelt's New Deal. Herbert Hoover and the Republican Party came to represent rich Americans in the minds of many Hispanics. Republicans were responsible for the crash that affected their families. Many Hispanics therefore credited Roosevelt's New Deal with pulling them from dire straits. They found employment in federal job programs, which "deepened their attachment to the Democratic Party."

Hispanic partisan realignment was therefore part of a broader national political realignment, but, except on particular issues in particular places, it would be an overstatement to say that Hispanics were a powerful force driving political change in the early twentieth century. Estimates of the size of the Hispanic population across time are unreliable, primarily because the Census Bureau was constantly experimenting with and changing the methods and terminology they used. But without a doubt, Hispanics represented a small percentage of the US population as a whole. Mexican Americans played an important political role in southwestern territories, home to the largest Hispanic communities in the United States. In Texas, California, New Mexico, and Arizona, government officials conducted their business in both Spanish and English. But that changed when territories became states, and the non-Hispanic white population eclipsed the Hispanic population. Hispanics lived in Florida and New York as well, but they had a greater impact on politics in Cuba and Puerto Rico than in the Southeast or Northeast.

The erosion of Hispanic political influence led to their skepticism toward US politics. For decades after the US-Mexico War, which annexed half of Mexico to the United States, Mexican Americans remained powerful leaders in states and territories throughout the Southwest. Immediately after the Spanish American War, when many Cubans and Puerto Ricans believed that the United States had helped liberate them from Spain, citizens of Cuba and Puerto Rico expected the United States to treat them as equals. But by the early twentieth century, Mexican Americans, Cubans, and Puerto Ricans were subordinates. Many remained prominent doctors, lawyers, businessmen, and politicians, but they were treated overwhelmingly as members of an inferior race.

Their skepticism had the effect of turning some Hispanics away from political involvement altogether. Many came to believe that political participation in the United States wasn't for them. A Republican governor running for reelection in the early 1960s asked his

Mexican American friend to stump for him. His Hispanic surrogate agreed to deliver a speech in Spanish, in which he told his *queridos conciudadanos*—beloved fellow citizens—that he remembered a time in the mining town where he grew up when the "members of our race never aspired to take part in politics." Politics, they believed, "was for the Americans," not them.

Moreover, until the 1950s, Hispanics mainly engaged in politics as members of groups. Mutual aid societies such as the Alianza Hispano Americana, founded in Tucson, Arizona, in 1894, or the League of United Latin American Citizens, formed in Corpus Christi, Texas, in 1929, were established to help Hispanics fight for inclusion and against discrimination. These organizations or the labor unions to which Hispanic employees belonged represented Hispanics politically. If a non-Hispanic politician needed Hispanic support, he approached the organizations that he assumed could influence the votes of their members. Sometimes the politician also approached the employers of Hispanics, who bought their votes, pressured them to vote for a particular candidate, or threatened them with consequences ranging from losing their jobs to violence if they didn't comply.

What changed by the 1950s? Hundreds of thousands of Hispanics had served in World War II. They fought for freedom and democracy at home, so when they returned from the war they expected it at home. The skills they acquired in the military qualified them for better jobs than they had before they joined, and their GI Bill benefits paid for their education and helped them buy homes outside the often-segregated barrios where they had grown up. As the Hispanic surrogate for a Republican governor put it, Hispanics realized that they were "as American as any Anglo-Saxon." Groups such as the Alianza Hispano Americana and the League of United Latin American Citizens grew dramatically in the mid-twentieth century, and were joined by others, including the American GI Forum, the Community Service Organization, and the Mexican American Political Association. These groups and others—including labor unions or student organizations—still represented Hispanics and influenced Hispanic politics, but Hispanics also demanded rights and engaged in politics as individuals.

Postwar era demands for equality led to the political rise of some notable Hispanic Democrats—including Edward Roybal, a congressman from Los Angeles, Hector Garciá, the founder of the American GI Forum, and Vicente Ximenes, a founder of the Viva Kennedy clubs who worked for President Lyndon Johnson—which may help explain why historians have assumed that the war made Hispanics Democrats. But this didn't map out consistently across the Hispanic population.

Many Hispanic Republicans also described World War II as a pivotal moment in the evolution of their political views. One group of Mexican American businessmen, politicians, and lawyers at a conference in Washington, DC, in the late 1960s, "bonded" over their wartime experiences and decided to form the Republican National Hispanic Council—later renamed the Republican National Hispanic Assembly. They argued that the war was responsible for their upward mobility. It gave them skills that led to successful careers, paid for their education, and helped them purchase homes, often outside of the barrios where they grew up. Fernando Oaxaca, one of the men present at the meeting in Washington, who became known in GOP circles as "Mr. Republican," is a prime example. For him, the war was a "significant turning point" in his life.

Born in El Paso, Texas, in 1927, Oaxaca enlisted in the military toward the end of the war, and after his service, graduated from college with a degree in engineering that was paid for by GI Bill benefits. He started his own company and began to live a life of "affluence" that made him "politically conservative." He resented that the taxes he paid were redistributed to people who he claimed hadn't worked as hard as he had. To his way of thinking, the GI Bill may have opened the door to opportunity, but his individual talents and hard work made the real difference.

In the 1950s, the balance of power in the United States was also shifting from the Northeast to the Southeast and the West. Before mid-century, New York had the greatest number of electoral votes because it had the largest population. But during and after World War II, the population of the West exploded because of Sunbelt economic growth fueled primarily by the growth of the military-industrial complex. Soon states such as California, Texas, and Florida would have more electoral votes than, or almost as many electoral votes as, New York. These states with growing electoral influence also had some of the largest Hispanic populations. As the states with large Hispanic populations became more important nationally, Hispanics became voters to be recruited; the leaders of both parties increasingly believed that they had to compete for Hispanic votes to win elections.

Hispanic Republicans embraced their new role in partisan politics. They argued that the Democratic Party had begun to take them for granted, and therefore ignored them. Hispanics who were early converts to the Republican Party argued that Democrats paid attention to them only during the election season, but then ignored them when the election was over. Hispanic Republicans began to argue that all Hispanics would be better off if the parties saw them as swing voters who would vote for the candidate who better represented their interests, regardless of party. The United States had two major parties, they said, and Hispanics should participate in both of them.

Republican candidates in California understood these shifts earlier than others. A US representative from California, Richard Nixon, was running for a US Senate seat in 1950. He and California governor Earl Warren, who was running for reelection himself, recognized the importance of recruiting Mexican American voters in their state, especially East Los Angeles. They hired a local member of the Young Republicans named Stuart Spencer, who opened a campaign headquarters, put up signs, dispatched a roving "sound truck" to blast advertisements in Spanish, and recommended visits to Mexican Americans in their homes. The Republicans won a quarter of

Mexican American votes and deemed the result a great success. The 1950 effort to recruit Hispanic voters was the first by a Republican in California. The key had been showing up and presenting Mexican Americans with an option; going door to door in their communities to show them that the Republican Party would be a presence. It was a lesson that Republican candidates across the United States would learn as the twentieth century wore on.

It is important to note, though, that Nixon and Warren in 1950 appealed to Mexican Americans, not Hispanics. The concerns of Puerto Ricans, Cubans and Cuban Americans, or Mexicans and Mexican Americans were not yet defined as "Hispanic." It wasn't even during the 1950s that members of particular national groups really began to consider whether they were also members of a national ethnic group that joined all of them. Flores with his Latinos con Eisenhower group was an exception to the rule.

Widespread debates about what brought different national groups together—people of Cuban, Puerto Rican, and Mexican birth or heritage, for example—didn't take place until the 1960s. Hispanics then were still motivated more by issues particular to their national group—Puerto Rico's territorial status, the Cuban Revolution, and one-party rule in Mexico, or immigration—rather than issues that applied to Hispanics nationally. In some ways, that remains true today. But since the 1960s, Hispanics have wrestled with the idea that they're both representatives of particular national groups and also part of some larger ethnic group—Hispanics, Latinos, and so forth. Over time, the self-identification of Hispanics as Republicans became a result of both local and national politics.

HISPANICS WHO WERE POLITICALLY ACTIVE BY THE 1950S SAW THE 1952 ELECTION AS A consequential one, as did all Americans. The influence of the Soviet Union was spreading around the world, through Europe, Asia, and into the Western Hemisphere. Mao Zedong had taken control in China, the Soviets had tested an atomic bomb, and Communist North Korea invaded South Korea with Soviet backing, which had sparked the establishment of NATO. Eisenhower ran against the Democratic governor of Illinois, Adlai Stevenson, and positioned himself as the candidate better prepared to combat communism. He had seen its dangers firsthand. The most urgent battle of the time, he said, was between tyranny and freedom, collectivism and individualism. The Soviet Union and the United States represented diametrically opposed visions of the world and of human progress. Hispanics and all Americans had to take a stand.

Hispanics also had to decide whether they would continue to support Franklin Roosevelt's New Deal and the Democrats. Even though most Hispanics believed that the New Deal had benefited them, Republicans in the 1930s had called the New Deal a Bolshevik-inspired Communist plot. By the 1950s, they called the New Deal a Soviet-inspired Communist plot. It was antithetical, Republicans argued, to American values of freedom and individualism. Some Hispanics were persuaded, especially Hispanic veterans, professionals, and those who felt that Democrats took their votes for granted.

The Cold War had already taken root in the Americas, including the countries from which many Hispanics had immigrated to the United States. In 1950, for example, there had been several outbreaks of violence over Puerto Rico's status, both on the island and in the continental United States. The fight was between nationalists agitating for independence and Puerto Ricans who favored commonwealth status or statehood. Puerto Rican nationalists had waged a decades-long independence war against Spain that didn't end until 1898, and, as they saw it, the United States then stepped in as a new imperial power. Their struggle with the United States was part of this much longer fight against empire. Nationalists, therefore, weren't necessarily Communists, but that's how they were seen in the context of the global Cold War.

The recently elected governor of Puerto Rico was Luis Muñoz

Marín, and he supported a measure—Public Law 600—that would lead him to be remembered as the "Father of Modern Puerto Rico." He said the law would represent a new "compact" between the United States and the island. It wouldn't fundamentally alter their relationship, since Puerto Rico would remain a territorial possession. But for the first time since the Spanish American War, it would give Puerto Rico the ability to draft its own constitution, enabling the island to govern its internal affairs. Puerto Rico would become a commonwealth of the United States, an "Estado Libre Asociado" (Free Associated State). But as long as the US Congress could approve Puerto Rico's Constitution, independence fighters—Pedro Albizu Campos foremost among them—called Public Law 600 a "sham." Despite their protests, the law was signed on July 3, 1950, as the Puerto Rico Federal Relations Act of 1950.

Another group of Puerto Ricans, those who favored making Puerto Rico a state like any other, also saw the new law as a farce. Like the nationalists, statehooders believed that commonwealth status was akin to colonialism and did not recognize the full equality of Puerto Rico and Puerto Ricans. But this was the only thing nationalists and statehooders agreed on. As the supporters of the various statuses would get caricatured, independence and commonwealth were the statuses favored by radical leftists, or liberal and moderate Democrats, whereas statehood was the status favored by the Republican Party of Puerto Rico and the national Republican Party.

In the fall of 1950, just a few months after the passage of Public Law 600, nationalist "commandos" favoring independence, as one Puerto Rican scholar has called them, expressed their opposition by attacking police stations in multiple towns on the island. Albizu had been released from prison in 1947, the year before Muñoz Marín's election, and within three years had reestablished himself as the leader of Puerto Rico's independence movement. He received the blame when six nationalists shot up the governor's mansion in San Juan. Muñoz

Marín was their intended target. He was inside, but ducked when he heard bullets flying through the windows.

At the same time, two Puerto Rican nationalists attempted to assassinate President Harry Truman while he was napping at Blair House in Washington, DC. Long seen as the ringleader behind all nationalist uprisings, Albizu was arrested again, and this time the House Committee on Un-American Activities (HCUA, or HUAC) labeled him a Communist. Eisenhower wasn't anywhere near the shootings, but his home—the president of Columbia's house, on Morningside Drive in New York City—was placed under armed guard for twenty-four hours afterward.

EPISODES OF VIOLENCE IN PUERTO RICO AND WASHINGTON, DC, IN THE MID-TWENTIETH century were seen as early Cold War conflicts in the Americas, but Latin American immigrants who became Hispanics, and Hispanics who were born in the United States, had developed understandings of conflicts within and between Latin America and the United States for more than a century.

Latin America's nineteenth-century wars for independence from Spain were one reference point. Even when they didn't directly involve the United States, they provided a way for Hispanic Republicans to think comparatively about the histories of Spanish and US imperialism in the Americas. Other nineteenth-century conflicts—including the US-Mexico War (1846–1848), US adventurism in Mexico and Central America (1850s), and the Spanish American War (1898)—also became referents. So did twentieth-century conflicts, such as the Mexican Revolution (1910–1920), numerous US military occupations in the Caribbean basin spread across the early twentieth century, the Cuban Revolution of the 1930s (the backdrop of Desi Arnaz's story), and World War II. They all led Latin Americans and Hispanics to think critically and not so critically about the role of the United States in the hemisphere, and Latin American histories of resistance and cooperation.

Seen from a distance, there were similarities that merged these conflicts in the minds of many—they resulted in the expulsion of Spain from the Western Hemisphere, the rise of the United States as an imperial power, and the exodus of many thousands of Latin Americans from their home countries, many of whom settled in the United States—but they often had distinct causes, were different kinds of conflicts, featured diverse actors (elites, peasants, intellectuals, military officers), and contained a range of lessons for the emigrants and refugees of violence who settled in the United States. In hindsight, as the product of historical memory, Latin American immigrants and Hispanics referred to these episodes as ways of explaining their evolution as conservatives and their affinity for the Republican Party of the United States. They weren't part of the same group in Latin America, and it took time for them to come together in the United States, but they would eventually articulate similar ideas about their admiration for law and order in the United States, its defense of capitalism and individualism, and their previous connections to the colonial government, military, and businesses of the United States.

Even though Hispanic Republicans became loyal to the United States—against the Spanish crown or the Mexican government, for example—a connection with their Spanish heritage became an important part of their conservatism. During the civil rights era, when many Hispanics embraced identities as members of a brown race and articulated their solidarity with indigenous peoples, Hispanic Republicans wrote counterhistories of their community that foregrounded their Spanishness. Manuel Machado, for example, wrote what he described as a corrective to the "bleeding heart liberal" version of Mexican American history, titled *Listen Chicano!*, with a foreword by the Republican senator from Arizona Barry Goldwater. Machado claimed that the "militancy" of Chicanos and their lionization of "Mexico's Indian heroes" were behind the denigration of "Spanish glory and achievement." He and other Hispanic Republicans revered what they considered to be Spain's record of accomplishment in the Americas; how the Spanish empire had civilized and modernized many different regions.

Hispanic Republicans also proudly recalled the period when their families arrived in the Americas from Spain. Their Spanish ancestors, they said, were concerned primarily with politics in their old home and earning money in their new one. Rita DiMartino, a prominent Puerto Rican Republican, said her maiden name, Dendariarena, was Basque. Her family kept the name when they moved to Bayamón, Puerto Rico. Meanwhile, the family of her compatriot Luis Guinot Jr. moved from Barcelona in the 1850s, like hundreds of other Catalonians. He said that Catalonians had a reputation within Spain for being both industrious and independence-minded, so when they moved to the Americas the Spanish crown wanted them to move to the poorer colonies, where they wouldn't cause trouble. It is interesting that DiMartino and Guinot claimed a lineage that stretched back to some of Spain's most separatist regions. Nevertheless, a shared Spanish identity became a way for Hispanics from different backgrounds to blur the distinctions between them. Their common past united them across borders between Europe, Latin America, and the United States.

While many Latin American immigrants who became Hispanic Republicans highlighted their Spanish heritage, they also bucked the anti-Yankee, anti-imperialist views represented in the late nineteenth and early twentieth centuries by the Cuban writer José Martí, the Uruguayan writer José Enrique Rodó, and the Mexican writer José Vasconcelos. The Latin Americans who became Hispanics who became Republicans admired the economic force, religious freedom, political ideals, and modernity of the United States. Latin American admiration of the United States stretched back to the early nineteenth century, when Latin Americans fighting for independence from Spain cited the American Revolution as their inspiration, and the Mexican writer Lorenzo de Zavala, the so-called Mexican Tocqueville, wrote a narrative of his travels through the young nation, titled *Journey to the United States of America*, first published in Paris in 1834.

Such admiration was often expressed in contrast with their acquired distaste for Spain, and here we see how Latin American immigrants and Hispanics could distinguish between their Spanish heritage and Spanish governance. Evidence of US progress compared with Spain's decline provided part of the rationale for their desire to leave the empire. Some of the Latin Americans who opposed Spain sought political inclusion in the United States because they believed that the democratic nation, and the Republican Party in particular, would defend their freedom.

The Afro-Puerto Rican José Celso Barbosa, for example, "acquired a deep admiration for the U.S. Republican Party," and for the same reason that African Americans did, because it was the "party of emancipation." Barbosa was the first Puerto Rican to receive a medical degree in the United States, from the University of Michigan in 1877, and he was a founder of the Partido Republicano de Puerto Rico (the Republican Party of Puerto Rico), established in 1902. He became known as the father of the prostatehood movement on the island. Zoraida Fonalledas, who is today a proponent of statehood and also the National Committeewoman of Puerto Rico's Republican Party, has a picture of him hanging in her office, to remind her, she says, of "his quest for the full equality of all Puerto Rican-American citizens."

Many other Latin Americans in the late nineteenth and early twentieth centuries also admired the United States and the Republican Party. After centuries of domination by Spain and the Catholic Church, many Cubans, for example, joined other Christian sects and developed business ties that created great wealth in Cuba and the United States. Many sent their children to school in the north. Some of them stayed, while others returned to their Latin American home countries and became part of the intellectual, business, and political ruling class.

Importantly, these things weren't considered conservative at the time. Many, in fact, were liberal, since they were rejections of the conservative and, as they saw it, backward-looking, regressive Spanish Empire. The subset of Latin American immigrants and Hispanics who saw things this way thought of Spain as the conservative oppressor and the United States as a liberal emancipator. Patriotic fervor, religious devotion, and free enterprise became seen as conservative, and Republican, only later in the twentieth century. These complicated associations are part of the story of how partisan identities flipped between the nineteenth and twentieth centuries, part of the difference between conservatism in Latin America and the United States, and part of why it's dangerous to look into the past for expressions of admiration for free enterprise, or the embrace of Christianity, as markers of an evolving Hispanic conservatism.

Hispanic Republicans would also cast the nationalist rebellions of the early twentieth century as plots carried out by the radical left, regardless of their historical differences, including whether they were civil, religious, or military conflicts, or whether their protagonists were liberal or conservative. Henry Ramirez, who served in the Nixon administration, for example, recalled how his family had fled Mexico because of the "militant secularism" of the 1917 Mexican Constitution, signed in the middle of the Mexican Revolution. The so-called Cristero War of the late 1920s was a rejection of the revolutionary constitution, when Catholics in the country took up arms to defend against attacks on the open practice of their religion. As a counterreaction, the Mexican government, Ramirez wrote, waged a "bloody French-Revolution style persecution of the Catholic Church."

The Cristero War was also one of the episodes that led to the formation of Mexico's Partido Acción Nacional (PAN, or National Action Party), which Hispanic Republicans during Ronald Reagan's presidency called the Republican Party of Mexico. Its founders, like Ramirez, opposed the secularism of the Constitution, radical reforms adopted by Mexico's postrevolutionary presidents, and one-party rule in the country. It was a refuge for Mexican Catholics who remained in Mexico after the Cristero War, just as the United States became for Mexican Catholics who chose to leave. As soon as the Ramirez family arrived in California, his parents enrolled him in a Catholic school, where he could practice his religion without government interference.

Arnaz's story offered a similar tale about conflicts in the Caribbean during the 1930s, and so did the story of nationalism's rise in Puerto Rico decades before the violence of 1950. The election of Albizu in 1936 as president of the Puerto Rican Nationalist Party impelled moderate and conservative Puerto Ricans to look to the United States for help resisting his influence. Under his leadership, the nationalist advocates of independence engaged in bloody conflicts with police and the military. In response to Albizu's agenda, the Republican Party included in its 1940 party platform a statement supporting statehood, calling it the "logical aspiration of the people of Puerto Rico." Puerto Rican Republicans have always seen Albizu as a violent radical supported by a small minority of Puerto Ricans and have believed that most Puerto Ricans desire closer alliances and greater equality with the United States.

Through their experience of violent conflicts waged, as they understood them, by radical leftists in Mexico, Puerto Rico, and Cuba, immigrants from Latin America and the Caribbean saw the United States, by contrast, as the symbol of democracy and freedom in the world. To be sure, the United States had its own authoritarian tendencies, including the impulse to control labor activism, to maintain laws that made it impossible for nonwhites to become truly equal, and to possess far-flung territories. Yet Hispanics whose families had immigrated from Latin America and the Caribbean saw the early decades of the twentieth century as a dangerous and uncertain time and believed that the United States would provide stability. Their home countries were engaged in a struggle between authoritarianism and democracy, and it wasn't clear which side would win. Hispanic Republicans placed their faith in the United States, the best hope for order, peace, and prosperity. And in 1952, they placed their hopes in Dwight Eisenhower.

THE SO-CALLED RED SCARE WAS ABOUT MUCH MORE THAN HEMISPHERIC RELATIONS, of course. After World War II, the House Un-American Activities Committee (HUAC) questioned Hollywood filmmakers about their Communist affiliations, Whittaker Chambers exposed Alger Hiss as a Communist, and Senator Joseph McCarthy of Wisconsin accused hundreds of members of Congress of being Communists. This atmosphere of suspicion took a toll on Hispanic communities by forcing Hispanics to publicly declare their loyalty to the United States, and by forcing more than a million immigrants out of the country.

One of the most important pieces of Cold War legislation was the McCarran-Walter Act, signed in June 1952, in the middle of Eisenhower's and Stevenson's presidential campaigns. Named after its sponsors, the Democratic senator from Nevada Pat McCarran and the Democratic representative from Pennsylvania Francis Walter, the law excluded foreigners, deported immigrants, and denaturalized US citizens deemed to be subversives.

President Truman vetoed the law, arguing that it was discriminatory and would "intensify the repressive and inhumane aspects of our immigration procedures." When Congress overrode his veto, it became a "major campaign issue," according to the *Baltimore Sun*. Eisenhower's Republican backers in the Senate supported the president's veto. Eisenhower himself said that he would ask for the law to be revised, and, after his election, expressed the generally proimmigrant sentiment that the United States needed to "strike an intelligent, unbigoted balance between the immigration welfare of America and the prayerful hopes of the unhappy and oppressed."

Hispanic views of the McCarran-Walter Act shifted from mild support to strenuous opposition. The Alianza Hispano Americana and the League of United Latin American Citizens at first toed the

Cold War line that they were also worried about Communists entering the United States. They represented Mexican Americans who believed that Mexican workers, whether they entered illegally or legally,
competed for their jobs. This was also the position of the National
Farm Workers Union, which supported the McCarran-Walter Act.

But after the law went into effect, many Hispanics came to recognize
its discriminatory consequences. Mexican immigrants frequently got
patted down at the border, were asked about their political beliefs, and
were searched for Communist propaganda, while anyone of Hispanic
descent who said anything critical about the United States—whether
they were here illegally or not—risked being labeled a Communist.
Patriotism, capitalism, and anticommunism defined Hispanic Republican politics, but so did their opposition to discrimination against
immigrants and Hispanics in general.

Although Hispanics were not a powerful political force in 1952, the
campaign and election were in many ways a proving ground for them.
Into the 1950s, they were active mainly locally, in Mexican American
communities in the Southwest and Puerto Rican communities in the
Northeast. Hispanics represented only 2 percent of the US population as a whole, and an even smaller percentage of the American electorate. Of those who did vote, only a small percentage—perhaps 5 to
10 percent—voted for Eisenhower. He had won the election handily,
capturing 55 percent of the popular vote and the electoral votes of all
states except nine in the upper and lower South, where Democrats
still dominated. He didn't rely on the Hispanic vote, but his campaign laid some of the groundwork that later Republicans built upon.

The Latinos con Eisenhower group set a precedent for the Viva
Nixon, Latinos con Goldwater, Viva Ford, Viva Reagan, and Viva
Bush groups. As a result of their organizing in 1952, Eisenhower's
Hispanic supporters also began formulating a rationale for why the
Republican Party and its leaders should pay attention to them. The
Republican Party was the defender of freedom and liberty around
the world; the Democratic Party took Hispanics for granted; and the

Hispanic population, small at the time, was growing. They had served their country loyally in World War II and were patriots. These arguments resonated with Hispanic Republicans for decades to come.

Unfortunately for Eisenhower's Hispanic supporters and the Republican Party generally, the 1952 election was also the first time that Hispanic Republicans felt ignored after they'd given their time, energy, and money to help elect the Republican candidate.

IN JANUARY 1953, EISENHOWER LEFT NEW YORK, WHERE HE WAS STILL PRESIDENT OF Columbia University, to be inaugurated as president of the United States. Once Eisenhower was in the White House, Flores—hoping that Hispanics, and especially he, might gain influence in the new administration—argued that Hispanics possessed unique skills that could help the United States win the Cold War; namely, they spoke Spanish, were personally and intimately familiar with Latin American cultures and traditions, and had relationships with anti-Communists south of the border who could become important allies. They had also gained valuable experience abroad during World War II, so they knew something about diplomacy and international relations.

Flores also argued that Hispanics had an important role to play in domestic politics. He said that his organization was made up of Democrats who left the party because, after decades of supporting Roosevelt and Truman, they hadn't been accorded the "proper representation which our large Spanish speaking vote and population deserves." The idea that the Democratic Party ignored them and took their votes for granted became a rallying cry for Hispanic Republicans. Flores claimed that there were some 25 million to 30 million Hispanics in the United States (this estimate was almost ten times greater than the US census figure of 3.2 million), and 5 million were eligible to vote. They could become a bloc of great importance.

Hispanics had also spent a considerable sum to get Eisenhower elected, and Flores argued that they deserved something in return for their efforts. Flores claimed that Latinos con Eisenhower contributed $10,000 to the campaign, so he thought Hispanics like him should be given opportunities to work in the government. He asked to be considered for the position of deputy assistant secretary for inter-American affairs. His request was denied, but he wrote the president again to ask that he be sent to South America with Eisenhower's brother Milton, whose counsel the president relied on in all matters related to Latin America. Flores explained in incomplete, telegram-style prose that the president should appoint him because of his "personal acquaintance with many officials who will be visited" and because his "knowledge of language, customs, traditions, economics and politics of those countries, will be an asset in my evaluating present trend of thought and opinion there toward official Washington."

Eisenhower never took Flores up on his offers, and Flores complained to anyone who would listen—Arizona senator Barry Goldwater, former Colorado governor Dan Thornton—that the president was ignoring him. Flores wrote Eisenhower himself to say that Latinos con Eisenhower was withdrawing its "backing and support" for him. The members of his group had hoped "to serve and represent our U.S.A. in positions of honor and trust having to do with Spanish speaking affairs, countries and peoples." They wanted to serve their country, as they had done on battlefields in Europe and the Pacific.

Flores was in fact expressing a frustration shared by Hispanics in both parties, that the United States sent ambassadors to Latin America who weren't Hispanic, didn't speak Spanish, and knew little about the region. But several Republican leaders considered Flores to be an overly ambitious, self-promoting pest and that that more than anything else explained why his requests were ignored. Moreover, it wasn't necessarily the case that a Mexican American from the Southwest knew any more about Latin America than a non-Hispanic would. The countries of Latin America had different histories and cultures, so it didn't always follow that a Mexican American could claim expertise on Argentina, Brazil, or even Mexico.

Flores congratulated Eisenhower for making a greater effort among African American voters and for appointing several to his administration. "We are happy for them," he wrote, adding that "we don't begrudge them this advancement and representation." Yet speaking for all Hispanics, he wrote, "We too merit and deserve this same consideration."

The White House forwarded Flores's letters to the Republican National Committee (RNC). If he had campaigned for Eisenhower, the White House didn't know it. Because Flores claimed connections in Arizona, the RNC asked Goldwater to do some digging. Goldwater reported back that "Flores has no standing in the Republican Party," and his organization was "practically non-existent." Just as Hispanic Republicans claimed that Hispanics were ignored by Democrats, Flores's letters to Eisenhower were representative of how Hispanics also felt insufficiently recognized by Republicans. This, too, was a theme reprised by later generations of Hispanic Republicans, even in the years when they experienced greater inclusion in the party. Their broad agreement on many of the Republican Party's positions, their ongoing frustrations with Democrats, and the Republican Party's promises, even when they were unfulfilled, inspired their evolution as Republicans.

BETWEEN EISENHOWER'S ELECTION IN 1952 AND HIS REELECTION IN 1956, THE COLD WAR at home and abroad continued to dominate national politics. And again, developments in the Americas shaped Hispanic politics in the United States.

Soon after Eisenhower became president, Joseph Stalin died, giving Eisenhower hope that the next Soviet premier would help usher in a period of reduced Cold War tensions. It wasn't to be. Continued Soviet aggression in Europe made the situation across the Atlantic increasingly tense, just as more trouble in the Americas appeared on the horizon. The conflicts closer to home were domestic struggles over economic and political inequality, but Eisenhower saw them through the lens of the international struggle against communism. He believed that they posed a grave danger to the country and to world order.

As populist nationalists took control in several American nations, Eisenhower and his inner circle saw their rise as more evidence that communism was on the move. Leaders such as Juan Perón of Argentina, Getúlio Vargas of Brazil, and Jacobo Árbenz of Guatemala sought to redefine the economic and political relationship between their countries and the United States, from one of subordination and dependency to one of equality and modernization. Their ideas, Republicans believed, threatened the United States, and Red subversion was thought to be behind them. These nationalist movements led Eisenhower and a succession of presidents after him to believe that the Cold War in the Western Hemisphere was a top priority. They also became touchstones for Hispanic Republicans, who came to see them as signs of the dangers of the radical left.

Cold War violence in the Americas struck close to home when, in 1954, Lolita Lebrón, a Puerto Rican nationalist based in New York, traveled to the capital city and fired twenty-nine bullets into a gathering of the US House of Representatives, injuring five congressmen. The day after the shooting, investigators in New York said they discovered a "cache of Red propaganda" in the apartment where she and four others planned the attack. Muñoz Marín said that "all Puerto Ricans" opposed the "outrage," but the *Los Angeles Times* said that the nationalists who committed the violence showed no remorse. Lebrón was labeled a Communist.

The percentages of Puerto Ricans favoring different territorial statuses changed over time. There were those who favored commonwealth status (such as Muñoz Marín), independence (such as Albizu), or statehood (whose main proponents were Miguel Angel García Méndez and Luis A. Ferré, who cofounded the Partido Estadista Republicano, or Prostatehood Republican Party). But in the 1950s, Muñoz

Marin's party reigned. It was called the Partido Popular Democrático (PPD, or Popular Democratic Party). This was the low point of the statehood movement—the "dark years," as one prostatehood Puerto Rican later called it. Even Eisenhower backed Muñoz Marin because he had brought political stability, democratic leadership, and relative prosperity to the island, primarily through tourism and an economic development initiative called Operation Bootstrap.

Nevertheless, the establishment of the commonwealth, or Estado Libre Asociado, in 1952 and the Prostatehood Republican Party in 1956 drew battle lines that separated those who argued for commonwealth status, independence, or statehood into the twenty-first century. In these early decades, the status issue also divided Puerto Ricans who identified as Democrats from those who identified as Republicans. By and large, Democrats favored commonwealth status (maybe even independence), while Republicans favored statehood.

It hadn't been three months since Lebrón fired on Congress when violence broke out in Guatemala, in June 1954. With backing from the US military and the CIA, Colonel Carlos Castillo Armas forced the resignation of Jacobo Árbenz Guzmán, who had won a democratic election in 1951. Árbenz continued the liberal social reforms initiated by his predecessor. He expropriated and redistributed uncultivated lands, including hundreds of thousands of acres owned by the banana-producing United Fruit Company, whose US-based owners argued that large swaths of its land had to remain uncultivated to rotate crops and hedge against natural disasters such as hurricanes.

Árbenz ignored them and expropriated 400,000 of the 550,000 acres in Guatemala owned by United Fruit. In the eyes of Hispanic Republicans, his violation of the rights of property owners was no different from what had transpired during the Mexican Revolution of 1910 and the Cuban Revolution of 1933. Guatemala had offered the company's executives $3 per acre to buy the land back, but they demanded twenty-five times that amount. The expropriation angered United Fruit and the company's representatives in the US government, but the presence of Communists in the Guatemalan government, and Árbenz's acceptance of Soviet arms shipments, crossed a line for Eisenhower, who began to mobilize against him.

The United States sent arms and airplanes to Armas, who had been living in exile in Honduras. The CIA trained his troops, and US warplanes bombed Guatemala City. Inspired by Árbenz, the Argentine medical student Ernesto "Che" Guevara moved to Guatemala to help defend against the attack by the United States. After Armas took control, Guevara fled to Mexico. The operation against Árbenz was known as PBSUCCESS, and it went against decades-old principles of nonintervention agreed upon by American nations including the United States.

Critics maintained that Eisenhower was a zealot who was quick to label any popularly elected liberal in Latin America a Communist. They also claimed that protecting lands owned by United Fruit was the real cause of the intervention (in point of fact, the company had its lands restored as soon as Armas was installed). Regardless, the motivations behind US involvement in the coup "cannot be separated from the political economy of the Cold War," as one historian has put it. It was a clear sign that Eisenhower would take extreme measures to fight communism in the Americas. After Armas took control, Eisenhower sent Vice President Nixon to Guatemala. Nixon and the anti-Communist, right-wing military dictator stood together on the steps of the presidential palace. A "great throng" of Guatemalans shouted its "enthusiastic approval": "Viva Nixon" and "Viva Armas."

The crowds may have cheered, but the military intervention had heightened the tide of anti-American, anti-imperialist sentiment in Latin America. In an age defined more by decolonization movements around the world, it was out of step with the times, and out of character for a nation that claimed to stand for "self-determination." But Hispanic Republicans were inspired. To them, the interventions ending in regime change showed that the United States would be the defender of freedom, democracy, capitalism, and private property. More often than not, they sided with the interventionists.

AS THE UNITED STATES INVADED GUATEMALA, THE COLD WAR AGAIN TOOK A TOLL ON HISPANICS at home. In the spring of 1954, the supporters of the McCarran-Walter Act argued that the so-called wetback problem needed to be solved. Unauthorized Mexican immigrants were called "wetbacks" because some of them swam across the Rio Grande to enter the United States. The binational labor agreement known as the Bracero Program, which the United States and Mexico had entered into during World War II, managed the flow of migration between the two countries. But the demand for immigrant labor was at least as great as the supply, and the desire of Mexican immigrants to work in the United States was greater than the number of Bracero contracts available. It was also greater than pressures to keep them in Mexico, so many chose to enter the United States illegally.

The number of unauthorized Mexicans apprehended in the United States increased throughout the 1940s and early 1950s, and by 1954 many Americans believed that something had to be done to restrict their entry. In June of that year, President Eisenhower directed the new head of the Immigration and Naturalization Service (INS), General Joseph Swing, to initiate Operation Wetback. The Border Patrol officers in his charge raided farms, ranches, hotels, restaurants, and other businesses that employed unauthorized immigrants to round them up; place them in detention centers; and load them in buses, trains, or planes for deportation. Swing, one historian has written, promised it would be a "spectacular show of U.S. immigration law enforcement." By 1955, more than a million undocumented immigrants had been deported.

Many Americans supported the initiative. The wartime demand for labor had subsided, they said, so US citizens and legal immigrants could meet the country's needs. An article in the *San Antonio Express-News* claimed that the "wetback invasion" also "endangers public health, increases the crime rate, overly burdens welfare agencies, hurts the general economy, depresses social standards, risks subversive infiltration, displaces worthy domestic labor, and traffics in human misery." Such arguments became the bread and butter of the anti-immigrant movement later in the century. The major difference was who made them.

In the 1950s, Democrats representing labor unions and Hispanic organizations whose members included Mexican American workers were the main opponents of unauthorized immigration, whereas liberal Democrats concerned with the exploitation of workers and Republicans who represented businesses that relied on immigrant labor did not support the crackdowns. For their part, Border Patrol officers argued that they were protecting immigrant workers from the abuses of farmers, and saving them from laboring in conditions they compared to slavery.

The language of the Cold War was front and center in debates over Operation Wetback. Senator McCarran, one of the authors of the McCarran-Walter Act, said there were "subversive dangers inherent" in the growing numbers of unauthorized immigrants entering the United States and that the border was a "direct threat to the internal security of the Nation because it is an open door for Communist agents." He endorsed Operation Wetback as a noble effort to establish an "Iron Curtain about the borders of the United States" to keep out the "militant Communists."

Soon enough, it wasn't just unauthorized Mexican immigrants who were suspected of being Communists, but Hispanic citizens of the United States as well. Once Cold War hysteria subjected them to discrimination, Hispanics adopted different views of the era's repressions. Historians have argued that the cautious response to Cold War–era discrimination by many Hispanics was strategic. Some who were Communists or fellow travelers may have found refuge in radical labor unions, but most distanced themselves from communism and even sounded anti-Communist alarm bells, either to gain inclusion or to avoid more extreme forms of exclusion. They chose sides in the Cold War in part because life in the United States could be made difficult for them if they didn't choose or if they chose the wrong side.

Hispanic Republicans sympathized with the plight of both undocumented immigrants and the businesses that hired them, and for these reasons they generally supported the comparatively lenient immigration policies favored by the Republican Party. But like their fellow Republicans, they also worried about the infiltration of the United States by Communists. For them, anticommunism wasn't a strategy; it was who they were. They wanted to wipe out communism abroad and from their new home. Some of them had come from countries in Latin America and the Caribbean where, they said, radicalism was rampant. The ills of leftist governments were why they moved to the United States in the first place, and their preference for the tough stand against those governments taken by Republicans was an important part of what drew them to the Republican Party.

THE COLD WAR IN THE AMERICAS, BOTH ABROAD AND AT HOME, PROVIDED THE CONTEXT for Eisenhower's reelection campaign in 1956. In anticipation of the election, Flores attempted once more to sit down with Eisenhower. In September 1954, shortly after the summer of immigrant roundups, Flores heard about Eisenhower's upcoming visit to Los Angeles. He wrote the White House to request a few minutes with the president, to impress upon him the importance of paying attention to Hispanics. Whether Eisenhower paid attention or not, Flores argued, would "have much to do with whether or not he and [the] GOP receives a large part of the Spanish speaking vote." Eisenhower didn't meet with him, and, as Flores had promised when he felt slighted by Eisenhower after the 1952 election, he didn't support the president in 1956.

Another Mexican American from California replaced Flores as

the main Hispanic vote seeker for Eisenhower. Manuel Mesa, the "State Chairman for Southern California of the Eisenhower-Nixon Committee's Latin-American Division," ingratiated himself to the Eisenhower team more than Flores ever did. Mesa sent eighty-three thousand mailings, delivered twenty-five thousand pamphlets, and handed out fifteen hundred "ME GUSTA IKE" buttons. In particular, the campaign targeted "doctors, small businessmen, lawyers, and churches of different denominations, labor organizations and women's organizations."

Like Flores's group, Mesa's focused on disaffected Democrats. Mesa estimated that 80 percent of their effort was spent on Democrats and only 20 percent on Republicans, whose number was small and whose votes they already had. Whatever they did, RNC chairman Leonard Hall thought their efforts were a success, even though Eisenhower again won less than 10 percent of Hispanic votes. Hall sent Mesa a letter after the election that said his support of the president had given him hope that Eisenhower's reelection was the "first step toward making the Republican Party by 1960 the dominant party in the nation." Republican leaders already saw the potential in recruiting Hispanics to their side.

After the election, in which Eisenhower defeated Stevenson by an even wider margin—in their rematch, the president lost only seven states, all of them, once again, in the lower and upper South—he again focused on the Cold War in the Americas. In 1958, he sent Nixon to Latin America for a second time, but the vice president wasn't welcomed as kindly as he had been in 1955. Latin Americans remembered the US intervention in Guatemala just a few years earlier. In April and May, Nixon spent eighteen days in Bolivia, Colombia, Ecuador, Peru, Uruguay, and Venezuela to talk with heads of state, students, labor leaders, and others about trade and US policy toward the region. The Latin American tour was one phase of his world travels to Africa, Asia, and Europe, where he also aimed to secure Cold War allies.

The trip wasn't all bad. He again heard "Viva Nixon" chants, this time at a soccer match in Ecuador with ten thousand people in attendance. Thousands flocked to him. He gave them ballpoint pens with his name on them, signed objects his admirers put in front of him, and hugged children, "letting himself be kissed." But in addition to those who praised him, he was greeted by the protests of students and workers, who yelled "Fuera Nixon" (Nixon Out!), "Down with Nixon," "Death to Nixon," and "Yankee imperialist."

Nixon had stated that his approach to diplomacy was simple: if they "like me," they "like my country." Assuming that this was his mind-set when he embarked for Latin America, he would have accurately perceived that Latin Americans were deeply ambivalent toward him and the United States. The reception was more or less cordial depending on where he went. In some countries, his supporters drowned out his opponents. In others, support and opposition split down the middle. In Peru and Venezuela, in particular, Nixon found trouble. Protesters stoned his car, spat at him, burned posters with his face on them, and menaced him with death threats. "Without any doubt," Nixon said, they "were Communists."

The vice president cut the Latin America trip short. He was in Venezuela when he decided to head home. Eisenhower had reached him by phone and told him that he was worried for his safety. Newspapers in the United States later claimed that Nixon had a near-death experience. He evacuated the country with an escort of "more than 10,000 Venezuelan soldiers and police." It was a "full-scale military operation," one newspaper reported. On his way back to Washington, he stopped in Puerto Rico, where the leaders of the island scrambled to welcome him properly. Governor Muñoz Marín arranged a dinner in his honor, and Nixon and his wife, Pat, spent the night in the same mansion where Puerto Rican nationalists had attempted to kill the governor just a few years earlier. Recognizing the island territory as his turf, Nixon said when he landed in San Juan, "I am glad to be home again."

The journalist Walter Lippmann said the "whole South American tour was misconceived" because those who planned it did not understand political dynamics in the countries visited by Nixon, where nationalist and internationalist, pro-Communist and anti-Communist, pro-American and anti-American forces struggled for supremacy. The conflicts in Latin America were caused by vast social and economic inequalities, demands for land reform, and disdain over the accumulation of wealth by the few. Growing resentment of the United States was only partially to blame. Because the US government didn't know what it was getting into, or the depth of anti-American feeling in the region, Lippmann wrote, Nixon was left vulnerable to the "diplomatic Pearl Harbor" he encountered in Latin America.

In the weeks and months after Nixon's travails in Venezuela, Eisenhower attended to Latin America with renewed urgency and, some have argued, sensitivity. Regime changes such as the one in Guatemala no longer seemed like unqualified successes. The United States should be wary of dictators and more supportive of democratically elected leaders, they concluded. As Nixon famously put it, there would be a "formal handshake for dictators; an *embraso* for leaders in freedom." (Nixon probably meant *abrazo*, the Spanish word for hug, which may have been too close to "embrace" for him to avoid the slipup.)

The United States wouldn't reverse its fundamental positions on economic, political, and military support. It still favored private investment and free-market capitalism above nationalized economies, but it became more willing to prop up particular commodities, such as coffee, that could destabilize the country if they were to crash. Plan A was to firm up relationships with democracies such as Puerto Rico. Plan B could still include military intervention.

The United States couldn't continue pushing capitalism and anticommunism without also working to improve economic and social conditions for the multitude of Latin Americans if they wanted to avoid the storms churning on the horizon, the "surging, swelling, revolutionary demand," as Eisenhower's brother Milton put it, "not just for aid, but for rapid social revolution in country after country." The strategies of military intervention and economic aid ebbed and flowed over the years, and between the two parties.

By the time Nixon returned from Latin America, Flores's Latinos con Eisenhower group was defunct, and true to his word, Flores had stopped supporting the president. But that didn't stop him from recommending himself for assignments. After Nixon's rough treatment, Flores wrote Eisenhower to offer his assistance based on his connections in the region. He said he could get a "direct, intimate, personal report from the Presidents of Chile, Peru and Venezuela giving the 'why' and 'wherefore' of the Nixon incidents." Flores again argued that Hispanics were valuable Cold War assets, and the White House once again ignored him.

Flores's letter to Eisenhower may have been another vanity-driven fool's errand to get close to power, but he also tried to show the Republican Party how Hispanics like him could be useful to them. In the fearful years of the Cold War, when international conflict seemed imminent, they were Spanish-speaking citizens of the United States and loyal Americans who could support the country's aims while improving its standing in the hemisphere. For Hispanics such as Flores, the Cold War in the Americas seemingly presented an opportunity to rise in a party they agreed with ideologically but to which they still didn't have a secure connection. Developments in the hemisphere were perilous, but Hispanic Republicans thought they could use the instability to their benefit.

Some Hispanic Republicans at the local level had already figured out how to benefit from their alliance with Republican leaders. In addition to Mexican Americans in California, who had supported Warren and Nixon, Puerto Ricans in New York became an important part of Governor Nelson Rockefeller's coalition. They generally embraced the moderate conservatism of Republicans such as Eisenhower and Rockefeller, as opposed to the stark conservatism of Goldwater or Ohio senator Robert Taft.

Puerto Ricans helped elect Rockefeller in 1958. "Rocky" spoke Spanish, had business connections in Latin America, and campaigned in their communities on a platform of low subway fares, the need for more affordable housing, and ending police brutality. He threw a "Rockefeller fiesta" in Spanish Harlem, on 108th Street, just days before the election. Some of the three thousand Puerto Ricans in attendance hoisted him onto their shoulders and "bore him to and fro amid wild cheers." They cast 30 percent of their votes for him. Confirming their sense that Rockefeller's election could have economic benefits for Puerto Rico as well, his brother Laurance, shortly after Nelson was reelected, announced that he would extend his hotel empire to the island by building the Dorado Beach Hotel and Golf Club in San Juan.

Soon after Nixon's fateful Latin America tour and Rockefeller's reelection, the Eisenhower administration's nightmare scenario came true in Cuba, when the young lawyer Fidel Castro, his brother Raúl Castro, Che Guevara, and a few dozen other rebels managed the iconic revolution that many in the United States perceived as an existential threat. For Hispanics motivated by anticommunism—who viewed developments in Puerto Rico, Guatemala, and Venezuela with apprehension and US interventions with approbation—the Cuban Revolution sealed their impressions of the dangers posed by leftist governments in the Americas.

BEFORE THEIR SUCCESSFUL REVOLUTION ON NEW YEAR'S DAY IN 1959, WHEN THEY RODE triumphantly into Havana, the guerrillas—called *barbudos*, or bearded ones—had launched two earlier attempts to take the island. The first, in 1953, was a failure that resulted in many rebel casualties. The Castro brothers were imprisoned, but soon were released in what one historian has described as an act of "dictatorial benevolence" by President Fulgencio Batista. It would prove to be a mistake. Fidel and Raúl relocated to Mexico City, where they connected with Guevara. In a little over a year, the trio assembled a force of about eighty fighters. They sailed toward Cuba on an old yacht called the *Granma*, which landed in December 1956 at Playa Las Coloradas, on the southern side of the island.

Local peasants tipped off Batista loyalists, who killed seventy of the eighty-two men on board. The Castro brothers and Guevara were three of the twelve who survived. They took refuge in the mountains of eastern Cuba, and over the next two years gained popular support that delegitimized Batista. The United States had supported Batista for years, since he became president in 1952. But the relationship turned cold when Batista lost influence, so Castro might have become a palatable alternative.

At first, Castro's revolution found widespread popularity, even among some of the Cubans who later moved to Florida and spent the rest of their lives opposing him. After only a few months in power, Castro had made his Communist leanings clear. His speeches were laden with references to Marxism. He followed the Mexican and Guatemalan revolutions by nationalizing Cuban industries and expropriating and redistributing the landholdings of wealthy Cubans and Americans. When he visited New York in 1960, he met with Malcolm X, a black nationalist who was deeply critical of the United States, and gave a four-hour speech at the United Nations about the ills of US imperialism.

Things got worse from there. Cuba bought oil from Russia, and the US-owned oil refineries in Cuba refused to process it. When Castro nationalized the refineries, the United States stopped purchasing Cuban sugar and embargoed all trade with Cuba. The United States trained Cuban exiles to invade the island, and Soviets furnished Cuba with nuclear missiles. By the early 1960s, Cuba had become ground zero of the Cold War in the Americas.

Facing all the Cold War conflicts in the hemisphere during the 1950s, Mexico fashioned itself as a nonaligned country that could mediate between the United States and the rest of Latin America. It was a difficult position, since Mexico had to balance doing nothing that would jeopardize economic and political stability in its relationship with the United States and solidarity with Latin American countries that also saw the colossus to the north as their greatest threat. But instead of remaining neutral, Mexico successfully managed conflicts while remaining deeply involved in regional affairs. Mexico was both sides of the Cold War at once.

On the one hand, E. Howard Hunt and William F. Buckley Jr., two American spies working for the CIA in Mexico City in the early 1950s, monitored Communist activities in the country and developed relationships with anti-Communist informants. On the other hand, Cuban revolutionaries launched their 1956 attack against Batista from Mexico, with support from Mexicans.

Mexican president Adolfo López Mateos reminded Latin Americans that he was the first leader in the region to recognize Castro, praising him for implementing agrarian reforms in Cuba that were similar to the ones in Mexico after the Mexican Revolution. But López Mateos also forbade former president Lázaro Cárdenas and thousands of members of the Student Front in Defense of the Cuban Revolution from traveling to Cuba to support Castro. Mexico was able to walk the line between right and left in part because of one-party rule. Claiming to hear the critiques of anti-Communist conservatives in the PAN, as well as leaders of prolabor, anti-imperialist groups, the Partido Revolucionario Institucional (PRI, or Institutional Revolutionary Party) nevertheless imposed its will. The Cold War between the United States and Latin America—from Puerto Rico to Cuba to Mexico—became the main filter through which Hispanic Republicans viewed not only international relations but also the formation of partisan political identities in the United States.

THE EIGHT YEARS OF EISENHOWER'S PRESIDENCY TOOK PLACE DURING A PARTICULARLY tense period of the Cold War between the United States and Latin

America. Indeed, it was a period of global tensions that forced Hispanics in the United States to take sides. Some were radicals charged with bringing communism into the United States, while others were Cold Warriors themselves, fighting to eliminate communism at home and abroad. Both groups became more politically involved during the 1950s. But the decade remained difficult for them. Hispanics such as Flores felt like they were being ignored. Others suffered more violent forms of discrimination, including deportation.

When they looked back on the 1950s, Hispanic Republicans said they supported the politics of the Republican Party nationally. They agreed with Eisenhower's positions on free enterprise, his fight against communism, and the idea that the United States had an important role to play in spreading democracy around the world. But their Republican identities also looked different on the ground, depending on where they stood.

A Californian himself, Vice President Nixon knew that winning the votes of Mexican Americans there required attention to their particular interests. When he ran for president in 1960, he again enlisted the services of Stuart Spencer, who opened a community center "right in the middle of the barrio" that offered free polio shots, hosted job fairs, brought in lawyers to help community members with "immigration problems," and organized seminars on health care and college scholarships. Spencer recognized a fundamental problem: politicians asked Hispanics for their votes right before an election, but then disappeared. Spencer wanted to show Mexican Americans that the Republican Party would be there year-round, whether there was an election or not. That would be critical to winning their loyalty.

In New Mexico, meanwhile, Nixon had to focus on other issues to win Mexican American support. The rancher and longtime Republican Abe Peña—whose father, Pablo Peña, had been a leading Republican in the state at the time of the territory's constitutional convention in 1912, which made New Mexico the forty-seventh state in the Union—was also concerned about communism and sought to protect individual liberties and free-market capitalism. Born in San Mateo in 1926, Peña attended college at New Mexico State University, served in the army, and spent time in Australia on a Fulbright scholarship before returning to New Mexico. He ran Peña & Peña Ranch with his father in the small town of Grants, about two hours west of Albuquerque, surrounded by several Indian reservations. He told a friend that he had dedicated his life to fighting the "erosive and insipient design of collectivism, welfare state advocates, naïve and warped Americans who wish to destroy our basic tenets." New Mexico was still a "refuge for rugged individuals," he wrote, but only connections forged among New Mexico's businessmen would help them remain free.

In addition to communism and anticommunism, Peña, like other New Mexicans, was also interested in the relationship between environmental protection and economic development. He wanted to protect the environment of the Land of Enchantment, but only if doing so wouldn't interfere with the development of the state's ranching, timber, and mining industries. As an outspoken advocate for economic development, he became a leader of the local Chamber of Commerce and the state Republican Party. He donated cattle from his ranch to feed hungry attendees at Chamber of Commerce fund-raisers and gave money to Republican candidates whenever they asked. Later in life he led the Peace Corps in Honduras and Costa Rica, and the US Agency for International Development in Paraguay and Bolivia.

Meanwhile, in New York, the key to success for Republicans was providing economic opportunities for the city's large Puerto Rican population, and also developing relationships with Republicans in Puerto Rico. They didn't vote in US elections, but Republicans knew they influenced Puerto Ricans on the mainland.

By 1960, Luis Ferré was their leading representative. He had become vice president of Puerto Rico's Partido Estadista Republicano— the Prostatehood Republican Party. He visited New York and other places with large Puerto Rican populations to campaign for Republicans such as Rockefeller. During a visit to the White House, he also managed to convince Eisenhower to back him for governor of Puerto Rico instead of Muñoz Marin, the founder of the commonwealth. Eisenhower and Ferré were both Republicans, and Eisenhower had come around to supporting statehood instead of commonwealth status if that's what Puerto Ricans themselves wanted.

Ferré also became nationally known in the Republican Party. He traveled to the Republican national convention in Chicago to endorse Nixon and promote statehood. Seeking to capitalize on the fact that Alaska and Hawaii had become states in 1959, he made the case for why Puerto Rico should be the next addition. It came down, once again, to the politics of the Cold War in the Americas. Statehood for Puerto Rico would assure that the United States had a strategic outpost in the Caribbean. The island, he said, "would serve as a natural bridge between the United States and Latin America, and would be beneficial in every aspect." His words must have comforted the Republicans assembled in Chicago, who were nervous about developments in Cuba.

As Ferré spoke, Cuban exiles were beginning to stream into Florida. Many of them had supported Castro's revolution at first. They had grown tired of Batista's corruption and cronyism, and many agreed with Castro that the United States exerted too much control over the island. But after the revolution took a Marxist turn—expropriating lands, nationalizing businesses, limiting the freedom of the press— they turned against Castro and made the same decision to move to Miami that a smaller number including the Arnaz family had made during earlier upheavals. They staked out different positions on the best way to topple Castro and heard more from CIA spies like Howard Hunt than from Democrats or Republicans.

The Cold War in the Americas—abroad and at home, from the attacks by Puerto Rican nationalists to the military intervention in Guatemala, and from Nixon's tour in Latin America to the Cuban Revolution to Red-baiting in Congress and against immigrants—highlighted how the US role in the Americas and debates over the best way to contain communism shaped the thinking of Hispanic Republicans about their role in American politics. They also narrated their political evolution as the result of their individual and family experiences throughout the first half of the twentieth century. They were indebted to the country that gave them freedom and opportunity after their families escaped lives of poverty and other forms of oppression in Latin America. Capitalism made possible the fulfillment of their American dreams. The Democratic Party took their votes for granted.

The success Republicans found among Hispanics locally taught them they could win Hispanic votes. Republicans carried this lesson into the 1960 election. Because both parties were experiencing tremendous flux, they sought new votes wherever possible. African Americans flipped from Republican to Democrat as the South flipped from Democrat to Republican. But Hispanic Republicans across the United States had not yet figured out which issues united them and which increased their influence within the Republican Party, let alone what it meant for them to be a voting bloc of national significance. This would be their work in the future.

CHAPTER 2

A NEW GENERATION

Arizona senator Barry Goldwater's manifesto *The Conscience of a Conservative*, published in April 1960, argued that conservatism had lost its way. Goldwater's supporters commissioned the book in 1959, hoping it would launch his bid for the presidency in 1960. Ghostwritten by William F. Buckley Jr.'s brother-in-law and vetted by Goldwater adviser Stephen Shadegg, the book called for an end to the old conservatism—represented by moderates such as Dwight Eisenhower and New York governor Nelson A. Rockefeller—and the rise of a new conservatism, or, rather, a return to conservative principles that had been tested by time but observed in the breach. Goldwater didn't enter the race against Nixon, whom he openly endorsed when the speculation about his own run reached a fever pitch, but his backers got their wish in 1964, when he entered the race and became the Republican Party's nominee.

In *The Conscience of a Conservative*, Goldwater claimed that the people of the United States were fundamentally conservative, but their representatives—both Democrats and Republicans—had failed them because they had accepted the New Deal state, had abandoned fiscal restraint, and had grown comfortable coexisting with communism, with the result that they'd bargained away true freedom. He criticized the state of American politics in chapters on subsidizing farmers, compulsory union membership, welfare programs, income taxes, civil rights, and the threat of communism, from the Soviet Union to Cuba, which had fallen in the year the book was commissioned and the year before its publication.

Much of Goldwater's message of purity in the pursuit of fiscal restraint, limited government, and the defeat of communism abroad and at home resonated with a "new generation" of Hispanic Republicans who, according to a Mexican American Republican and Goldwater supporter from Arizona, were ready to abandon their decades-long fealty to the Democratic Party. Goldwater found greatest purchase among Mexican Americans in the Southwest and Cubans in the Southeast. Puerto Rican Republicans in New York still supported Rockefeller, who had some clear advantages given his experience in Latin America, the fact that he spoke Spanish, and the fact that a significant minority of Puerto Ricans had already voted for him in statewide elections. New York's small Cuban exile community was a different story. They were more conservative than the Cuban exiles in Florida, immersed in Buckley's writings, and in agreement with Goldwater that Rockefeller was too moderate.

Hispanics in 1964 saw politicians paying greater attention to their communities. Many had started to believe that their participation in electoral politics offered them the best hope for upward mobility, inclusion, and an end to discrimination. Like African Americans, they fought for change through the courts and labor unions, and now they would fight at the ballot box. Many of the most politically active Hispanics were middle-class professionals or military veterans of World War II and the Korean War who were ready to do their part, or already had done their part, to curb the spread of communism. In the early 1960s, there was some daylight between Hispanic Republicans and Hispanic Democrats on issues including immigration, civil rights, and US–Latin American relations, but not a lot. Both sought greater political power and increased representation in government.

Beyond the general message of Goldwater's book, Hispanic Republicans, in particular, were most interested in what Goldwater had to say about civil rights and communism.

The civil rights issue had inflamed the nation's passions throughout the 1950s, and *The Conscience of a Conservative* poured fuel on the fire. The book came out six years after the *Brown v. Board of Education* decision, in which the Supreme Court declared separate public schools for black and white children to be unconstitutional, thus beginning the long and incomplete road toward integration, as well as massive resistance against the decision by Southern states in particular. It was published three years after Eisenhower had sent federal troops to Little Rock Central High in Arkansas to secure the entry of nine African American students. In both instances, Goldwater argued, the Supreme Court, and then President Eisenhower, had gotten it wrong.

Not only did the Constitution not require integrated schools, but it also did not "permit any interference whatsoever by the federal government in the field of education." In the *Brown v. Board of Education* case, the justices had asserted their own will. In doing so, he claimed, they delivered a message that "what matters is not the ideas of the men who wrote the Constitution, but the *Court's* ideas." The precedent was dangerous, Goldwater wrote, because "if we condone the practice of substituting our own intentions for those of the Constitution's framers, we reject, in effect, the principle of Constitutional Government: we endorse a rule of men, not of laws." The same logic applied to the conflict at Central High, in which the president abused his constitutional authority by sending the Arkansas National Guard to intervene in a fight that wasn't theirs.

Goldwater carefully argued that his personal feelings—against segregation, for equality under the law—didn't matter. "I believe that it is both wise and just for negro children to attend the same schools as whites," he wrote, and that "to deny them this opportunity carries with it strong implications of inferiority." Yet he wouldn't "impose" his views on other states or individuals. What they did was "their business, not mine."

His words marked a pivotal moment in the turning away of African Americans from the Republican Party, and contained challenging ideas that Hispanics were forced to reckon with as well. For a long time Hispanic Republicans had argued that federal civil rights initiatives ignored them, while African Americans were the intended beneficiaries. Many also saw themselves as members of the white race, so the protection of African American civil rights may have seemed inconsequential. But many others felt the need to have their civil rights protected, since they were discriminated against just like African Americans were: through segregated schools, redlining practices that barred them from certain neighborhoods, or social norms that excluded them from other public places. Even those who supported the African American movement for civil rights, though, sometimes disapproved of the tactics used by African Americans, their loud protests and demonstrations and their radical demands for change.

Many Hispanics supported the *Brown v. Board of Education* decision, in part because Mexican Americans in California had fought and won their own desegregation battles a few years earlier. The 1947 *Mendez v. Westminster* ruling, which held that Mexican American and non-Hispanic whites in Southern California would go to school together, in fact served as precedent for the Supreme Court in 1954. But more in alignment with Goldwater's position, many Hispanics also opposed integration through busing, because they wanted to be able to send their children to schools near their homes, which put them at odds with most African Americans.

Nevertheless, the views Goldwater expressed about civil rights contributed to his extremist reputation and to the suspicion of many Hispanics that he just didn't like them. This was an idea that Goldwater's Hispanic supporters struggled to overcome.

On communism, Goldwater used the example of the Cuban Revolution to make a point about freedom, which was really a warning against the treacheries of unfreedom. In *The Conscience of a Conser-*

vative, Goldwater asserted that "The child born in America and the child born in Cuba are created equal—but because the Cuban child is born to tyranny, he cannot enjoy the freedom in which that basic equality will be respected and in which he will have the opportunity to strive for self-fulfillment." The Cuba comparison served at least two purposes for Goldwater. First, it appealed to Cuban exiles by focusing on the negative effects of Castro's dictatorship. Second, it allowed him to sidestep the fact that, if he focused on the United States instead, he would have to reckon with the same tyranny at home as abroad. To careful readers, *The Conscience of a Conservative* offered a primer for how Goldwater would discuss civil rights and anticommunism during his presidential campaign.

LYNDON JOHNSON'S CIVIL RIGHTS ACT WAS ENACTED ON JULY 2, 1964, IN THE THICK OF the campaign. After John F. Kennedy's assassination the previous fall, Johnson vowed to pass a flurry of liberal legislation that would continue the dead president's agenda. The Civil Rights Act was one of the first measures he took up. Because of how sweeping and controversial it was, civil rights, said an article in the *Dallas Morning News*, had become "the No. 1 issue" in the campaign.

Goldwater had honed his ideas about civil rights over the course of years, but he gave his first big campaign speech on the matter at the Conrad Hilton Hotel in Chicago just two weeks before the election. He stressed that the freedom to not associate was as important as the freedom to associate and that "forced integration" was "just as wrong as forced segregation." Again, the answer for Goldwater was moral leadership that encouraged individuals to change their beliefs, and this was something that the federal government could not accomplish. The core of his critique of the Civil Rights Act was therefore about preserving individual freedoms, whether that meant an individual's decision to associate with a member of another racial or ethnic group, or a decision to remain apart. He gave a version of the speech ten days later in Cleveland, as the election drew closer. Racial tensions had rocked the city the year before.

To better understand what voters on the ground in Texas believed about civil rights, the *Dallas Morning News* sent a "task force of reporters" to all corners of the state to interview whites, African Americans, and "Latin Americans," who in all probability were of Mexican heritage. Texas was a western extension of the southern states that gave Goldwater his biggest bloc of support, and one of the states he couldn't win without. Texas also had large African American and Hispanic populations, so members of both communities could be interviewed about their views of the election and civil rights. The interviews would help shed light on a matter that might determine the outcome of the election.

Reporters heard and reported a range of opinions. Some non-Hispanic whites who had been Democrats their whole lives planned to vote for Goldwater because they opposed the Civil Rights Act. Meanwhile, a "Negro nightwatchman" told one reporter, "Every Negro that can walk or crawl to the polls will vote for LBJ." Mexican Americans were more divided. Some organizations, such as the Political Association of Spanish-Speaking Organizations (PASSO), supported Johnson and the Civil Rights Act. The League of United Latin American Citizens (LULAC) supported it as well, but as the president of the organization wrote, they fought for civil rights "intelligently in a calm and collected manner," not through "sit-in's or picketing or other outward manifestations."

The views of the LULAC president were echoed by other Mexican Americans. A teacher from Austin told a *Dallas Morning News* reporter, "My people have always been sympathetic to the Negroes' struggle for civil rights, and we're Democrats." But "some Latin Americans," they continued, believe their "demonstrations and lawlessness are going too far." This particular individual may have still voted for Johnson—the reporter didn't follow up after the election—but such hesitation was exactly the sentiment the Goldwater campaign tried to exploit in order to pry Democrats from the Democratic Party. Any difference in opinion between Mexican Americans and African Americans over the Civil Rights Act was part of a longer history. Mexican Americans sometimes chose not to participate in African American protests, even if they had the same grievances, because they viewed them as disruptive and counterproductive.

Some Mexican Americans even tried to convince fellow Hispanics that Goldwater was the real civil rights champion. One in California argued that Goldwater had employed plenty of Hispanics at his Phoenix department store; helped desegregate Phoenix schools and Arizona's National Guard; and voted for the Civil Rights Acts of 1957 and 1960. In fact, he had only voted against one Civil Rights Act: Johnson's in the summer of 1964, in the thick of the campaign. And for this Goldwater had good reason, since it "took away the liberties of the majority in order to bestow them on the minority."

A Mexican American in Arizona made a similar point. Whereas the Johnson administration's version of civil rights meant giving jobs to African Americans by taking them away from Mexican Americans, Goldwater's would mean jobs for all. So Goldwater's opposition to the Civil Rights Act found some support among Hispanic Republicans, but it probably didn't afford him an opportunity to expand his support from Hispanics beyond those who saw civil rights as he saw them. Those who were with him were with him, but the majority of Hispanics who supported civil rights took pause.

Despite protests from moderate Republicans, including Republicans of African American and Hispanic descent, the 1964 Republican Party platform removed the civil rights plank that had appeared in the 1960 platform. The 1964 platform barely affirmed that Republicans would uphold the bill, with just a couple of sentences on the matter. When it came to civil rights, some Hispanics whispered a belief that became more pronounced over time, that gains for African Americans necessarily came at a cost to Hispanics, that Democrats worked for the civil rights of African Americans only, and that this was one of the reasons Hispanics ought to become Republicans.

If Goldwater's Hispanic supporters agreed with him that the Civil Rights Act bowed to the pressures of African American activists, they differed with him on the important immigration question of the early 1960s: whether to renew the Bracero Program, a binational labor agreement between the United States and Mexico that sanctioned the importation of some 4.5 million Mexican guest workers between 1942 and 1964. "Braceros" was a Spanish word that meant those who work with their arms, and it was the name applied to the guest workers brought in under the auspices of the program (as opposed to "wetbacks," a slur against undocumented workers).

In December 1963, Mexican Americans had pressured Congress to terminate the Bracero Program. It would expire in twelve months, at the end of December 1964. The announcement caused a panic among growers in California, Texas, and other areas that relied on Mexican agricultural labor. One historian called it a "legislative success" because Mexican American opposition to the program brought Mexican Americans "to the attention of liberal leaders."

Johnson's Mexican American supporters in his home state were gratified that the president "went all out" to support their effort. He didn't worry about alienating powerful growers because he believed their needs could be met through regular immigration laws instead of special programs. Growers, though, feared their crops would rot without Mexicans to harvest them, so conservatives from the Young Republicans to Goldwater himself favored the extension of the guest worker initiative. Republicans stated that the Bracero Program worked not only for American growers but also for Mexico and Mexicans, since workers sent home money that aided their communities. Hispanic Democrats, however, said Republicans supported the Bracero Program mainly because growers wanted them to.

The greatest pressure against the Bracero Program came from labor unions, including those headed by Mexican American organizers such as Ernesto Galarza and César Chávez. Democrats sympathized with the braceros, acknowledging that they were seeking jobs that paid better than ones that may have been available in Mexico. But they also argued that braceros and undocumented Mexicans, who could be hired cheaply, lowered wages for American citizens. They believed that domestic laborers would do agricultural work if growers guaranteed them a minimum wage, housing, transportation to the fields, and regular work schedules. Finally, they argued that the protests of growers were a ruse. There was no chance that their crops would go unpicked. They only wanted to secure their supply of cheap Mexican labor.

While the Republican Party line was support for the Bracero Program, Hispanic Republicans disagreed. On the Bracero Program and other avenues of legal migration, including the thousands of Mexican commuters who crossed the border every day to work in the United States, Hispanic Republicans held views that were more like those of Democrats—that Mexican workers took jobs from American citizens and lowered their wages. One of Goldwater's staunchest supporters, Robert Benitez Robles of Arizona, proposed advertisements in Spanish and English that criticized the Johnson administration for its "open door" policy allowing one million foreigners to enter the United States every year. Johnson ultimately buckled to such pressure and canceled the program.

The nature of the disagreement between the Republican Party and Hispanic Republicans over immigration would change over time—from supporting the Bracero Program or not, to charges by Hispanic Republicans of xenophobia and racism later in the twentieth century—but their differing views of the Mexican guest worker program were a signal moment of Hispanic Republican dissent. It foreshadowed how immigration would continue to be a point of tension between Hispanic Republicans and other members of their party, one that grew increasingly bitter over time.

GOLDWATER'S ANTICOMMUNISM WAS AS CONSEQUENTIAL FOR HISPANICS AS HIS OPPOSI-tion to the Civil Rights Act. To Goldwater, federal civil rights legis-lation and communism were two sides of the same coin. He argued that the Civil Rights Act was a government imposition that limited the freedom of states and individuals to make decisions about racial inclusion for themselves; similarly, communism was an affront to the freedoms Americans took for granted. Communism and anticommu-nism were central themes in *The Conscience of a Conservative*, and they would be core concerns of Goldwater's presidential campaign as well. Even though Goldwater chose not to run in 1960, he followed the election closely, and in the years after Kennedy's narrow victory—he defeated Nixon by less than one percentage point—he became in-creasingly vocal in his critiques of how the Democratic president had made a mess of the fight against communism.

Given that Americans in the 1950s had witnessed conflict after conflict between enemies, allies, or ambivalent partners of the United States in Puerto Rico, Guatemala, Venezuela, Cuba, and elsewhere—and that they sensed more trouble on the horizon—the Latin Ameri-can Cold War was a recurrent theme for both Nixon and Kennedy. Nixon assumed that Cuba and the fight against communism would be strong suits for him. He had earned a national reputation as a thirty-five-year-year-old congressman from California, only two years into the job, when, in 1948, he helped expose Alger Hiss as a Com-munist sympathizer before the House Un-American Activities Com-mittee. That reputation led Eisenhower to select Nixon as his running mate.

Nixon's crusade against communism defined his years as vice president as well. During his campaign against Kennedy, he touted Eisenhower's (and his) accomplishments in Latin America. Seemingly forgetting his reception in 1958, he focused on how there had been eleven dictators in the region in 1952, but by 1960 there were only three, including Castro. Just a few months after Castro took power, Nixon traveled to Moscow to attend the American National Exhibition, where he famously showed the Soviet leader Nikita Khrushchev the sort of kitchen he would find in many households in the United States. It was supposed to demonstrate the supremacy of American capitalism and modernity. His world travels made it possible for him to claim a wealth of international experience combating communism.

But Kennedy didn't concede that Nixon was the anti-Communist candidate. He recognized the dangers of communism as well, but said his solutions would be more effective. Kennedy reminded voters of Eisenhower's failures in Latin America, especially the fact that a neighboring island in the Caribbean was apparently lost to communism. Castro was so aggressive, he argued, because Eisenhower and Nixon had backed vicious dictators such as Fulgencio Batista, who killed tens of thousands of his opponents. Only following Nixon's disastrous tour did Eisenhower turn to economic assistance in addition to sanctions and military intervention, whereas loans and development programs to combat poverty, he said, would be the foundation of his approach.

Nixon and Kennedy sparred on the campaign trail and during four televised debates in September and October 1960—the first in history—over who had the greater resolve to defeat communism and keep peace in the hemisphere. In every debate, they clashed over Cuba, the subject that gave rise to some of their most heated exchanges. They disagreed over whether the United States should intervene militarily, what economic sanctions had accomplished, and how to win Latin American allies willing to help the United States isolate the island.

Kennedy stirred controversy when he suggested he would support the efforts of Cuban exiles and anti-Castro Cubans to overthrow the dictator. Nixon feigned shock at what he called Kennedy's dangerous suggestion given the nonintervention pact agreed to by the members of the Organization of American States—including the United States—even though he had suggested a covert plan for exiles to invade Cuba that Eisenhower had already set in motion. Kennedy revised his statement. He said he supported nonintervention, too, but wanted to make sure Americans knew that as soon as Castro stepped out of line, he would respond, perhaps with military force. Goldwater watched and thought the stances of both men were too moderate; they shouldn't be so cautious about using the military in Cuba or any other Communist country.

Kennedy kept up the anti-Communist drumbeat in his inaugural speech on January 20, 1961. He said that the United States would "pay any price, bear any burden, meet any hardship, support any friend, oppose any foe to assure the survival and success of liberty." He referred not only to the Soviet Union and the Vietnam War, to which a growing number of US troops was committed, but also to conflicts in Latin America. A little over a month into his presidency, he gave a speech that established the Alliance for Progress, an economic assistance program that was often referred to as a Marshall Plan for Latin America. At the same time, he was making final preparations for the armed invasion of Cuba he inherited from Eisenhower. It would be Kennedy's first armed strike against communism in the Americas, but it wouldn't be his last. In his brief time in office, Kennedy attacked Cuba, Haiti, the Dominican Republic, and Guatemala. His public promises of economic assistance ran alongside these covert military maneuvers, demonstrating that he and Eisenhower didn't differ substantially; the Republican and Democratic presidents ended up approaching countries that did and did not stand for democracy and freedom in much the same way.

Goldwater said his approach would have differed substantially. He supported the use of the military to oust Castro, but used a bungled attempt to invade Cuba as the basis for his critiques of Kennedy and, after Kennedy's assassination, Lyndon Johnson. The failed invasion fueled Goldwater for years, and it would fuel the Republican Party's success among Cuban Americans, in particular, for decades longer.

On April 13, 1961, more than a thousand Cuban exiles, backed by the CIA, left Guatemala, Nicaragua, and Panama by boat and by plane. They were headed for Cuba to launch an attack against Fidel Castro. Planning began in March 1960, when Vice President Nixon suggested the plot to Eisenhower. After recruiting the exiles, the CIA began training them in Central America and at US bases including Fort Knox in Kentucky, Fort Benning in Georgia, and Belle Chasse Naval Air Station in Louisiana.

The recruits formed Assault Brigade 2506, the serial number assigned to a comrade who died during training. They made landfall on April 17. Their assault on Playa Girón and the Bay of Pigs, a little more than ten miles apart on the southern side of the island, was in almost every way a failure. The CIA had charged the Cuban Revolutionary Council, made up of leaders from the Cuban exile community in Miami, to form a "government in exile" that would run Cuba after they overtook the island.

They never got the chance. The landing area hadn't been surveyed properly. Boats sank in shallow coral reefs or were turned back by the Cuban Air Force. Counterrevolutionary groups on the island planned to back up the invaders, but the CIA never said when. Finally, at the last minute, Kennedy called off US air cover for the exiles without telling them. As a result, more than a hundred exiles died and more than eleven hundred were captured.

Cuban exiles back in Miami huddled together in churches and parks, awaiting news of the invasion. When they learned what happened, they begged Kennedy to intervene. When he didn't, they felt betrayed. The invading rebels were held in Cuba for almost a year and a half, until December 1962, when they were ransomed and returned to the United States in exchange for $53 million worth of baby food, powdered milk, pesticides, and medicine and medical equipment. Other incidents continued to inflame US-Cuban relations during the Kennedy years, including the Cuban Missile Crisis of October 1962, a thirteen-day standoff between the United States and Cuba caused by the discovery of Soviet missiles on the island, and Operation Mongoose, which authorized covert attacks against Castro subsequent to the Bay of Pigs. Yet it was the Bay of Pigs that cast the longest shadow over Cuban exile politics.

The failed invasion became a rallying cry and call to arms for Cuban exiles and Republicans, who used the episode as an object lesson in the dangers of communism's spread in the hemisphere, and the Democratic Party's ineptness in dealing with it. Exiles took their first steps toward the Republican Party. Leaders such as Manuel Giberga and José Manolo Casanova became increasingly involved in politics. Giberga was celebrated as "Washington, DC's ambassador to Cuban exiles." He earned widespread acclaim for securing visas for more than twenty-five thousand exiles around the world who were granted entry into the United States. In time, Casanova became a leader of the Republican National Hispanic Assembly.

The failure of the Bay of Pigs operation, the withering criticism of Castro, and the failures of Democrats to deal with the Communist leader became some of Goldwater's most consistent talking points in 1964, both in the primary race against Rockefeller and in the general election campaign against Johnson. His political career, from the time he became a member of the Phoenix City Council in 1949, had always had an anti-Communist bent. As a presidential candidate, he latched on to Cuba as a target of his anticommunism.

Perfectly echoing the Cuban exiles, Goldwater said that Kennedy had betrayed them by not sending the promised backup at the Bay of Pigs. Moreover, Kennedy's concessions during the Cuban Missile Crisis were a "complete bungling" of the "Cuba situation." The president may have avoided a nuclear attack, but in doing so he negotiated away his ability to respond aggressively to the continued buildup of Soviet troops and MiG fighters, which were permitted according to the agreement between Kennedy and Khrushchev because they were "defensive" rather than "offensive" weapons. Goldwater scoffed at the distinction.

Once Johnson took office after Kennedy's assassination, Goldwater charged that he, too, was soft on communism. Johnson underestimated the threat posed by Castro, but Goldwater said he recognized that the island was a "cancer eating away at the security of the entire hemisphere." He demanded a Senate investigation of the "disgraceful" Bay of Pigs episode and backed proposals to set up a Cuban government to rival Castro's on the US naval base at Guantánamo. He said he would "train refugees and mercenaries" to carry out another "assault on the island," and he couldn't promise that he wouldn't drop an atomic bomb on Cuba. These were the kinds of statements that led Goldwater's critics to claim he was too extreme to become president of the United States. He might launch a nuclear attack, or cede decision-making authority to trigger-happy generals who would launch one for him. But for Hispanic Republicans, his extremism was the reason for their support.

Hispanics whose political views were shaped by Cold War anticommunism—which is to say, many of them—were taken with Goldwater's tough rhetoric against leftist leaders in Latin America and the Caribbean, especially Castro. Goldwater capitalized on the Bay of Pigs debacle, which helped the Republican Party dominate among Cuban Americans for a generation because they blamed the fiasco on Kennedy and the Democrats. These were both lessons that future Republicans would learn from Goldwater's Hispanic campaign.

THE 1964 ELECTION WAS THE FIRST RACE BETWEEN TWO WESTERNERS WHO CLAIMED A personal connection to Hispanics. Johnson was from the Texas Hill Country and had been a schoolteacher in the small Texas town of Cotulla, south of San Antonio, in the heavily Mexican Rio Grande Valley. Meanwhile, Goldwater was from the Arizona borderlands. Exaggerating by a hundred miles or so, he liked to say that he "was born on the Mexican border" and that he "spoke Spanish about the same time that I spoke English." Their personal connections led both Johnson and Goldwater to believe that they were well positioned to receive Hispanic support.

By the time of the 1964 campaign, Republicans and Democrats had come to believe that it was important to appeal to Hispanic voters. The Hispanic population was small but growing, and the political balance of power in the country was shifting, not only from the East to the West, but to the Southeast as well. Texas and California became battlegrounds as Texas trended Republican and California trended Democratic. Florida became increasingly important as Cubans reshaped politics there.

Because of these broader political shifts, both parties had reasons to want to extend their minority base beyond African Americans; having lost their hold on the South, Democrats in other parts of the country had begun to forge multiracial coalitions of supporters, while Republicans, having lost support from African Americans, needed to find it from other minorities. One historian has described the Goldwater campaign's decision to pursue Mexican American voters as a "Southwestern Strategy," a necessary counterbalance to the Republican party's pursuit of white voters and abandonment of African American voters in the South.

The Kennedy campaign in 1960 in some ways provided a model for success. Part of Kennedy's success among Hispanics stemmed from his personal appeal, including his military service, which resonated with Hispanic veterans; his Irish background, which, he suggested, made him familiar with the plight of immigrants and ethnics; and the facts that his wife spoke Spanish and he was Catholic. But it also stemmed from the fact that a group of Mexican Americans in the Southwest joined forces with Puerto Ricans in New York to court Hispanic votes for him.

When they headed to the 1960 Democratic national convention in Los Angeles, Hispanic Democrats didn't know they would leave united behind Kennedy. Hispanic attendees included a Democratic senator from New Mexico, Dennis Chávez; San Antonio City Council member (and soon-to-be congressman) Henry B. González; American GI Forum founder Hector P. Garcia; and Los Angeles City

Council member Edward Roybal, who soon became the first Hispanic congressman from California since the nineteenth century. They arrived at the convention divided between Johnson, who was the choice of Hispanics in Texas, and the establishment candidate Adlai Stevenson, who had been Eisenhower's opponent in 1952 and 1956.

While there, they cornered Kennedy's brother and campaign manager, Robert Kennedy, to discuss a "voter turnout effort" for Mexican Americans, as one historian has described it. Kennedy's brother liked the idea, and mentioned the possibility of high-level appointments for Mexican Americans if they won the election with significant Hispanic support. That was enough for them to shift their support from Johnson or Stevenson to Kennedy. The Viva Kennedy clubs were born of this meeting between Kennedy's brother and Mexican American leaders, and only later did they include Puerto Ricans from New York. After the convention, the leaders of the major Hispanic organizations all got involved, attending a conference in October 1960 at the Waldorf-Astoria in New York to solidify their partnership. In the end they helped Kennedy win some 85 percent of the Hispanic vote.

Nixon didn't have a strong Hispanic campaign in 1960, but Goldwater could learn from Kennedy. When Nixon campaigned in the Southwest, he heard the same "Viva Nixon" cries that he had heard in some Latin American countries. He also campaigned in New York, where he had help from Puerto Rico's leading advocate for statehood, Luis Ferré, who stumped for him there. He also had his Cuban American friend Charles "Bebe" Rebozo, who could have given him an inroad in Florida. But the Cuban population there in 1960 was still small, most Cubans hadn't become citizens, and Florida still had only ten electoral votes. To be sure, Nixon had his Hispanic supporters, but they didn't come together as Kennedy's had.

Kennedy's Hispanic supporters had great expectations of what he might do for them in return for their support. They expected patronage in the form of economic programs that would benefit their communities and appointments to domestic agencies and diplomatic posts. Many of Kennedy's Hispanic supporters made the same arguments that John Flores had made to Eisenhower, that Hispanics should be assigned to Latin American countries because they were familiar with their languages and customs. Some opportunities did come their way, but many did not, which again opened the door for Republicans to charge that the president (and Johnson after Kennedy) had failed to keep his promises to them.

Still seeking his opportunity for political influence, Flores, the founder of the Latinos con Eisenhower group, inserted himself into the 1960 and 1964 elections. In 1960, he supported Johnson in the primaries, but later claimed to be the "national cochairman" of the Viva Kennedy clubs. A leader of the Viva Kennedy campaign said that he didn't recall Flores working with them. Then, after Kennedy's victory, Flores issued a "seventeen-point grievance proclamation," which stated that Kennedy had let down "millions" of Mexican American supporters by failing to appoint enough Mexican Americans to his administration. He said he had rallied Hispanics throughout the Southwest to express their disappointment with Kennedy's "New Frontier," which he found to be not so new after all. Flores also noted that Mexican Americans had fewer opportunities for government employment compared with other minorities, especially African Americans. One of Flores's allies in Colorado said that Kennedy's emphasis on "Negro recognition" aggravated the "problems faced by Latin Americans." Hispanics "used to be second class citizens," he said, but "now we're being relegated to third class status."

Flores had backed Eisenhower before he didn't, then backed Kennedy before he wouldn't again. In 1964, he formed a group called Latinos con Goldwater. His Hispanic critics—there were plenty of them—called the group a "phony publicity stunt" and said Flores was a self-serving scam artist. Maybe so. He had switched sides many times, and apparently claimed credit for things he hadn't done. He had also listed multiple addresses and job titles, from California to Washington, DC. All of these things called into question his focus, commitment, and even his identity. Another possibility, though, is that Flores was playing the field, courting support for Hispanics wherever he could find it. This would mean that his goals weren't all that different from the goals of other Hispanic Republicans, who wanted Democrats and Republicans alike to fight for their votes.

The Flores drama aside, Goldwater recognized the success of the Viva Kennedy movement and aspired to run a similarly effective Hispanic campaign. Instead of launching a truly national effort to recruit Hispanic voters, Goldwater focused on two groups in two regions: Cuban Americans in the Southeast and Mexican Americans in the Southwest. The Nationalities Division of the Republican National Committee—the purpose of which, according to the *Washington Post*, was to recruit naturalized immigrants from "countries under Communist rule or nations threatened by Red Subversion, such as Latin America"—appointed a Cuban American as the head of a "Cuban-American Affairs" committee in the Southeast, and a Mexican American as the head of a "Hispanic Division" in the Southwest. Stanley Ross oversaw their efforts as the division's "coordinator of Spanish-speaking sections." Ross himself was not Hispanic, but he had written several books on Latin America; was the editor of Spanish-language newspapers in New York, including *El Tiempo*, *El Diario de Nueva York*, and *La Prensa*; and had organized several Amigos de Goldwater clubs.

Fernando Penabaz led the Cuban-American Affairs committee. Born in Baltimore in 1916, Penabaz spent the early decades of his life moving back and forth between the United States and Cuba. Upon settling in Florida after Castro's revolution, Penabaz became a "writer in exile" for several newspapers. Virtually all of his columns were about Cuba, US–Latin American relations, the failures of the Democratic Party, and communism and anticommunism. Cuban exiles like him, he wrote, were baffled by American politics. They left the island believing that all Americans were anti-Communist, but upon arrival they learned to their "great dismay" that the United

States was, in fact, "dominated" by liberals who were "almost generally pro-Communist." He therefore organized "teams" of exiles—many of them housewives—to make phone calls, mail pamphlets, and hand out campaign literature to encourage fellow exiles to vote for Goldwater. Penabaz was the ideal Cuban American surrogate.

Out west, the Californian Manuel Ruiz served as the national chairman of the Hispanic Division. Born in Los Angeles in 1906, Ruiz was part of an older generation of Hispanic Republicans that was already politically active by World War II, when he criticized the Mexican American participants in the Zoot Suit Riots of 1943, during which Mexican American youth were subjected to harassment and violence for wearing clothing that was seen as subversive. By 1964, Ruiz had earned a reputation as the "dean of the Latin-American attorneys," having worked for US lawyers whose clients did business in Mexico, Central America, and South America, to help them understand the fundamentals of the law there, and for Latin American lawyers with clients who did business in the United States. Republican candidates in California solicited his help recruiting Hispanic voters. For the Goldwater campaign, he would organize Mexican American voters in California, Texas, Arizona, New Mexico, and Colorado.

Ruiz had Robert Benitez Robles working with him, as the Hispanic Division's main representative in Arizona. Robles was from Yuma, in the southwestern corner of the state, where Arizona, California, and Mexico met. Like Penabaz, he was a lawyer by training, with a law degree from the University of Arizona in Tucson. He had served in World War II, spent two years toiling in copper mines, worked as a teacher, and at the time of Goldwater's campaign sold insurance for New York Life. Goldwater chose Robles over the Tucson department store owner Alex Jácome Jr. As he explained in a letter to a friend in Mexico, he chose Robles because Jácome was a "rich man" who didn't "represent the average paisano." Goldwater still had high regard for Jácome, but said that he needed someone who was "nearer Bob's level." Robles stumped for Goldwater throughout the

Southwest; wrote Spanish-language advertisements; and helped the leaders of the Nationalities Division formulate strategy.

Robles was perhaps Goldwater's leading Hispanic advocate. In many ways he was a more dogged supporter than Ruiz, who had flirted with Rockefeller before signing on to work with Goldwater. In a speech at Knott's Berry Farm, the Southern California theme park owned by the prominent Republican Walter Knott, Robles proved that his anti-Communist vitriol matched Goldwater's. His speech was on the subject of "extremism." Anyone who criticized Goldwater, he said, was a "left wing, Communist-inspired" propagandist. Goldwater's critics hacked away at the "mighty Goldwater tree of human decency," but, thankfully, Goldwater's supporters refused to be "brainwashed." The Soviet Union forced people into slavery and held them in prison camps. Closer to America's shores, Fidel Castro was an assassin, a torturer, and a fierce dictator whom Johnson allowed to install Soviet missiles and military bases just off the Florida coast.

Goldwater also had support from a Cuban American in California named Tirso del Junco, a doctor who had settled in the Los Angeles area about a decade before the Cuban Revolution of 1959. He attended the University of Havana with Fidel Castro and became increasingly active in Republican politics after the Bay of Pigs invasion. He preached all across Southern California about the ills of communism, forming a group of Goldwater supporters called Republicans of Latin Extraction, which operated alongside Latinos con Goldwater and other groups that supported the senator. Two decades later, del Junco became a leader of the Hispanic Republican movement.

Goldwater sought support from Puerto Ricans who largely favored Rockefeller, but he didn't launch the same coordinated effort among them that he did among Cuban Americans and Mexican Americans. Puerto Ricans had predicted in 1960 that Rockefeller could have won 45 percent of the Puerto Rican vote there if he had been the party's nominee instead of Nixon. In 1964, when Rockefeller competed against Goldwater in the primaries, they claimed that

Rockefeller could have won 50 percent if he were to face Johnson in a general election. But conservatives within the Republican Party had spoken; they wanted someone who stood against civil rights and organized labor, and who would punish the left in Latin America. That wasn't Rockefeller, but it described Goldwater perfectly.

Knowing how influential Puerto Ricans on the island could be with their compatriots on the US mainland, Goldwater saw his strongest play for Puerto Rican support not in New York, but rather in his effort to secure Puerto Rico's five convention votes. After Goldwater learned that Puerto Rico had pledged these votes to Rockefeller, he announced that he, too, favored Puerto Rican statehood "when its residents desire." It worked. The head of the Republican Party of Puerto Rico, Miguel García Méndez, who was Luis Ferré's brother-in-law, said that Puerto Rican delegates would not commit their votes before the national convention. If Goldwater could get their votes, he believed, that might help him with Puerto Ricans in New York.

Goldwater's appeal to Puerto Rican Republicans on the island was just one instance of his effort to seek support from Hispanics and Latin Americans who were unable to actually cast a vote in the election. An expat in Mexico who was friends with Goldwater offered the rather limited perspective that Mexicans on both sides of the border supported the Arizonan because they were against handouts and didn't like "dirty people." Goldwater had developed personal and business relationships with Mexicans and Mexican Americans in the Southwest and northern Mexico, and he hoped that his associates there would encourage friends and family members who were US citizens to vote for him. His Mexican friend also observed that Mexico, with the election of Gustavo Díaz Ordáz as president, had become a "right of center" country, so if Goldwater won, the two countries would "see closer relations than ever before."

This last point caught Goldwater's attention, because throughout his career, he sought closer relations with Mexico and other Latin American countries. He saw friendship and cooperation as the best tools against communism. If the people "on this great land mass of North and South America" could "be combined in unity of purpose and feeling," he wrote, "it could well be the bulwark that would end communism around the world."

Goldwater appealed to Cuban exiles as well, even though they hadn't become naturalized and were therefore unable to vote. Like Puerto Ricans on the island, they held great sway among the Cuban Americans in Florida who had naturalized, so Goldwater believed that exiles would help funnel them in his direction.

GOLDWATER FELT HE WOULD CLINCH THE NOMINATION WITH SUPPORT FROM HISPANICS, and that he would make a good showing among them in the general election. He didn't come close. Despite Goldwater's appeal to fellow anti-Communists, his upbringing in the US-Mexico borderlands, his facility with Spanish, and loyal supporters such as Robles, he won single-digit support from Hispanics; even less than Nixon had won in 1960. Johnson more than neutralized Goldwater's appeal to Hispanics with his Viva Johnson clubs, which had been modeled on the Viva Kennedy clubs. He also touted his support for civil rights and other Great Society programs. The World War II veteran and American GI Forum leader Vicente Ximenes headed the pro-Johnson effort, and by the fall some three hundred Hispanic groups supported him.

Even in the thick of the campaign, Robles, the Mexican American Goldwater supporter from Arizona, made some clear-eyed assessments of how Goldwater might fare. In his conversations with fellow Hispanics, he said he encountered an "alarming number" who believed that "Goldwater dislikes Latins," and added that he was doing "all within my power to change this image." He confirmed his suspicion that Democrats were ignoring Hispanics because they felt they could count on Hispanic votes, but he found that Republicans ignored them as well. He had invited many Hispanics to a meeting in Texas to discuss "issues of importance to Latins." But only two showed up, compared with twenty "Anglos." His conclusion was that there was only a "slight chance to win over some of the Latins this year."

Toward the end of the race, Goldwater's Hispanic supporters already looked to the future, and found reasons for optimism. Disregarding the precedent set by the Viva Kennedy clubs, Manuel Ruiz wrote in a letter to the prominent New Mexican Carlos Sedillo, "The Republican Party is the first Party that has given us an opportunity to open channels upon a national level with relation to the Hispanic segment of our citizenry." Ruiz proposed that "as soon as the elections are over and things get back to normal" they should get together with other Hispanic Republicans to share their experiences with the Goldwater campaign and plan their "future activities."

Goldwater's Hispanic supporters did what they could for him. Their radio ads highlighted his promises to lower taxes, stop wasteful spending abroad, end the Vietnam War, and defend the United States against communism. Goldwater was a true friend to Hispanics. He was the candidate who "knows and loves" them. But their efforts weren't nearly enough. The election was so lopsided that a better showing among Hispanics hardly would have mattered. Johnson won the popular vote by the largest percentage since the uncontested election of 1820 (61 to 39), and the electoral college with more votes than anyone since Roosevelt in 1936 (486 to 52).

Even though Goldwater lost, the election of 1964 was an important milestone for Hispanic Republicans. For the first time ever, a Republican candidate had done better among Hispanics than among African Americans. That would become the norm. Also, Goldwater may have won a small percentage of the Hispanic vote, but he inspired loyalty among those who did vote for him, and his followers became part of the Hispanic Republican base going forward. Sedillo said he would "go anywhere in this Country and Puerto Rico, at my own expense, and speak to gatherings of Americans of Cuban ancestry in Florida, Puerto Ricans in New York, Mexican-Americans in California, Texas," and "anywhere in the Nation where the Gold-water candidacy may need us."

A couple of weeks after the election, Robles wrote a postmortem analysis of how Hispanics had voted and why. His main takeaway was that Republicans couldn't expect to succeed if they tried to gather support only in the last couple of months of a campaign. Going forward, Robles argued, they should make a sustained effort to let Hispanics know they would be there "day-in-and-day-out." He sounded a lot like Stuart Spencer, who helped Nixon in 1950 and 1960. The Hispanic Division was a start. It was created for the 1964 campaign, but remained active afterward with an eye toward linking Hispanic Republicans nationally and convincing them to "jump the fence and vote Republican" in 1968.

Robles offered a playbook for Republican recruitment efforts. Their events didn't have to be anything fancier than a "simple wiener roast," but organizers had to be sincere. Republicans had to "act natural" and "beware of phony gestures." After making initial contact, they should go to the homes of Hispanics and invite them to their next gathering; "show an interest in their everyday living"; "give them precinct work to do"; "ask for their opinion"; and "make them feel important and wanted." Once Republicans accomplished these things, they could talk with Hispanics about issues, and when Hispanics grasped the issues, they would study them like they "study their catechism." They would "preach" Republican positions because they were a loyal and "most appreciative people." That had been part of the problem.

Kennedy and Johnson received a great deal of credit for recruiting Hispanics through their Viva Kennedy and Viva Johnson campaigns, Robles believed, and because Hispanics were a loyal people they were already in the bag for Democrats in 1960 and 1964. In a letter to Goldwater after the election, he offered a history lesson to prove his point. Hispanic support for Democrats stretched back to the New Deal, when Roosevelt "fed them during the depression." Out of reverence and appreciation, they hung photos of him next to images of La Virgen de Guadalupe. They repaid Democrats with their blind devotion.

Over time, Mexican Americans began to discriminate against fellow Hispanics who identified as Republicans. In a letter to a group of Mexican American Democrats in Arizona who had invited Robles to speak at one of their meetings, he criticized them for saying that they "wouldn't do anything to help a 'Mexican Republican.'" When he declined their invitation, he also chastised them. "You discriminate against your own people," he said, "simply because their political philosophy may differ from yours."

But according to Robles, these Hispanic Democrats weren't loyal to the Democratic Party because they believed Kennedy and Johnson had done so much for them. In fact, Robles, like many Hispanic Republicans, felt that Kennedy and Johnson and all Democrats were talk without action. They appointed few Hispanics to posts in their administrations, and they delivered on their promises only during campaign season, when they were looking for votes. (Over time, they would accuse Republicans of the same thing.)

Nevertheless, Hispanic Republicans had to reckon with Goldwater's "shellacking," as one Goldwater adviser put it. The "ethnic campaign" didn't work, but the Republican Party still had to find new ways to find new voters. They had no choice, because they were losing African American votes. Robles put it bluntly: Republicans needed "the Latin vote" to "counteract the loss of the Negro vote." The sentiment was widespread. Manuel Ruiz also wrote, "The Hispanic American vote is needed to neutralize the Negro vote which will vote democratic." African Americans were slowly but surely turning away from the Republican Party. Goldwater's opposition to the *Brown v. Board* decision, busing, and the Civil Rights Act of 1964 hastened their exit. But Hispanic Republicans weren't like African American Republicans. They didn't reject the Republican Party wholesale in the 1960s, and they wouldn't in the future.

AFTER HE DEFEATED GOLDWATER HANDILY, JOHNSON CONTINUED WITH THE WAR ON POV-
erty initiatives he began after Kennedy's assassination, including the
establishment of the Office of Economic Opportunity's Jobs Corps,
Head Start, and Community Action Programs. Liberalism seemed to
dominate, and the conservative movement seemed vanquished.

Hispanic Republicans had mixed views of Johnson's Great Society. One group in California recognized its potential to "elevate"
Mexican Americans who "struggle in the morass of poverty to a position of increased self-reliance." But the group's members also viewed
the Great Society as a "boondoggle" designed to further enrich already wealthy bureaucrats. They didn't recommend ending the War
on Poverty, but rather called for the hiring of more Mexican Americans as War on Poverty administrators in the areas its programs were
meant to benefit.

In addition to the Great Society programs, the Immigration and
Nationality Act of 1965 became an important piece of the legislative
deluge that was the Eighty-Ninth Congress, which had a Democratic
supermajority in both chambers. Largely seen as a liberal law, the
roots of which could be found in post–World War II arguments for
pluralism and calls for racial equality during the Cold War, it did
away with the discriminatory national origins quotas in the Immigration Act of 1924. But it also placed the first restrictions on immigration from Western Hemisphere countries including Mexico, which,
as a result of the law, would be able to send a total of 120,000 immigrants to the United States per year, and no more than 20,000 from
any one country.

The law passed with bipartisan support in the Senate, by a margin
of 76 in favor, 18 against, and 6 abstentions. Democratic supporters
of the Western Hemisphere quotas, which were squeezed in at the
last minute, argued that they reflected the equality inherent in the
law. It wouldn't be right to allow unlimited immigration from Western Hemisphere countries while restricting immigration from others.
The Western Hemisphere quotas also placated Democrats who had

argued for a long time—at least since debates over the McCarran-Walter Act—that Mexican immigration needed to be curbed as the number of undocumented Mexicans arriving without papers increased, and threatened to rise even more with the termination of the Bracero Program.

Republicans, including Hispanics, had differing opinions. Growers and other employers of undocumented immigrants had always enjoyed the benefits of cheap Mexican labor, so the Republicans who represented them in Congress argued against restrictions, including the Western Hemisphere quotas. Nativist Republicans also argued against the law, but on different grounds. They objected to its family reunification provisions because they foresaw—accurately, it turns out—that the numbers of immigrants from Asia, Africa, and especially Latin America would increase sharply. Naturalized immigrants could bring their closest relatives, who wouldn't count against the overall allotment. Hispanic Republicans generally sided with other Republicans who opposed restrictions, but they also opposed nativists in their own party.

On October 10, 1965, exactly one week after Johnson signed the new immigration law at Ellis Island, Castro announced that he would allow Cubans to leave. The number of Cubans in the United States jumped. In response to the new arrivals, who overwhelmed the immigration bureaucracy, Congress, on November 2, 1966, passed an amendment to the 1965 law that said Cubans who had been present in the United States for two years could apply to become an "an alien lawfully admitted for permanent residence." The date recorded on their applications would be "thirty months prior to the filing of such an application," to expedite their status adjustment. This was the Cuban Adjustment Act.

Airlifts from Havana to Miami had begun shortly after Castro's announcement, and by January 1968 two flights per day, five days per week, were bringing planeloads of Cubans to Miami, where public and private agencies helped them resettle. They arrived at a rate of 860 per week, or 44,720 per year. The Western Hemisphere quotas were to take effect on July 1, 1968, and in their final report to the president and Congress, the Select Commission on Western Hemisphere Immigration recommended that Cubans not be included in the 120,000 quota ceiling because of their special status as refugees. Five members of the commission wrote a dissenting minority report but were overruled. A powerful lobby of Cuban exiles fought for decades to help Cuban refugees maintain their special status.

EVEN IN THESE YEARS OF DEMOCRATIC ACTIVISM, LAWMAKING, AND POWER, THE GOLD-water supporter Bob Robles was confident that Hispanics would play an important role in the revival of the Republican Party. They were spread across the country and still had little connection to one another, but Robles believed they could come together to support Republicans by the time of the midterm elections of 1966, and certainly the presidential election of 1968, if "Republicans merely invite them into the fold and give them work to do and treat them as though they are needed and appreciated." Inviting Hispanics into the fold was already working at the state level, especially in New York, Texas, and California. The national party had only to catch up.

In New York, the two-term governor, Nelson Rockefeller, took a hiatus from presidential politics and was running for reelection against the Democratic president of the New York City Council. Rockefeller had already done well among the city's Puerto Rican voters in earlier campaigns, but in 1966 he won a third to a staggering 50 percent of their votes, depending on the precinct. The secret to his success was his embrace of a liberal agenda that included support for Medicaid, the first state minimum wage law in the country, and affordable housing. He also promised to wage war against increasing drug use by African American and Puerto Rican youth, whose parents wanted to rid their communities of narcotics. That is, Rockefeller hewed to the moderate conservatism that seemed to fall by the wayside when Goldwater won the 1964 nomination.

Also in 1966, Senator John Tower's Hispanic supporters had successfully come together as the "Mexican American Republicans of Texas." The Lone Star State hadn't had a Republican senator since Reconstruction, but that changed with his election in 1960. Before his victory, liberal and conservative wings of the Democratic Party jockeyed for position with Hispanic voters. To Hispanics, "supporting a Republican was anathema," one historian wrote. But Tower began making inroads among Mexican Americans, especially in each of his reelection campaigns.

The key had been on-the-ground organizing in Corpus Christi, San Antonio, and El Paso. Echoing Robles after Goldwater's loss, Hilario Sandoval, the Mexican American chairman of the El Paso Republican Party and future Nixon administration official, said Republicans had to show Mexican Americans that they "really care." Terms such as "liberal" and "conservative" didn't mean anything to them; Republicans had to make a more personal appeal. A little Spanish on the campaign trail wouldn't hurt, but not if spoken by a non-Spanish speaker who had been coached to say a few words. Republicans also needed to remind Mexican Americans that their lives hadn't improved under Democratic rule. Their "standard of living" was "lower than the downtrodden Negro community," Sandoval said, and urged them to return to the party they supported from Reconstruction until the New Deal.

With the help of Sandoval and Celso Moreno of Corpus Christi, Tower won 30 percent of the Mexican American vote in 1966. His margin improved in 1972 and 1978. Hispanic Republicans for decades to come—especially, but not only, in Texas—recognized him as one of the figures who first brought them into the Republican Party.

Finally, in California, Ronald Reagan successfully courted Mexican Americans during his 1966 gubernatorial campaign. Reagan had been one of Goldwater's surrogates during the 1964 campaign. The "Time for Choosing" speech he delivered on Goldwater's behalf launched him to national political fame. He repeated Goldwater's themes of individual rights and freedoms in his own run for office.

No Democrat had been elected as the governor of California between the 1890s and the 1950s. As one historian has put it, "California was a solidly Republican state in all of its politics and was known as such." But that changed with the election of Pat Brown in 1958. From that year forward, the state's Republican leaders, including Reagan, ran against what they considered to be the tide of liberalism that had taken over their state, with its support for organized labor, big government, and student activism.

Reagan appealed to middle-class Mexican Americans in particular by saying that the Republican Party would protect their hard-earned property rights. Like other conservatives in the state, Mexican American Republicans sought to overturn the Rumford Fair Housing Act, passed in 1963 to end housing discrimination. They didn't want anyone to be able to tell them whom they could rent or sell to. Reagan also promised to defend their right to attend bilingual neighborhood schools instead of getting bused to schools outside of their communities. That was how he cleaved Mexican Americans from African Americans, who supported integration through busing.

What seemed like a period of liberal dominance between 1960 and 1966 gave way to a period seen as one of liberal decline. In the midterm elections of 1966, Republicans picked up forty-seven seats in the House of Representatives, three seats in the Senate, and eight governorships.

Puerto Rican Republicans were also on the move. In 1967, Puerto Ricans participated in the island's first plebiscite to gauge how they felt about the status question. The prostatehood position lost, but garnered more support than expected. This led to a split between the two leaders of the statehood movement, Miguel García Méndez, who remained the leader of the Partido Estadista Republicano (the

Republican Statehood Party), and Luis Ferré, who founded the Partido Nuevo Progresista (PNP, or New Progressive Party). As the PNP and Ferré became the leading proponents of statehood, the Partido Estadista Republicano folded. The period was also a high point for the proindependence movement, which had the support of activists on the island and the mainland. But those who favored independence were a small number compared with those who favored statehood or commonwealth status.

Republican success stories in particular places didn't have clear lessons for the Republican Party nationally. Rockefeller, Tower, and Reagan were different candidates, and New York, Texas, and California were different states. Victory in one state wouldn't necessarily translate to victory in another, let alone success across the country. Moreover, if any one of their paths to success was embraced as the best path forward nationally, it would have taken the Republican Party in different directions, toward moderation or conservatism, or toward a northeastern or a southern or a western strategy. Still, they demonstrated again that Republicans could find success among Hispanics, so Hispanic Republicans and the Republican Party endeavored to figure out how their experiences could be like individual markers that, when seen from above, offered a legend for how to recruit Hispanic Republicans wherever they were.

Over time, Mexican American, Cuban American, and Puerto Rican Republicans began to articulate a set of ideas that brought them together. They were sympathetic to many of the issues that Hispanic Democrats fought against, including high dropout rates, employment discrimination, unaffordable housing, and lack of access to government programs. They grew up together in the same communities, and many considered themselves to be civil rights activists. Still, they criticized the Democratic Party for ignoring Hispanics. Hispanic Republicans didn't come up with this argument themselves. Leading Democrats helped them.

Edward Roybal, a Democratic congressman from Los Angeles,

grew frustrated with Johnson. He cited the lack of diplomatic posts in Latin America and scant opportunities for Hispanics to get government jobs. The Equal Employment Opportunity Commission itself needed to "practice equal employment," he said. Roybal stated that the president began to ignore Hispanics as soon as the polls closed in 1964.

He became particularly enraged when no Hispanics were invited to a White House conference on civil rights. Some Hispanics called it the "White House (all Negro) Civil Rights Conference." When Roybal was told that the conference wasn't "set up to discuss the problems of the Spanish-speaking," he was careful to emphasize that he didn't have any "hostility toward the advances made by the Negro community," but that he had voted for Johnson's civil rights bill because he thought it applied to "all Americans regardless of race, color, or creed."

In part as a response to Roybal's criticisms, Johnson established the Inter-Agency Committee on Mexican American Affairs in 1967, which sponsored a civil rights conference that same year in El Paso that brought together Hispanics in the Johnson administration with the leaders of national Hispanic organizations. Demonstrating how Chicano activists looked down upon Hispanic politicos, César Chávez had a supporter read a statement outside the convention hall that called the participants inside "perfumed sell outs." Chávez and other critics saw the conference as more symbol than substance.

Hispanic Republicans savored the infighting among Democrats. They clipped articles from newspapers and filed them away, then used the stories contained therein as a wedge to pry Hispanics away from the Democratic Party.

As the 1968 election loomed, Hispanic Republicans from across the country—members of the new generation that Robles talked about—began meeting about the possibility of playing a larger role in Republican Party politics. Some banded together to form the Republican National Hispanic Council, which became the Republican

National Hispanic Assembly. Their names were Martin Castillo, Benjamin Fernandez, Manuel Luján Jr., Fernando Oaxaca, and Francisco Vega. All of them were Mexican American, and except for Vega, all were from the Southwest. It didn't take long for the Republican Party to notice them. One was elected as the new congressman from New Mexico on the same day that Nixon reclaimed the White House for Republicans. Several went on to work for the president, whose flurry of Hispanic appointments symbolized his commitment to making Hispanics loyal Republicans.

PART II

INFLUENCE

CHAPTER 3

NIXON'S HISPANICS

The hundred or so Hispanics who visited the White House on November 14, 1971, all received a memento: a photo of President Richard Nixon, "smiling, relaxed," seated in the Oval Office next to Romana Acosta Bañuelos, his nominee to become the new treasurer of the United States. Just beneath the photo was a crisp one-dollar bill with her autograph. The event was part of Nixon's effort to win reelection with Hispanic support. In a year's time, he would pull off one of the more surprising developments in the history of Hispanic politics when he became the first Republican candidate to capture a third of the Hispanic vote. In November 1971 it was hardly certain that he could do it.

Nixon had nominated Bañuelos as treasurer in September, and she was confirmed in December. She had entered the United States as a child with her Mexican parents. With meager earnings from jobs in El Paso and Los Angeles, she started a tortilla factory in California that became an extremely profitable food distribution company. Later she helped found the Pan American National Bank in East Los Angeles, which served primarily Mexican and Mexican American customers. Bañuelos was one of several Nixon appointees who became symbols—tokens, critics would call them—of Hispanic influence. Over time they became Nixon loyalists who wouldn't turn their backs on the president, they said, because he never turned his back on them.

The photo of Nixon and Bañuelos was given out at the first ever "Sunday service for the Spanish-speaking people" hosted at the White House. In addition to Hispanic leaders, Supreme Court justices, Lyndon and Lady Bird Johnson, and other dignitaries were there. The archbishop of Boston delivered a twenty-five-minute sermon in Spanish in the East Room. Not everyone understood him, but that wasn't the point. As one of Nixon's Hispanic cabinet appointees later described it, the mass was a symbolic celebration of "national cultural inclusion." It was a way to demonstrate that Nixon was sensitive to Hispanics; that he acknowledged their culture in the most prominent way, in a most iconic place. After the mass, Nixon and Johnson flanked the archbishop in the State Dining Room, shaking hands for nearly an hour—twice as long as he had preached.

Nixon needed to increase his support among Hispanics because the 1968 election had been a bruising affair. The nation was deeply divided over the Vietnam War, student protests, and civil rights. After Johnson signed the Civil Rights Act, the story circulated that he believed the law would cause Democrats to lose the South for a generation. Republicans and Alabama governor George Wallace, who ran with General Curtis LeMay as an independent and split the conservative vote with Nixon, made sure this would be the case. Nixon also knew that he wouldn't find much support from African Americans, who resented his hostility to the Kerner Report's conclusion that systemic discrimination was rampant, his "racially charged" law-and-order rhetoric, and his "back room dealings" with the likes of Strom Thurmond, who increased Nixon's support among Democrats in the South.

President Johnson also introduced chaos into the election when he announced in March 1968 that he would not seek reelection, largely because of US failures in Vietnam. He and John F. Kennedy had gained the widespread admiration of non-Cuban Hispanics (recall the Bay of Pigs), but after Johnson bowed out, it wasn't clear whom Hispanics would support. Texas governor John Connally was a possibility, but many Mexican Americans in his state blamed him for the fact that agricultural workers hadn't received the minimal pay hike they demanded. Robert Kennedy gained momentum in California, riding the wave of his brother's popularity and earning support from César Chávez. Some 75 percent of Mexican Americans in the vote-rich state—which now claimed almost as many electoral votes as New York—had planned to vote for Kennedy when a bullet killed him in Los Angeles, right after he won the California primary.

Johnson's vice president, Hubert Humphrey, ultimately secured the nomination. He had the backing of some leading Hispanic Democrats, including Texas congressman Henry B. González, a former Viva Kennedy member. Gonzalez formed Arriba Humphrey (Up with Humphrey), while others formed Viva Humphrey and Ciudadanos para Humphrey (Citizens for Humphrey) groups. The vice president claimed to be a Kennedy and Johnson disciple and embraced Robert Kennedy's Hispanic agenda, including support for greater representation in higher education, organizing rights for agricultural workers, and federal government jobs "from the White House on down."

Hampering Humphrey's chances for success among Mexican Americans, in particular, was the formation of a third party in the Southwest called La Raza Unida. It was the political arm of the Chicano civil rights movement, operating alongside the agricultural, student, and antiwar movements. Hispanic youth all over the United States were articulating new notions of ethnic nationalism, calling for Puerto Rican independence and a new homeland for Chicanos in the Southwest, which they called Aztlán. La Raza Unida tapped into this momentum and directed it toward increased political participation. They registered Mexican American voters and urged them to only support candidates who represented "La Raza"—the race. Like Hispanic converts to the Republican Party, they, too, had grown frustrated with Democrats who promised inclusion but failed to act, even though Democrats had ended the Bracero Program and signed the Civil Rights Act.

Hispanics weren't ready to shift their support to the Republican Party, either, even though Nixon's Hispanic campaign was more organized than Humphrey's (Nixon had learned this lesson after 1960), with more paid workers and volunteers in key states. They found persuasive Humphrey's argument that Nixon and the Republican Party opposed the New Deal programs that had benefited them. Nixon countered by enlisting high-profile surrogates such as the Cuban American performer Desi Arnaz, who was the national chairman of the Viva Nixon group. Arnaz traveled across the country before the 1968 election, trying to convince Hispanics to support Nixon.

Like other Republican campaigns in the previous two election cycles, the Nixon campaign believed that Texas would be critical. It could be a swing state, but was trending toward the Republicans, joining the southeastern states they hoped would form a solid South. Arnaz sought to help him win enough Mexican American votes in the state to beat both Humphrey and Wallace. He opened Viva Nixon clubs throughout the state, in Austin, San Antonio, Corpus Christi, Victoria, and Del Rio.

The entertainer spent several days in Texas in late October and early November. In San Antonio he made light of a serious situation, joking that Fidel Castro had ruined the "barber business" in Cuba because everyone there wore their hair long, on their heads and their faces. There were "no haircuts." He delivered another stand-up routine in El Paso. His gray hair peeked out from underneath the sombrero he wore as he cracked jokes at Humphrey's expense. He also told Mexican Americans that the election would be up to them. Before leaving, Arnaz tuned a guitar and sang a song in Spanish he had written to praise the "character and qualities of Richard Nixon." If Nixon won, he said, Hispanics would have a friend in the White House. He dashed toward his plane and said, "Don't forget to go out and vote."

In the end, Nixon won the election by less than 1 percent—largely because he won California, Florida, New York, and Illinois—but his margin of victory would have been larger if Wallace didn't win a plu-rality of the popular vote, and all of the electoral votes, in Arkansas, Louisiana, Mississippi, Alabama, and Georgia. The Democrat Hu-bert Humphrey edged Nixon in Texas, also because of Wallace, who won 19 percent of the vote statewide, and an even greater percent-age in the counties along the state's eastern border with Louisiana. Since Nixon and Wallace split the conservative vote, Humphrey won the state's 25 electoral votes by a mere 1.2 percent. Nixon couldn't get enough of the Mexican American vote in Texas. The state, and the ability of Republican candidates to find Hispanic support there, would loom increasingly large in determining the outcome of presi-dential elections.

With Johnson's withdrawal from the race, Robert Kennedy's as-sassination, the formation of La Raza Unida, and their reluctance to vote for a Republican, Hispanics didn't have a clear candidate in 1968. But Nixon would remedy the situation by 1972, convincing a significant minority of them that they had a home in the Republican Party, and a sympathetic president in the White House.

EARLY IN NIXON'S FIRST TERM, REPUBLICAN OFFICIALS ARGUED THAT HISPANICS HAD TO be an important part of his strategy for reelection. Barry Goldwa-ter, who had just reclaimed his Senate seat, warned Nixon that His-panics could not be ignored. A Republican congressman from the Central Valley of California also made the point clear. In July 1969, he wrote that Nixon had done better among Mexican Americans in California than among Hispanics in other parts of the country. He won a "healthy" 20 percent of their votes compared with single dig-its nationally, and he could increase that share in 1972. Nixon also had done much better among Hispanics than among other "minori-ties" in California (he meant African Americans). At the same time, Governor Ronald Reagan maintained a "good rapport" with Mexi-can American Republicans in the state, and the Viva Nixon groups remained active even after the election, continuing to support Nixon while he was in office.

But in the months after the election, many Hispanics had grown "despondent." Those who voted for Nixon complained that African Americans "were treated better" and that Mexican American Democrats were getting all of the "positions and recognition." Mexican American Republicans felt like they were again relegated to second-class status. The California congressman said Hispanics were "too loyal to neglect and too important to offend." Nixon had to make more of an effort if he hoped to turn things around by 1972. He had to win back the support of his Hispanic followers in California and across the United States.

Nixon did more than turn things around. Between 1968 and 1972 he was able to reverse his fate among Hispanics by making political appointments, creating new financial programs, and forming cabinet-level committees whose goal was to connect leaders in Washington, DC, with Hispanics across the United States, to ensure that Hispanic communities benefited from federal programs. It was classic patronage politics, and it was an effective way of building a Hispanic Republican base.

The president had plenty of help from a loyal group of Hispanic Republicans called the Brown Mafia. They were some of the Hispanics who served in his administration, including Henry Ramirez, chairman of Nixon's Cabinet Committee on Opportunities for Spanish Speaking People (CCOSSP); the economist Benjamin Fernandez, chairman of Nixon's National Economic Development Association (NEDA); and Philip Sanchez, the head of Nixon's Office of Economic Opportunity (OEO). He also had the support of a newly elected congressman from New Mexico, Manuel Luján Jr., who, as the lone Hispanic Republican, was outnumbered by five Hispanic Democrats in the Ninety-First Congress (1969–1971).

Nixon's success in 1972 was transformative for the party. In almost every presidential election afterward, Republican candidates won a percentage of the Hispanic vote nearer to the third that Nixon won in 1972 than the single-digit support he earned in 1968. He established a new normal, and developed a national strategy that future Republicans sought to duplicate. Earlier Hispanic campaigns had gestured toward a national strategy but didn't pull it off. In Nixon's efforts to recruit Hispanics, he also pushed the boundaries of the law, as Americans discovered in the first year of his second term.

NIXON HAD LEARNED THE IMPORTANT LESSONS OF THE SIXTIES. JOURNALISTS HAD WRITten his political obituary after his loss to Kennedy in the 1960 presidential election, as well as his loss to Pat Brown in California's 1962 gubernatorial election. Even in 1968, things hadn't gone smoothly for him. Reagan received more primary votes than Nixon did. Reagan had spent his first years in the governor's mansion railing against student activists and supporting growers in their struggles against César Chávez's United Farm Workers union, even as he appointed more Hispanics to posts in his administration than any California governor before him. Fending off Reagan became increasingly difficult for Republican contenders. For the moment, though, Nixon had rebounded.

In his first term, he improved his position considerably. He worked to replace lost African American votes, and tempered the influence of African American civil rights activists by aggressively recruiting Hispanics. It was a lesson Nixon learned from *The Emerging Republican Majority*, written by the strategist Kevin Phillips, who had worked on his 1968 campaign. One of Nixon's takeaways from reading the book over Christmas in 1969 was that it would be a waste of time to fight for the votes of African Americans. Their votes were already lost, he came to believe, so he had to forge a different coalition of voters to rally around him. It had to include Hispanics, who he believed would be critical to his success.

Nixon's Hispanic supporters argued that his deep connection with them since childhood had motivated his concern for them once he was in office. They recounted how, at age nine, Nixon moved with his family to the Murphy Ranch in Whittier, California, about midway between East Los Angeles and Yorba Linda, where he was born. His father opened a grocery store and a gas station, and Nixon worked at both family businesses while attending Fullerton Union High School, an hour away by bike. He transferred to Whittier High School for his junior year to be closer to home. Nixon helped his parents with tasks related to the operation of their businesses, but to hear Hispanic Republicans tell it, he toiled with them. "Richard Nixon was once a poor white man who worked alongside poor Mexicans in the orchards and fields," one wrote.

Nixon was a lot like them, they believed. "He got to know us in the hot, dusty, and dirty fields and orange groves where hard physical work was requirement number one," claimed one Mexican American from the area. He "got to know us while we lived our culture in accord with our values: God, family, hard work, respect for the law, and an aversion to handouts," he said. Nixon also was the first president born in California. He wasn't like the others from back East, who didn't know anything about Mexican Americans. It was a nice bildungsroman of Nixon's ties with Hispanics, but it relied on the sweat of his Hispanic supporters to rally their communities to his side. They were the leaders of an emerging movement of Hispanic Republicans, and they said Nixon was the first president to bring them into the Republican Party.

One of the first things Nixon did as president was figure out what to do with a committee that Johnson created: the Interagency Committee on Mexican American Affairs (ICMAA). The title of the committee singled out Mexican Americans, perhaps because they were the largest Hispanic group, or at least the one with which Johnson, a Texan, was most familiar. Yet the documents that established the committee said it was also meant to "reach and involve other Americans of similar ethnic or cultural background such as Puerto Ricans, Cubans, and other Spanish Surnamed Americans."

In Johnson's memorandum establishing the ICMAA, he expressed sympathy with Mexican Americans' "search for equal opportunity and first-class American citizenship." Johnson claimed that Mexican American leaders had praised his administration's efforts to help them during his first three years in office, but he felt he could do even more to focus on the needs of their communities.

Johnson appointed Vicente Ximenes as chairman of the new committee. Ximenes had worked on the Viva Kennedy and Viva Johnson campaigns. At the time of his appointment, he was serving as head of the Equal Employment Opportunity Commission. His job with the ICMAA would be to convene the heads of several government agencies: the OEO and the Departments of Labor, Agriculture, Housing and Urban Development, and Health, Education, and Welfare. These were the individual agencies that made up the interagency committee. Their purpose was to make sure that programs of the federal government benefited Hispanics, and to establish new programs if existing ones didn't meet their needs. They brought together public funding, private investment, and Hispanic initiative to find "workable solutions" to urgent problems, including poverty, the language barrier, unemployment, and educational inequality. Improvement in these areas, Ximenes said, would lead to "more productive citizenship."

Nixon was determined to continue Johnson's organization in some fashion, because he wanted to show Hispanics that he would do as much for them as Johnson had done. Following custom, Ximenes resigned shortly after Nixon's inauguration and Nixon replaced him with Martin Castillo, a lawyer and registered Democrat from Los Angeles who nevertheless campaigned for Nixon and worked closely with Hispanic Republicans. Sometimes Democrats such as Castillo were as important to Nixon as Republicans, because they were examples to other Democrats considering switching sides. The ICMAA had a new chairman, and now Nixon had to figure out how the mission and broader constituency of the organization would match the goals of his administration.

In the early weeks of his presidency, Nixon made clear how his civil rights strategy would differ from the strategy of Democrats, as caricatured by his supporters. Specifically, he said that economic uplift would be the central plank of his appeal to Hispanics, African Americans, and other minorities. Because of Nixon's attention to entrepreneurship, Hispanic Republicans exaggerated that Nixon "introduced" them to capitalism. This was a "historical fact," a Mexican American in his administration proclaimed, even though Hispanic merchants, entrepreneurs, and business owners had embraced capitalism for centuries before Nixon's election. His point, though, was that Nixon's approach would be different. Republicans argued that handouts to poor people were the foundation of the Democratic Party's antipoverty programs, but Nixon sought to use the tools of government in a limited way to help them help themselves.

On March 5, 1969, Nixon announced Executive Order 11458, which established the Office of Minority Business Enterprise (OMBE) within the Department of Commerce. In later decades it became the Minority Business Development Agency, and its purpose was to coordinate the efforts of state and local governments, and institutions such as public universities and private foundations, to make sure individual communities were benefiting from the federal government's economic development programs.

In addition to establishing and funding the OMBE, Nixon directed the resources of the Small Business Administration (SBA) toward minority communities, especially through loans to minority-owned businesses. The SBA ran what one of Nixon's Mexican American supporters described as the "tremendously successful" 8(a) program, which required federal government procurement agencies to purchase goods and services from minority-operated businesses. The African American population was twice as large as the Hispanic population, so one might assume that it made sense for African American businesses to receive the lion's share of 8(a) funding. But to Hispanic Republicans, the disparity seemed to confirm their complaints that the government's main civil rights priority was placating African Americans. Nevertheless, millions of dollars flowed to Hispanic businesses and, in theory, Hispanic communities as well. These sorts of programs were the crux of what Nixon supporters called his "black capitalism" and "brown capitalism" initiatives to promote entrepreneurship within minority communities.

THE ICMAA WOULD BE CENTRAL TO NIXON'S EFFORTS TO SERVE AND RECRUIT HISPANICS, but he wanted to make some changes. First, the committee needed new personnel. Johnson's committee, one Republican congressman wrote, had been "staffed with unfriendly militant Democrats to the exclusion of friendly, nonviolent Republicans." Nixon needed more of these friendly, nonviolent Republicans. The ICMAA also had really involved only Mexican Americans, even though it was supposed to include all Hispanics. This, too, would have to change. Nixon got an earful from the leading New York Republicans, Governor Nelson Rockefeller and Senator Jacob Javits, who had large Puerto Rican constituencies. His administration embraced a new name—the Cabinet Committee on Opportunities for Spanish-Speaking People—as the umbrella held over Puerto Ricans, Cubans, and "all Americans of similar ethnic or cultural background."

Nixon signed the bill creating the CCOSSP on the last day of 1969, and in remarks he prepared for the occasion he made clear that the goals of the committee and his brown capitalism initiatives were aligned. "Too many members of this significant minority group have been too long denied genuine, equal economic opportunity," he said, and "in signing this bill, I reaffirm the concern of this government for providing equal opportunity to all Spanish-speaking Americans—to open doors to better jobs and the ownership and management of business." He emphasized that the CCOSSP would work closely with the recently created OMBE to help Hispanics "launch their own businesses" through a combination of government assistance, investments by "private enterprise," and the "proven drive and talent of the Spanish-speaking peoples." His closing words—the cherry on top, a little flourish in Spanish—were, "I sign this bill *con gusto*—with the enthusiasm and determination to make equal opportunity a reality in these United States."

The central plank of the CCOSSP was its classically Republican, probusiness, limited-government emphasis, pitched at the middle class. The Nixon administration would "encourage wider participation by private industry and organizations so that the burden of reform is ultimately lifted from the shoulders of government and returned to the democracy of the people," he said. The described functions and aims of the CCOSSP became an early blueprint for the Nixon administration's efforts to uplift Hispanics economically, through public-private partnerships that would lead to the improvement of Hispanic lives by giving them the tools of capitalism and free enterprise.

The CCOSSP would define the needs of the Hispanic community, then propose ways of meeting these needs. Together, the CCOSSP and the president came up with the so-called Sixteen Point Program, which was a systematic effort to increase federal government employment for Hispanics. Nixon appointed the New Mexican Fernando E. C. De Baca as the director of the Sixteen Point Program. As part of their effort, the White House also directed the Census Bureau to count Hispanics as a unique category of Americans for the 1970 census. Instead of the criteria they relied on earlier, such as birthplace, language, or surname, they now asked about whether a respondent was of "Spanish origin," which they believed would offer a more accurate measure of the population. They hired Spanish-speaking enumerators, published questionnaires in Spanish, and urged Hispanic participation through advertisements in Spanish-language newspapers, on the radio, and on television. Being counted was a crucial step toward solving their economic woes, because it could help determine where the government directed resources.

The CCOSSP worked closely with government bureaucracies in charge of helping to create business opportunities for Hispanics and other Americans. The main engines driving Nixon's brown capitalism agenda were the SBA and the Department of Commerce. These two government offices came together to fund NEDA. The head of the CCOSSP said the Department of Commerce's OMBE was their "gateway to capitalism," and NEDA "developed Hispanic capitalists."

Established in 1970, NEDA began publishing monthly reports that offered brief biographies of the organization's leaders. According to this literature, members of NEDA's board of directors were successful businessmen who owned restaurants, started media companies that grew into "multimillion-dollar" businesses, and ran banks that served Hispanics where they lived. Their work for NEDA became an important line on their résumés that opened doors to other professional opportunities, including appointments to government posts throughout the Americas, as the "Peace Corps Director in Colombia" or as employees of the OMBE. Finally, they believed, their skills were being used for a good purpose. The reports were emblazoned with a drawing of a conquistador's sword and armored helmet, suggesting that NEDA's leaders were like the Spanish conquerors who paved the way for capitalist ventures like their own.

Benjamin Fernandez, one of the founders of the Republican National Hispanic Assembly, became NEDA's first chairman. He was the son of poor Mexican immigrants, and he went on to attend college at the University of Redlands and business school at New York University before starting his own management consulting firm in California. NEDA's basic objective, as one Mexican American administration official described it, was to promote the free-enterprise system among Hispanic business owners. Instead of loaning them money directly, NEDA helped Hispanic entrepreneurs develop business plans and raise capital, so they could open their own commercial banks and savings and loans in Hispanic communities that were underserved by giants such as Wells Fargo. NEDA established relationships in local communities, which helped it identify potential business owners and business opportunities for Hispanics in general.

In its first year, NEDA helped Hispanic businesses raise $50 million. In its second year, NEDA almost tripled that amount, to more than $140 million, and NEDA had opened offices in Albuquerque, Chicago, Los Angeles, Miami, New York, San Antonio, San Juan, Puerto Rico, and other cities with large Hispanic populations.

Well-respected Hispanic journalists nevertheless criticized Nixon's brown capitalism and NEDA in particular for failing to make a difference in Hispanic communities. At the end of August 1970, just a couple of days before he died in the aftermath of an anti–Vietnam War protest when a gas canister fired by a police officer struck him in the head, the *Los Angeles Times* reporter Ruben Salazar wrote, "In the barrios Chicanos immediately started calling NEDA *NADA*, which in Spanish spells 'nothing.'" They wanted more spending on education rather than assistance for business owners. Hispanic Republicans objected. "It almost appeared that Spanish-surnamed writers wore socialism on their sleeves and held capitalism in opprobrium," one wrote. Didn't Salazar know that "there is no Santa Claus" in socialism?

If Nixon's NEDA came in for criticism, his CCOSSP wasn't faring too well, either. Castillo resigned as chairman after only a year, in part because he felt that the White House wasn't paying Hispanics the attention they deserved. After Castillo resigned, Nixon left the CCOSSP without a chairman for almost eight months. By statute, the CCOSSP was supposed to have a chairman, an Advisory Council, and quarterly meetings for the delivery of reports on their progress. The Nixon administration had done none of it by the summer of 1971.

In early 1971, the CCOSSP was also plagued by competition between Mexican Americans and Puerto Ricans. As soon as there was a government agency with the express purpose of representing Hispanics nationally, Puerto Ricans sought greater inclusion. They generally complained that the CCOSSP continued to represent the interests of Mexican Americans only and that the Mexican Americans on the committee wanted it that way. The nine-member Advisory Council was supposed to include a mix of Mexican Americans, Puerto Ricans, and Cuban Americans. The number of representatives from each group was to be in proportion to their percentage of the Hispanic population.

Mexican Americans technically were underrepresented, since they counted for 60 percent of the Hispanic population and their four members held only 45 percent of the seats on the council. Puerto Ricans and Cubans technically were overrepresented, since their 15 percent and 6 percent of the Hispanic population, respectively, translated into 33 percent and 22 percent of the council, or three seats and two seats. Central Americans and South Americans represented 16.5 percent of the Hispanic population, while "other" groups represented 11.6 percent. But together they claimed exactly zero seats on the council.

Cubans seemed to have been okay with their allotment, based on the absence of critical letters to the White House. They were just getting involved in politics in the United States. But Puerto Ricans and their representatives in Washington and Puerto Rico pressured the CCOSSP to appoint more Puerto Ricans to the council and to do more for them in general, especially in New York and New Jersey, where most of them resided.

When Congress debated funding for the CCOSSP, Republican senator Clifford Case of New Jersey grilled his colleague Democratic senator Joseph Montoya of New Mexico about why the committee had done so little for Puerto Ricans. They were a people who were "just as Latin as yours are," he told the Mexican American senator, yet the CCOSSP had "accomplished nothing" for them.

Luis Ferré, who had recently come out on top in his struggle against Miguel García Méndez to become the leader of Puerto Rico's prostatehood faction, weighed in on behalf of Puerto Ricans as well. Ferré was elected as governor of Puerto Rico in 1968, as the candidate of the New Progressive Party, which he founded. In the eight months between Castillo's resignation and the appointment of his successor, Ferré wrote letters to Nixon to urge the appointment of particular individuals to the Advisory Council. He recommended prominent Puerto Ricans living in the United States, including Jorge Luis Córdova; Manuel Casiano; Jaime Pieras; and Luis Guinot Jr., a Puerto Rican lawyer living in Washington, DC, who had been friends with Ferré's daughter when he still lived on the island. When Nixon failed to appoint the Puerto Ricans Ferré recommended, Ferré sent more letters with more recommendations. These letters from Hispanics such as Ferré prompted members of the Nixon administration to consider the future of the CCOSSP.

Before he resigned, Castillo described the ideal CCOSSP chairman and Advisory Council member, which helped the White House consider who his replacement should be. They should have a "measurable commitment" to Hispanics, a "tenacity in pursuit" of their commitment, an understanding of the needs of Hispanics, "stability of judgment," and an ability to "relate to all vertical levels of Community leadership," from the bottom up. In May 1971, Henry Ramirez, who at the time worked for the US Commission on Civil Rights—where his coworkers, he claimed, held political opinions ranging from "very liberal to Marxist"—chimed in with a similar list. It would become his own job description. Nixon appointed him the new chairman of the CCOSSP just a couple of months later.

SHORTLY AFTER RAMIREZ'S APPOINTMENT, NIXON SUDDENLY NEEDED A NEW TREASURER because Dorothy Andrews Kabis died of a heart attack on July 3, 1971, while visiting her father's grave in Sheffield, Massachusetts.

Nixon's chief of staff, H. R. Haldeman, asked his aides to identify as potential replacements the "best qualified women" who were also Mexican American. The criticism Nixon was hearing both about the effectiveness of his programs and his handling of the CCOSSP worried him, and those around him, that he wouldn't receive Hispanic support in 1972. The appointment of a Hispanic to a high-level position such as treasurer of the United States might help turn things around.

William "Mo" Marumoto, a special assistant to Nixon, was put in charge of the search. Less than ten days after Kabis's death, Marumoto reported that he had identified some promising candidates. He had consulted Mexican American leaders, including Benjamin Fernandez, Henry Ramirez, and Philip Sanchez, and had come up with the names of eight women—seven from California and one from New Mexico. His sources in Colorado, Arizona, and Texas reported that there were no "qualified women" there. The surname had to be Spanish, so Gilda Bojorquez Gjurich had a last name that "excludes her from consideration." Maria Concepcion "Concha" Ortiz y Pino might have worked, but Marumoto judged her to be too old. She was a "wonderful old gal who has seen her day," he said. Blanche Gomez, chairperson of the Los Angeles Housing Authority, was first runner-up, and Bañuelos was the top pick.

Bañuelos was born in Miami, Arizona, a mining town where her father worked. She attended elementary school in Chihuahua, Mexico, before settling with her parents in El Paso at age fourteen. She had crossed the border with only thirty-six cents in her pocket. But hard work and determination, she said, helped her become the millionaire owner of her own business. Her message, the administration believed, would inspire any immigrant. "If you don't have money you have to . . . work more than the ordinary person," she said. That was what it took to get ahead. She worked at a sewing factory, ice cream factory, and ice cream and sandwich shop before she and her aunt started a tortilla factory in a converted garage. They earned their first dollar in 1949, but by 1971 Ramona's Mexican Food Products had grown into one of the largest Mexican food distributors in the West, with three hundred employees, a net income of more than $5 million per year, and a wide variety of products to sell, including their most popular food, the burritos of "both beef and potato and beef and beans variety."

Bañuelos seemed to embody everything the president hoped to convey about how free-market capitalism and hard work were the keys to prosperity. When her business was still growing, Bañuelos woke up before dawn, headed to the factory, and made the tortillas herself. Then she returned home to get her kids ready for school before returning to work to get going with the rest of her fourteen-to-sixteen-hour workday. Now her company satisfied the Mexican food cravings of more than ninety thousand people, who ate almost two hundred thousand of her tortillas every single day. Ramona's was headquartered in Gardena, California, just south of Los Angeles. But Bañuelos lived with her husband, Alexander, in the upscale neighborhood of Los Feliz, in an English Normandy–style home at the base of Griffith Park, not far from the Hollywood sign. It was twelve miles and a world away from East Los Angeles, where Ramona's first opened for business.

Bañuelos still inhabited that world, but now as the chairwoman of Pan American National Bank. Located in the heart of East Los Angeles, at 3626 East First Street, Pan American, established in 1964, was adorned inside and out with Aztec imagery that was typical of the Chicano murals on the facades of many buildings in the community. The logo on their letterhead contained an Aztec pyramid that resembled Teotihuacán. Bañuelos was one of the forty individuals who founded the bank, and she became chairwoman in 1968, by which point Pan American had millions in assets and had become a fixture of an institution that served primarily Mexican American—but also Jewish and Japanese American—customers. She was the first woman to head a commercial bank in the history of California. The other executives at the bank were all men.

In addition to her business qualifications, the White House believed Bañuelos was perfectly situated to convince Hispanics to vote for President Nixon. She had already done so much for Mexican Americans in Los Angeles, and her views were in sync with Nixon's. Bañuelos had set up Ramona's Mexican Food Products Scholarship Foundation, which helped Mexican American students at Roosevelt, Garfield, and Lincoln High Schools attend college in the area. Just a few years earlier, Mexican American students at these and two other area high schools—Belmont and Wilson—had walked out of their classes to protest high dropout rates and the lack of Mexican American teachers and curricula. The students became some of the leaders of East LA's Chicano movement for civil rights, the sorts of radicals whom Reagan gained a reputation for criticizing from the governor's office in Sacramento. Through her scholarship endowment, Bañuelos offered a more positive vision for how to uplift Mexican Americans.

The president was drawn to Bañuelos precisely because she was not a supporter of, nor a participant in, the Chicano movement. When a reporter asked her if she had "played any part" in the Chicano movement, she pursed her lips, shook her head, and said no. But she did "help the community," she said, especially through Pan American, which, for Mexican Americans in East Los Angeles, was the "only bank they've had." Nixon had found himself a Mexican American who opposed Chicano activism but nevertheless contributed to the uplift of other Hispanics through financial services. He had also found a woman who believed in the advancement of women but who was not part of the women's liberation movement. She thought that women should be able to do anything they wanted, but she didn't pursue her goals by taking to the streets, or by going to "bars" and doing all of "the bad things." She was too good to be true.

Bañuelos and her husband flew to Washington, DC, just days after Marumoto recommended her to Malek, to interview for the position. Malek reported back to Haldeman on how the meetings had gone. Malek told him that they had "thoroughly romanced" Bañuelos, with limousine service and lunch at the White House. Administration officials met her at the airport and whisked her to meetings with Marumoto, Sanchez, Ramirez, and others. At the end of her meetings, after Bañuelos had boarded her TWA flight back to Los Angeles, Malek said he found Bañuelos to be "articulate," "personable," and "attractive." She had been a Republican for ten years and was a "real believer in the President." She was offered the job immediately.

After just a couple of days attending to business back home, Bañuelos flew to DC again, this time to meet with Nixon in the Oval Office. It was a prime photo op, an occasion that underscored Nixon's "commitment to Mexican-Americans and his concerns that qualified women be appointed to high-level posts." Their meeting lasted fifteen minutes, which gave them plenty of time to smile for the cameras and exchange pleasantries. Nixon handed her a blue felt pen and a piece of paper so she could sign her name just as it would appear on American currency. The president joked that it was both neater and larger than his signature, and then turned to his Treasury secretary, John B. Connally, to ask, "You got room on the money?" Bañuelos later said that she had been "so nervous" at the White House that she needed to sign again for the actual bills. The White House decided that Nixon would wait until September to nominate her formally, so as to maximize the impact of the announcement by making it during Hispanic Heritage Week.

BEFORE NIXON ANNOUNCED BAÑUELOS'S NOMINATION, HE HAD TO PATCH THINGS UP WITH the members of the CCOSSP, who had been waiting for a new leader for months. Ramirez's term as chairman began on August 5, the same day that the CCOSSP met for the first time in more than a year.

Nixon and Ramirez posed to have their photographs taken in the Oval Office, and then moved into the Cabinet Room for the meeting of the full committee.

Ramirez described himself as an "ardent anticommunist Republican" and "civil rights warrior." He was the child of Mexican immigrants and had spent several years picking oranges, grapes, and walnuts in Nixon's hometown of Whittier, California. He became a beloved high school teacher, but first went to law school and pursued a graduate degree in theology, training to become a priest. He had moved from Southern California to Washington to work for the US Commission on Civil Rights. His deeply researched reports on Hispanics in the United States put him on the White House's radar. Ramirez served as chairman of the CCOSSP through the time of Nixon's resignation, and he remained a Nixon loyalist for decades to come.

Ramirez's loyalty to Nixon was stirred by the passion with which the president spoke about the needs of Hispanics. Staff member Robert Finch had given Nixon some fairly bland and generic talking points for the August 1971 meeting. Nixon sought to encourage "individual development" and "participation in industry and free enterprise." He saw the committee as recognition of the "prestige of the Spanish speaking people." The members of the committee were representative of all Hispanics, who had a "strong character built on religion, morality, strong family ties, and hard work." Etc., etc.

But Nixon went off script. According to Ramirez, the president spoke "feelingly and forcefully" on the need to improve the lives of Hispanics. Nixon told Ramirez during a meeting in the Oval Office, "We gringos have built an invisible wall of discrimination against you Mexicans, and you and I are going to knock it down." So far, the president said, he had not followed through on his promises. Deflecting responsibility from himself, he chastised his staff for letting the CCOSSP lay dormant. He told them they'd better not let him down again. They were stunned into silence.

As one of Ramirez's first acts as chairman, he axed two of the three Puerto Ricans on the Advisory Council and installed a Mexican American, Antonio Rodriguez, as executive director. The chairman of the CCOSSP was supposed to be Mexican American, but the executive director was to be Puerto Rican. When Ramirez replaced the Puerto Ricans with a Mexican American executive director, he elicited a scathing telegram addressed to Nixon from the National Association for Puerto Rican Civil Rights stating that Rodriguez was a "racist of the worst kind"—against whom and on what grounds, it did not say. He demanded Rodriguez's immediate resignation and the reinstatement of the two fired Puerto Ricans, for the "good of democracy."

Meanwhile, a Mexican American from California argued that the CCOSSP should appoint five Mexican Americans from California alone, and that all of the committee's meetings should be held either in California or Washington, DC. These intra-Hispanic tensions over the leadership of the CCOSSP foreshadowed later divides, especially as Cuban-American influence in the Republican Party increased.

The August 1971 meeting revived the CCOSSP, and in many ways inaugurated Nixon's reelection campaign among Hispanics. He had selected his new treasurer of the United States, had a Mexican American at the helm of the OEO, and appointed a new CCOSSP chairman. The next month, he formally nominated Bañuelos and sent Ramirez on the road to meet with Hispanic leaders across the country.

THE KICKOFF OF THE HISPANIC CAMPAIGN CAME AT A CRITICAL TIME IN NIXON'S PRESI-dency. His reelection promised to be tough because of the fervor of his opponents and a series of events that plagued his first term, including the leak of the Pentagon Papers; protests and shootings on college campuses; and the brutal suppression of a riot at Attica Prison, which housed many Puerto Rican prisoners from New York City. He needed Hispanic support more than ever before.

With a belly full of Ramona's burritos, which Nixon said he had eaten earlier in the day, he told the press corps and all Americans that his staff had "searched the country for a person of truly outstanding credentials and ability" to become the next treasurer of the United States. Bañuelos was the one. She had an "extraordinarily successful career as a self-made businesswoman" and had "displayed exceptional initiative, perseverance and skill."

Nixon neglected to mention that he directed his staff to find him a Mexican American woman, but reporters made that point for him. Nixon "didn't say, but might have," one wrote, "that in addition to her proven business acumen she is of Mexican-American descent, a group with which the President is not noticeably popular." Bañuelos herself rejected the idea that the president had chosen her because of her Hispanic heritage. She said the president "wanted someone of my qualifications and I just happened to have them and happened to be of Mexican-American descent." She asked, as if trying to convince herself, "So it's an accident, isn't it?"

From the moment of Nixon's announcement in September, Nixon launched a full-scale campaign to sell her to the American public and to Hispanics in particular. The *New York Times* reported that the confirmation of Bañuelos as treasurer was "expected to be routine." It would be anything but. Just days after her nomination, the Immigration and Naturalization Service (INS) raided Ramona's Mexican Food Products.

In the early morning hours of October 6, just before the beginning of their 6:00 a.m. shift, immigration officials burst into the factory to round up workers who didn't have proper documentation. The raid inflamed already tense debates about undocumented Mexican immigrants, whose number had increased as a result of the end of the Bracero Program and the implementation of Western Hemisphere quotas. More than seventy employees scattered, trying to escape through the back and side doors of the factory. Thirty-six of them were apprehended, including two dozen women who ran into the changing room to hide behind stall doors. They were bused back to Mexico that very day.

Bañuelos was at the factory when the raid took place. She insisted that she didn't know undocumented immigrants worked at her company. She added that she had done nothing wrong, since the law didn't require verification that her employees were permitted to work in the United States. Considering the raid on her company, the Spanish-language mass at the White House in November can be seen in a different light, as part of her penance for the negative attention she had brought upon the administration, and part of the public relations campaign that had become necessary to secure her confirmation.

Everyone involved pointed fingers at everyone else. Bañuelos blamed the Democrats who opposed her nomination. They were the ones, she insisted, who urged immigration officials to raid her factory. Reporters were on hand to cover the raid. How had they learned about it if they weren't tipped off by her opponents? she wondered. Some Californians wrote letters to the editor that blamed Bañuelos herself. The raid in the middle of her confirmation process was the sixth raid of her company in three years. Each time, undocumented immigrants were found to be working at the factory. She must have known that she employed undocumented immigrants. The INS blamed her as well. In a letter sent to the firm in 1969, they had given her the opportunity to stop hiring undocumented Mexicans on her own, which, they said, not only "encourages additional aliens to enter the United States" but also "deprives United States citizens and lawful resident aliens of necessary employment." Other letters to the editor blamed the immigrants themselves. "Violence to the law was done by the 36 Mexican citizens who were working for Mrs. Bañuelos," not the "California taco tycoon."

One editorial, as sarcastic as it was racist, said Ramona's was "in the habit of employing wetback taco-makers, and thereby allegedly keeping good American taco-makers out of jobs." Others advanced more serious arguments, including that Bañuelos was "unfit to hold any public office" because she was complicit in a system that exploited Mexican immigrants and pitted them against American workers.

The raid on Ramona's certainly didn't make Marumoto's job easier. He was responsible for the campaign on Bañuelos's behalf. He reached out to "veterans, educational, business, civil rights, community development" organizations—including the "Chicano, Puerto Rican, and Cuban" groups—to seek their endorsement, asking them to send telegrams to their senators and to the president. He had all of his "contacts" write editorials for Spanish-language newspapers and television and radio stations. He wrote all of Nixon's Hispanic appointees to ask them to "fully endorse the nomination both personally and by contacting their circle of acquaintances."

Marumoto urged the presidents of all the major Hispanic organizations, such as the American GI Forum, the League of United Latin American Citizens, NEDA, the National Puerto Rican Forum, and the recently formed Association of Cuban American Organizations, to hold a joint press conference to support Bañuelos. They complied, issuing their statement at a conference sponsored by NEDA. It was the "first time" that "representatives of the different Hispanic groups had met without friction," participants recalled. Even the Democrats there supported Bañuelos. They wanted to see Hispanics in "top positions, whoever appoints them."

At the same time that they endorsed Bañuelos, though, Hispanic leaders also began to speak of the need to hold employers responsible for hiring undocumented workers. "We urge the United States Congress to address itself to adopting a law which would make employers responsible for the verification of citizenship or legal residence of employees," they said.

The raid was an embarrassment for the administration since it drew unwanted attention from the press, but in the end Bañuelos's confirmation went smoothly. The Senate Finance Committee running the hearing investigated the raid and questioned Bañuelos about it, but ultimately decided that it shouldn't disqualify her. She was confirmed in December 1971. As the *New York Times* reported, it was "apparent" that "all sides had no objection." She was sworn in and her signature began to appear on all new paper currency.

Even though Bañuelos was confirmed, the raid on Ramona's had a lasting consequence: it was the genesis of public debates over employer sanctions. In part as a response to the Hispanic leaders who offered their endorsement of her, Congress began working on an immigration bill that made it an offense to "knowingly" hire undocumented immigrants. When he promoted the bill, Democratic congressman James O'Hara of Michigan mentioned the raid at Ramona's and another incident involving an undocumented gardener who worked at the Western White House in San Clemente. He joked that the bill should be called the "tortilla workers and White House gardener's protective amendment."

AS THE INS RAIDED RAMONA'S, RAMIREZ AND OTHER MEMBERS OF THE CCOSSP WERE embarking on their tour around the country for conferences in cities with large Hispanic populations. They started in Chicago in October; New York in December; Dallas in January 1972; and Denver, Atlanta, and San Francisco in February 1972. They called their tour Project Alpha, which would be a series of meetings between Hispanic leaders and the heads of regional offices of federal agencies.

Each place they went, the community's concerns were slightly different—employment in steel mills in Chicago, the need for health clinics in New York—but their complaints against the government were remarkably consistent. Very few Hispanics held high-level government jobs. Hispanics didn't have equal access to programs funded by the federal government. The government had shown insensitivity through its choice of the Hispanic representatives who met with them. For all of Nixon's talk about inclusion and attention to the needs of Hispanics, they saw very little of it.

At the meetings Ramirez offered a carrot: Nixon planned to "set aside" a certain number of federal jobs for Hispanics. These set-asides later led the conservative Mexican American Linda Chavez to call Nixon the "father of quotas." They were her primary reason for opposing a president that other Hispanic Republicans admired greatly.

Despite his offer, Ramirez and Nixon's Hispanic representatives failed to impress. Ramirez said he attended the meetings to listen, but many Hispanic leaders complained that he should have come with solutions. He thought his major accomplishment was to get White House officials such as Finch to travel to their communities to listen to their problems. But to the Hispanics they met with, this was hardly enough, and it demonstrated Ramirez's arrogance toward and lack of connection with the Hispanic community.

According to a Mexican American who reported back to the White House on the meetings, his sources complained that Ramirez had never "related" to the grassroots "community people" and would therefore continue to encounter "difficulty communicating with them." Hispanics in New York made an even harsher assessment of the White House's Hispanic representatives. Angel Rivera had "no connection with the New York Puerto Rican leadership." Edward Aponte had been chased from his job at a community college by students because he "does not have Puerto Rican sentiment—does not identify with the community." One New Yorker said that "all hell would break loose" if the White House was "depending on these two individuals" to represent the needs of the Hispanic community.

THE CCOSSP'S MEETINGS ACROSS THE COUNTRY CONTINUED INTO EARLY 1972, BY WHICH point the Nixon reelection campaign was in full swing. Nixon's main argument to Hispanics, simply stated, was that he had done more for them than any president before him. Whereas Kennedy and Johnson made promises, Nixon followed through. This wasn't always clear to Hispanics, even his supporters, but he had, in fact, collected historical firsts like trophies on a shelf. He was the first president to order the Census Bureau to count Hispanics as a distinct category of Americans. The set-aside programs Ramirez announced would be in proportion to their percentage of the population, so how could the federal government know how many jobs to set aside if it didn't know how many Hispanics there were?

Nixon established several government programs to create opportunities for Hispanic business owners. He held the first Spanish-language mass at the White House. He also appointed more Hispanics to government posts than any president before him, including the first Hispanic treasurer of the United States, the first Hispanic administrator of the SBA (Hilario Sandoval), the first Hispanic head of the OMBE (Alex Armendariz), and the first Hispanic head of the OEO (Philip Sanchez). Even if they didn't go exactly as he had hoped, he sponsored conferences across the United States designed to come up with meaningful solutions to the challenges Hispanics faced.

Nixon enlisted his Hispanic appointees to highlight his record of accomplishment. As they campaigned for him across the country, they said that economic progress had been a "third front" of the Hispanic civil rights struggles, alongside their more news-grabbing social and political activism. The president had worked hard to "generate jobs through minority business opportunities." His "brown capitalism" had been the "centerpiece" of his civil rights effort. Sanchez, the OEO head, said Nixon supported the organization because he "didn't want it to be another welfare agency." Hispanics had to help themselves. These were their talking points, the messages Nixon hoped they would get across to Hispanic communities.

They said Hispanics needed Nixon *Ahora Mas Que Nunca*—a Spanish translation of the campaign's "Now More Than Ever" slogan. Ads placed Nixon's face in the center of an Aztec calendar stone, surrounded by his Hispanic appointees. At one campaign stop in San Antonio, Marumoto said, "President Nixon has been the best President the Spanish-speaking has ever had." He "set a precedent for

Spanish-speaking involvement and achievement that will serve as a benchmark for future administrations."

Bañuelos became one of Nixon's most important surrogates. She logged more than eighty-five thousand miles between January and November 1972, campaigning for the president in more than a dozen states. She stopped in cities from Chicago to Los Angeles, often accompanied by fellow Hispanic appointee Philip Sanchez, who explained to one audience that Hispanics had voted for Democrats for more than twenty years "because of Roosevelt and for good reason." But it was time for a change. As a local Hispanic Republican official put it, Hispanic voters had to "quit being chloroformed by blind allegiance" to the Democratic Party. Not everything was "sweetness and light" with the Nixon administration, another said, but the important thing was that "we're in Washington and being heard." They were the members of the new generation of Hispanic Republicans Robert Benitez Robles wrote about after Goldwater's loss.

Bañuelos and Sanchez gave speeches in Spanish, were honored guests at fund-raising dinners, and attended community events throughout the Southwest, including one that served more than a thousand pounds of free barbecue and featured entertainment provided by mariachis, a Spanish-speaking disc jockey, and a performance of the hit song "La Bamba" by the Spanish-born singer and actress Lita Baron.

Bañuelos was the star of the show. She autographed bills with her name on them, then spoke powerfully "as a business woman, wife and mother." She said that Nixon's opponent, George McGovern, was a flip-flopper who couldn't be trusted. The president was ending the war in Vietnam and had already lessened tensions between the United States and China and Russia. Going forward, Hispanics would be increasingly influential in the administration, but they had to speak up and be heard. "We must be free to speak our minds, free to study the facts, free to vote for those officials who give us deeds

instead of words," she preached. "Our votes are tremendously important" because they "are the tools that can build the kind of country
we want."

Bañuelos told a crowd of Mexican Americans gathered in Texas
that she heard one question most frequently: "Why should the American of Mexican, Cuban, Spanish, or Puerto Rican descent vote for
Nixon?" Her answer was that Nixon was the first president to "give
recognition to the person of Latin descent," so they had to "help
re-elect him." After the events, she and other surrogates fanned out
into the community, visiting local banks and call centers where they
talked with volunteers and made calls for President Nixon.

Not everyone greeted Nixon's Hispanic surrogates warmly. Some
Chicanos called Bañuelos a *vendido*, or sell-out. The lieutenant gov-
ernor of New Mexico, a Democrat, said that Bañuelos and the other
Hispanic Republicans out on the campaign trail "do not represent
the kinds of opportunity needed by Spanish speaking Americans."
He called them "showcase Hispanos" before adding that "President
Nixon thinks he can get the votes of Spanish-speaking Americans by
putting a few mannequins in powerless show positions."

In California, Bañuelos was criticized at the end of the cross-
country bus tour she had taken with other leading Republican women.
They made more than fifty stops in the "New Majority People Ma-
chine." Their last stop was in Sacramento. As Bañuelos spoke, she got
drowned out by the throng of protesters, including many Hispanics
who were there to express their opposition to Proposition 22, which
failed but would have restricted the organizing rights of agricultural
workers. Governor Reagan called the protesters "riff raff" and said,
"Let's not let those bums stop us." McGovern campaign pamphlets
handed out at this and other gatherings warned Hispanics not to be
fooled by Nixon's rhetoric. They said that Nixon was the one who had
made promises he didn't keep, on jobs, housing, education, and many
other issues.

DESPITE COOL RECEPTIONS IN SOME PLACES, THE CAMPAIGN WAS A SUCCESS. IN THE days after the election, Bañuelos received letters from Hispanics expressing their thanks for her efforts, and expressions of how meaningful her travels had been. A Hispanic voter wrote a letter to Bañuelos that read, "In one day you won the respect and the hearts of many Hispanos." Ramirez's assistant at the CCOSSP similarly told her, "I have followed your travels and the speeches you have given and can assure you that they have been a direct cause of the high turnout of Spanish speaking voters in favor of the President." The praise continued to pour in. Everywhere she went, she had inspired Hispanic voters. The head of the reelection campaign in New Mexico told her that McGovern once had a comfortable lead among Hispanics there, but, on the eve of the election, polls showed Nixon up by two percentage points, forty-six to forty-four.

Nixon also reveled in his success. He had won the election convincingly, with forty-nine states and 60 percent of the popular vote. But he was especially gratified by how well he had done among Hispanics. He was the first Republican to win more than 30 percent of the Hispanic vote, setting a new bar for his successors. Hispanic Republicans believed that this accomplishment in particular enraged Democrats, who took their votes for granted. It "set the hair of the Democrats on end," CCOSSP chairman Ramirez wrote. It was a good thing, too, since Nixon won only 13 percent of the African American vote in 1972. That figure was up four points from 1968, when he won only 9 percent, but far short of the 20 percent his African American advisers hoped he would win.

From Air Force One, Nixon's counselor Robert Finch telephoned Ramirez to ask for his ideas about what the president might do to repay his Hispanic supporters. According to Ramirez, Finch said that Nixon wanted "to do something of such proportion and magnitude that he will be remembered by the Mexican Americans in a manner akin to their affection and remembrance for President Roosevelt."

Writing decades later, Ramirez said he immediately began telling Finch about his uncle Elías and the parishioners he had worked with at Sacred Heart Catholic Church in Pomona, California. His uncle worked at a steel mill and felt he could never return to his hometown in Mexico, out of fear that he wouldn't be allowed to reenter the United States if he left. "*Yo no tengo papeles*"—I don't have papers—his uncle told him. At Sacred Heart, Ramirez helped parishioners fill out forms for their baptisms, holy communions, confirmations, marriages, and funerals. He "came to know" them personally. They told him that they, too, didn't have papers. They told him of their "unfulfilled longings" that were haunted by "fleeting memories of youth" and by dreams that had vanished as time wore on.

Hearing of their plight, Nixon apparently decided on the spot: "The president wants to give them amnesty," Finch told Ramirez. "Get moving! And that's an order," he exclaimed. In his memoir, Ramirez placed himself at the center of what happened next. He said he began talking to members of Congress and working with Leonard Chapman, the new head of the INS. The goal was to figure out how to offer amnesty to undocumented immigrants.

Nixon had expressed his support for immigrants and immigration before, so his support for amnesty didn't come out of the blue. He had said in 1971 that he hoped "America will always be the land of the open door." His support stemmed from what he considered to be the positive impact immigrants had on the American economy, and how open immigration reflected American dynamism. As long as the "door is open," he said, "it means that this land will continue to grow and continue to prosper and continue to have that drive, which makes a great nation." If the initiative Ramirez and Finch discussed during their mile-high telephone conversation had come to pass, Nixon would have added to his trophy case by becoming the first president to grant amnesty to millions of Mexican immigrants. The so-called wetbacks were increasingly in the news, and so was Ramona's Mexican

Food Products when the INS raided the company again in December 1972, just after Nixon's reelection.

Ramirez recalled that momentum on immigration reform was picking up, but then, as he put it, the "dark clouds of Watergate crowded the skies." The *New York Times* profiled an individual who once was an "obscure congressman from the streets of Newark" but who gained increasing notoriety. His name was Peter Rodino. From 1971 to 1973, he served as chairman of the House Subcommittee on Immigration and Nationality. In January 1973, he proposed two bills that would have dealt with some aspects of the undocumented immigration problem, including increasing the number of work visas available to Western Hemisphere immigrants, and the employer sanctions that Hispanic leaders had called for after the October 1971 raids on Ramona's. His bills failed to go so far as the amnesty Nixon sought.

Rodino's attention to immigration soon got diverted. In late October 1973, following the notorious Saturday Night Massacre in which Nixon fired several Justice Department officials when they refused to dismiss special prosecutor Archibald Cox, Rodino found himself in the unenviable position of being the chairman of the House Judiciary Committee. His immigration proposals had gone nowhere, but the impeachment proceedings he oversaw eventually led to Nixon's resignation. Nixon's Hispanics were greatly dismayed. The scandal tarnished the president's reputation, threatening to erase all of the good he had done for them. Nixon's first term was greeted with a good deal of skepticism from Hispanics, but a significant minority of Hispanics became Nixon loyalists, and over time they would become loyal to the Republican Party as well.

CHAPTER 4

THE HISPANIC WATERGATE

Richard Nixon and Bebe Rebozo were together on a private island in the Bahamas on June 17, 1972, when they learned that five burglars had been arrested at the Watergate Hotel while fixing a bug they'd placed on phones at the headquarters of the Democratic National Committee (DNC). The following morning Nixon and Rebozo returned to Key Biscayne, Florida, where they had homes next to each other. After pouring himself a cup of coffee, Nixon read a headline on the front page of the *Miami Herald*: "Miamians Held in DC Try to Bug Demo Headquarters." Once the news sank in, Nixon, enraged, threw an ashtray against a wall and instructed his press secretary to state that the White House wouldn't comment on the "third-rate burglary attempt."

The president spent the next two years denying that he was involved, until the fateful summer of 1974, when the Supreme Court's unanimous decision in *United States v. Nixon* required the release of the so-called smoking gun tapes that led to his resignation. Americans already knew that individuals in the White House helped coordinate the cover-up. Details about their involvement at first dripped slowly but then gushed thanks to the reporters, investigative committees, congressional hearings, prosecutors, and judges who turned the faucet handle in the same direction. By the summer of 1974, most of Nixon's closest associates had resigned. Many were sentenced to prison.

Loyalists insisted on the president's innocence until his very last days in office. Others suspected that he knew about and consented to the bugging, redirection of the investigation, and payments of hush money to the burglars to buy their silence. Still, when Americans listened to the tapes, for the first time they heard their president committing a crime. Within two weeks, the waterlogged floor of Nixon's support caved in and he was forced to step down; the Supreme Court delivered its decision on July 24, and on August 9, Nixon waved his last good-bye from the South Lawn of the White House. He boarded Marine One, en route to San Clemente, California, to resume life as a civilian.

The Watergate affair has been called the greatest political scandal of the twentieth century, the standard against which all subsequent scandals have been judged. It caused many to lose faith in government, led to campaign finance reform because of Nixon's corrupt campaign spending, and drove Americans to demand greater transparency in politics, which led to broad transformations that reshaped the cultural and political landscape for decades to come. If Nixon had not resigned, he would have been the first president removed from office.

Although Watergate looms large in our understanding of America's recent political past—as the subject of hundreds of memoirs, histories, movies, documentaries, and podcasts—Americans have forgotten, or never knew, how hemispheric and Hispanic politics were central to the whole affair. If Nixon had his way, Watergate would have been dismissed as one more intrigue in a long line of intrigues that defined the Cold War in the Americas. For Hispanics increasingly involved in electoral politics, Watergate brought up the painful question of whether their votes could be bought or easily manipulated; whether they were part of someone else's plan or independent political actors. The Nixon campaign's approach to Hispanics demonstrated that Hispanics mattered, but not in the way they had hoped. Yet Hispanic Republicans felt included in the party as never before. They wouldn't leave Nixon's side.

IN THE DAYS AFTER THE BREAK-IN, THE BURGLARY AT FIRST APPEARED TO BE THE WORK of Cuban freedom fighters who had been trying to oust Fidel Castro since the Cuban Revolution of 1959. Rumors had circulated in Miami that Castro was funneling money to the DNC, to back Democratic candidates who supported him and were secret Communists. The Cubans at the Watergate were looking for hard evidence. There were three of them: Bernard Barker, the son of a Russian father and Cuban mother; Eugenio Martinez; and Virgilio Gonzalez. Their boss, E. Howard Hunt, gave them nicknames that suggested their intimacy. Barker was "Macho," Martinez was "Musculito" ("little muscle"), and Gonzalez was "Villo," an abbreviated form of his first name Virgilio.

All had been born in Cuba, but after the revolution that placed Castro in power, they moved permanently to Miami, like thousands of other Cuban exiles. By the time of the Watergate break-in, Cubans had begun to establish businesses and engage in politics both locally and nationally as members of Nixon's Cabinet Committee on Opportunities for Spanish Speaking People (CCOSSP). Yet they were still motivated primarily by the politics of the Cold War in the Americas, and the Cuban Revolution in particular. Cuban exiles found it intriguing that the unfolding national news story was about the issue they cared about most, and involved members of their own community.

Barker was a real estate agent. Martinez was one of his salesmen. They had done business with Rebozo's bank, Key Biscayne Bank & Trust. Gonzalez, meanwhile, was a locksmith who worked at the Missing Link Key Shop. His role in the break-in was clear. A fourth burglar, Frank Sturgis, was not Cuban himself, but was a mercenary who first fought with Castro against Fulgencio Batista, the Cuban

president seen as a puppet for US business inte̶
tion, Sturgis switched sides and became a fi̶
These and other Cuban exiles would stop at n̶
island from Castro's dictatorship. James McCord, ̶
was an outcast in more ways than one; he was the only mer̶
team without a connection to Cuba and, according to Hunt, was
committed to their cause and didn't have the "gonads" to do the job.

Martinez wrote an account called "Mission Impossible," which
chronicled the events that landed him at the Watergate. Like other
Cubans in Miami, he felt betrayed by Kennedy, who stranded his
compatriots at the Bay of Pigs. He wasn't there or at Playa Girón on
those fateful days in April 1961, but he participated in more than three
hundred missions in Cuba as an employee of the Central Intelligence
Agency (CIA). He wrote about the famous Eduardo, who approached
him about participating in a top-secret operation. They arranged a
face-to-face meeting on April 16, 1971, which Martinez recorded as
the ten-year anniversary of the Bay of Pigs (landfall at the Bay of Pigs
actually happened on April 17). As the location for their meeting, they
chose the Bay of Pigs monument on Calle Ocho in Little Havana.

Sitting there together, Hunt asked Martinez to join a White House
team to spy on traitors to the country who were helping the Soviet
Union. Hunt originated the story of how the White House learned
that Cuba was funneling money to the DNC, and it would be the job
of Martinez and the other Cubans to prove it.

To assemble the other team members, Hunt scrolled through
the Rolodex of friendships formed over decades of fighting communism in the Americas. At the time of the break-in, Hunt was working
for G. Gordon Liddy, a lawyer who was himself working for John
Mitchell—the "Big Man," Hunt called him—who had recently stepped
down as US attorney general to head Nixon's reelection campaign.
The main role Hunt played for Liddy was as one of his "plumbers,"
someone who would prevent secret information from leaking beyond
the White House. Liddy devised Operation Gemstone, a series of

—"Diamond, Ruby, Sapphire, Opal, and others," Hunt wrote— .igned to undermine Nixon's opponents. To the Cubans involved, ney were all part of their effort to defeat Castro and win the Cold War. Defeating Nixon's opponents was tantamount to waging war on Castro himself.

Some Gemstone plots involved tapping the phone lines of Democratic candidates. Another was to have Cubans working for Hunt tamper with the air conditioners at the Democratic national convention, which they hoped would "make the other party's tempers skyrocket in the hot summer climate," causing them to make mistakes. The "most audacious and salacious" of their plans, Hunt wrote, would have involved renting a houseboat near the Democratic convention, from which they could eavesdrop on conversations in convention hotel rooms and where they could "lure high-level officials into sexual excursions" with prostitutes. The break-in at the Watergate Hotel was another Operation Gemstone project, and Hunt was the perfect man for the job because of his wealth of experience running covert maneuvers for the CIA.

Missions throughout the Americas had led Hunt to the Howard Johnson hotel across the street from the Watergate, from which he watched the caper go awry. After graduating from Brown University and serving in World War II, he spent years circling the globe with his family, rooting down in countries sympathetic toward, or led by, Communists. Fluent in Spanish, he spent more time in Latin America than anywhere else, in Guatemala, Uruguay, Cuba, and Mexico, where he worked with the young conservative writer William F. Buckley Jr., who wrote *God and Man at Yale* in the same year that he moved to Mexico. The book became an early bible of the new conservative movement. In 1960 and 1961, Hunt helped plan the Bay of Pigs invasion. He traveled to Havana to gauge whether Cubans on the island would rebel against Castro after the landing of Cuban exiles, and to Mexico City to evaluate the potential leaders of a Cuban government in exile.

Harassment by the Mexican government forced his return to Florida, where he continued with the planning. In Florida, he met a "Cuban American CIA asset" named Barker, who "had ties throughout the Cuban community in Miami." Barker had been an "invaluable assistant in maintaining contact with various leaders and recruiting new members for Brigade 2506," the force that would land at the Bay of Pigs. Hunt didn't meet Sturgis during the Bay of Pigs preparations, but he had heard of him as someone who flew planes between Florida and Cuba to drop anti-Castro propaganda over Havana. The members of the team that Hunt assembled to break into the Watergate were all familiar to him from his days planning the Bay of Pigs invasion, and subsequent attempts by Cuban exiles, with CIA help, to remove Castro from power.

The rest is history: the men were caught red-handed in the DNC headquarters, then were arrested, convicted, and sent to prison. Martinez himself served fifteen months.

After the burglars were arrested, the Cubans lent credence to the idea that the break-in was part of their effort to liberate Cuba. In response to a question about his profession, Barker said he was an "anti-Communist." The others nodded in agreement; this was their professional identity, too. They maintained throughout the trial that they were at the Watergate as anti-Communist, anti-Castro Cold Warriors, that they had broken into the Watergate because they would do anything Hunt instructed them to do and anything that might help Cuba become free. Barker drew laughter when he responded to a question about what he thought of Hunt's instructions. "I wasn't there to think," he said, "I was there to follow orders." They were happy to dig up evidence that the Democrats supported Castro; they wanted to help Nixon if it would further their cause. Nixon had been an ally in their struggle, just as Goldwater had been. It was on Nixon's recommendation in March 1960, when he was vice president, that Eisenhower directed the CIA to begin planning the covert military maneuver against Castro. Moreover, Hunt was a hero to them. They were eager to join the team of the man that, for the past decade, they had known only as "Eduardo."

Once reporters began to pull on these threads of the case, it made perfect sense that the Cubans would have Hunt's phone number in their address books. If the Mexican Americans in the Nixon administration called themselves the Brown Mafia because of the aggressive tactics they used to court Hispanic support for Nixon, Cubans served as Nixon's Brown Mafia in a more literal sense. They were the crew that had ties to Miami mob bosses and Nixon associates, including Rebozo. They were eager to do the president's dirty work for him. They wouldn't even accept payment for breaking into the Watergate—only for travel to and from Miami, lodging in DC, and recompense for the wages they lost for missing work.

NIXON AND HIS INNER CIRCLE WERE THRILLED AT THE OPPORTUNITY TO BRUSH THE BREAK-in aside as the act of fanatical Cubans. They would "crank up" the Cuban story and stick to their line that the burglars had broken into the Watergate "for their own political reasons," as Chief of Staff Bob Haldeman put it. That would divert attention from the White House itself, they believed. They got the FBI's acting director, Patrick Gray, involved as well. One of the most damning details revealed by the tapes was how, in the Oval Office, Nixon, Haldeman, and White House counsel John Dean conspired to have Gray instruct his deputies at the FBI, including Mark Felt, who was unmasked as "Deep Throat" only decades later, to lay off the investigation of key Watergate witnesses.

Nixon's team wanted to make the Watergate investigation a matter of national security, which would have been the CIA's domain, rather than a domestic matter to be investigated by the FBI. There was plenty of overlap between international and domestic affairs during the Cold War. A non-Watergate example was the relatively new war on drugs, which Nixon made not only a matter of domestic policing but also a part of the Cold War in the Americas, as the United States sought to crack down on production and trafficking in Mexico and Colombia in particular. But at the outset of the Watergate investigation, the FBI and CIA understood their jurisdictions to be separate. In what Gray called the most shameful decision of his life, he at first went along with the White House's plan and instructed Felt and others at the FBI to mind their business and investigate no further. The tapes demonstrated that Nixon knew about and supported the plot for the FBI to go along with the Cuba story. As much as any single fact revealed in more than two years of investigation, this one led to Nixon's resignation.

But in the days after the break-in, Nixon's political instincts made him go beyond trying to deflect personal culpability; he also tried to devise a way to make the burglary work in his favor. It became a part of his plan to eviscerate his general election opponent, South Dakota senator George McGovern. If the "Cuban angle" became the way the story "starts to bounce," Nixon said, then he recommended reaching out to Rebozo, who could spin it into a way of raising funds from Cubans in Miami for his reelection, since they were "very much against" McGovern. To Haldeman, the plan demonstrated Nixon's political genius. It would "kill two birds with one stone," clearing Nixon while also firming up support from Miami's Cuban exile community.

The Watergate burglars and others actively organizing to overthrow Castro may have been extreme, but even the Cuban exiles who entered the political mainstream supported their cause. Their dislike for Castro united extremist and mainstream Cuban exiles; it defined them politically.

For a brief period, Nixon's spin worked. The arrested Cubans became powerful symbols. Once newspapers revealed that they had spent a decade traveling covertly between Miami and Cuba as anti-Castro activists, they became heroes. The *Miami Herald* interviewed Cubans who couldn't understand why the burglars at the Watergate were considered criminals. They thought they were working for a cause that almost all of them supported.

But the White House's effort to frame the break-in as another episode of the Cold War in Cuba quickly broke down as investigators uncovered White House involvement. Five burglars wearing suits and surgical gloves was enough to raise suspicions, but initials discovered in Barker's address book, found at the scene of the crime, first linked them to the White House: "HH" and "WH," which, the *Washington Post* reporter Bob Woodward ingeniously deciphered, stood for Howard Hunt and White House.

Woodward called the White House, asked to speak to Hunt, and was forwarded to the office of Nixon lawyer Charles Colson. Colson's secretary told him that Hunt also worked at the public relations firm Mullen & Company. That's where Woodward reached him and asked why Barker would have his phone number in his address book. "I have no comment," Hunt said. When Nixon learned of the address book snafu, he leaped to protect his best friend. "Is Rebozo's name in anyone's address book?" he asked.

THE CUBAN BURGLARS WITH CIA CONNECTIONS EMBODIED THE WATERGATE AFFAIR'S ENtanglement with the politics of the Cold War in the Americas, but a Mexican connection discovered in the days after the break-in suggested it was tied to US–Latin American relations more broadly, that the scandal became part of the web of connections that linked the United States to countries beyond its own borders.

The $100 bills that the burglars had in their possession when they were arrested—fifty-three bills, to be precise, along with electronic bugging devices and photocopying equipment—landed in Bernard Barker's Miami bank account via Texas, after they were laundered in Mexico by a prominent Mexico City lawyer named Manuel Ogarrio Daguerre. The money traveled some of the same routes—from Texas to Mexico to Texas to Florida—that financially had connected the United States and Latin America for centuries.

In the fateful Oval Office meeting of June 23, Nixon said the only way to explain the appearance of the Mexican money in Barker's possession was that the Cuban freedom fighters had approached Mexicans for help, and Mexicans agreed. Cuban and Mexican connections were part of the same Cold War story; they were working together to oust Castro. This again would make the Watergate break-in a matter of national security and therefore be in the jurisdiction of the CIA; it would throw the FBI off the scent of a trail that led to the White House.

The Cuban and Mexican connections could make the break-in at the Watergate seem like an effort coordinated by actors on the interfacing frontiers of the United States and Latin America that only happened to make its way to Washington, DC. But the truth was more damaging to Nixon.

Federal investigators traced the funds withdrawn from Barker's account to one of the biggest banks in Mexico City, Banco Internacional, SA. But that wasn't their point of origin either. In the months after the break-in, over the summer of 1972, investigators discovered that their path to Mexico in fact began in Texas. Three oil executives with Pennzoil and the Gulf Resources & Chemical Corporation—both companies based in Houston—served as the Texas heads of Nixon's reelection campaign finance team. They had raised hundreds of thousands of dollars from donors, including Democrats, who wanted their identities to remain anonymous.

The head of Gulf Resources, whose attorney in Mexico happened to be Ogarrio, said the most convenient way to preserve their anonymity was to route the money through Mexico. They sent Ogarrio $100,000. He converted it into four checks totaling $89,000, plus another $11,000 in cash. Then a young man "with a Mexican surname"—presumably one of Ogarrio's associates—returned the money to Houston in a "large pouch."

From there, the Texas oilmen flew the checks and a suitcase full of cash to Washington, DC, where they drove it directly to 1701 Pennsylvania Avenue, the headquarters of the Committee to Reelect the President (CRP, or CREEP, as critics called it). They handed the money to committee treasurer Hugh Sloan, who handed it to Liddy, who handed it to Barker, who deposited it in mid-April, two months before the break-in.

Ogarrio, of course, denied any wrongdoing. He was a "well-known" sixty-three-year-old, semi-retired lawyer who represented several American clients. It was routine, he said, for American companies to send cash to their American-owned Mexican subsidiaries, which was exactly what his clients had done in this case. Since his name on the back of the checks was typed, rather than written in ink, he denied that he had even signed them. He spoke little English and said he was in "delicate health" at the time, seemingly to avoid questioning from American reporters. He wanted nothing to do with them. But his son, who was also a lawyer, with a degree from Harvard Law School, confirmed that the signatures were his father's and said he found that fact worrisome.

Within days of the break-in, the FBI was planning to send investigators to Mexico to get to the bottom of how the transfer of funds had worked. As many government employees learned the details of the laundering, they suspected the effort to conceal donor identities came in response to a new campaign finance disclosure law (S.382, the Federal Election Campaign Act, signed by Nixon himself on February 7, 1972) that required candidates to provide lists of the names of individuals who'd given them money. It was exactly the kind of investigation that Nixon hoped to prevent by claiming that whatever had transpired was a matter of foreign relations and national security and should be handled by the CIA. But in fact, as a matter of campaign finance law, it was a domestic affair.

The campaign finance disclosure law took effect just days after the money, nice and laundered, returned from Mexico. Investigators wanted to know if the money was a personal donation, which was legal, or a corporate donation, which was not. The Texas oilmen contended that the money was their own personal contribution, but they had stated earlier that donors to Nixon's campaign didn't want to reveal their identities. If the contributions were personal, why invoke the desires of donors to remain anonymous? Also, had any of the money originated in Mexico, which would have constituted an illegal foreign contribution? Investigators wanted answers.

As soon as investigators discovered that the burglars had ties to the Nixon administration through shadowy figures such as Hunt and Liddy, as well as to the highest members of Nixon's reelection committee, including Maurice Stans and Sloan, the investigation focused increasingly on those in the president's inner circle and their associates. Reporters, politicians, prosecutors, and judges worked to answer the question made famous by Tennessee senator Howard Baker: "What did the president know, and when did he know it?" The thread that tied the Watergate scandal to the Cold War in the Americas was lost as investigators focused on Nixon's campaign finance team and how they spent their money, but another one linking it to Hispanic Republicans emerged.

ONCE IT WAS DISCOVERED THAT CRP MONEY WAS INVOLVED, NIXON'S WHOLE CAMPAIGN finance operation came under investigation. When that happened, Hispanics and their representatives in the federal government were caught in the cross hairs of the Senate Watergate Committee.

From the day after the break-in until Nixon's resignation two years later, one government committee after another formed to examine the whole range of so-called dirty tricks by Nixon's campaign team. The committees then determined exactly how the CRP raised and spent funds to bolster Nixon's candidacy and discredit his opponents, first Maine senator Edmund Muskie, who had been Hubert Humphrey's running mate in 1968, then McGovern.

The Senate Watergate Committee, established in February 1973, was the most famous of these committees. Their hearings were broadcast live for months on end in the spring and summer of 1973, which

became known as the "Summer of Judgment" due to the highly consequential revelations made by those who testified. A Democratic senator
from North Carolina, Sam Ervin, chaired the committee. The Republicans on the committee included Baker from Tennessee, Edward
Gurney from Florida, and Lowell Weicker Jr. from Connecticut. The
Democrats were Daniel Inouye from Hawaii, Herman Talmadge from
Georgia, and Joseph Montoya from New Mexico, a Mexican American who was well known to, and generally disliked by, Hispanic Republicans. The committee disbanded in June 1974, when they issued
their report on the break-in, cover-up, and "all other illegal, improper,
or unethical conduct occurring during the presidential election of
1972, including political espionage and campaign finance practices."

By November 1973, when key members of the Brown Mafia testified about their meddling in Hispanic politics, the Senate Watergate Committee had been conducting hearings for the better part
of a year. Americans who stayed tuned throughout the fall learned
the details of Hispanic involvement at the highest levels of patronage
politics, including how the president's closest Hispanic advisers, the
Hispanic beneficiaries of the CRP's crooked dealings, and potential
Hispanic voters, got caught up in the Watergate scandal.

Many observers argued that the activities of Nixon's Brown Mafia
were key to understanding Nixon's reelection campaign, and constituted abuses of power that, along with all the other evidence against
Nixon, could lead to the impeachment of the president.

If the earlier televised hearings included bombshells about the
White House's involvement in the cover-up, the testimony by the
members of Nixon's Brown Mafia revealed how Nixon's Hispanic
supporters tried to drum up support for the president by doling out
government grants and contracts, dangling appointments in front of
prominent Hispanics, cutting deals with some of the Chicano movement's leaders, and making efforts to suppress support for McGovern.
For many Americans, it may have been the first time they saw politically influential Hispanics on national television. For Hispanics,

however, the hearings exposed their political leaders as both power brokers who commanded the attention of the White House, and corrupt dealers toeing a line between personal ambition and community support.

Like many involved with the Watergate break-in, members of the Brown Mafia considered their sins to be political rather than criminal in nature. What was politics if not patronage intended to convince a group of people that you and your party represented its interests? Politics had worked that way for decades before Watergate. But it also became clear through the hearings that Nixon's team withheld or withdrew the benefits of federal programs intended for all Americans if their recipients did not, or would not, support the president. That crossed a line.

IN THE MICROPHONED, ECHO-FILLED CHAMBER WHERE THE HEARINGS TOOK PLACE, HIS-panic politics came under a spotlight as two members of Nixon's Brown Mafia took the witness stand on consecutive cool fall days. William Marumoto testified on November 7, and Benjamin Fernandez testified on November 8.

Marumoto was an "assistant to the president." He reported to Nixon aide Charles Colson and was the designated leader of the Brown Mafia. Sitting before the Senate Watergate Committee, answering their questions for hours on end, he represented the other members of the Brown Mafia: Alex Armendariz, head of the Office of Minority Business Enterprise and Hispanic liaison to the CRP; Henry Ramirez, chairman of the CCOSSP; White House aides Antonio Rodriguez and Carlos Conde; and Fernandez, chairman of the National Hispanic Finance Committee for the Reelection of the President. In his testimony, Marumoto detailed for the committee exactly how the Brown Mafia worked.

Marumoto nervously read a prepared statement. He anticipated that the committee might wonder why he, a Japanese American, was chosen to lead a group of Hispanics. He told the assembled panel that he had grown up alongside Mexican Americans in the "barrios of Southern California," and, like them, had experienced racism and a "lack of opportunity" in the United States.

When he was eight, the US government placed him and his family in an internment camp in Arizona. Despite such scarring mistreatment, he nevertheless learned what "opportunity and justice is in America." President Nixon, he said, had opened many doors to him and Hispanics, and he didn't do it only because he was seeking their votes, but because he believed it was "basically right, and basically just." Marumoto's love of the United States, loyalty to the president, and commitment to economic uplift and the creation of opportunities for Hispanics inspired his work in the White House.

Marumoto cited Nixon's achievements for Hispanics as the reason for his loyalty to the president. Nixon created government programs for them. He continued to fund the CCOSSP. He made the "Spanish-speaking people active participants in their government," Marumoto said, by "appointing top-caliber Hispanos to executive-level, policy-making positions." By the time he was called to testify, the president had made fifty-such appointments, which, Marumoto said, was eight times as many as the number that Kennedy and Johnson appointed combined.

Thanks to Nixon, Marumoto continued, Hispanics would no longer take a "back seat" to other Americans benefiting from government services, in particular the president's educational, health care, and economic initiatives designed to provide business and job opportunities. Echoing the rhetoric of black and brown capitalism, he said the administration "discovered" that Hispanics "wanted to be a part of the economic mainstream of this country, that they wanted to be businessmen, industrialists, developers" to "share in the nation's economic wealth." With Nixon as president, more Hispanics had participated in business and government than at any other time in US history.

Such noble goals and accomplishments, Marumoto insisted, motivated the Brown Mafia to help Hispanics take advantage of opportunities provided by the government. They informed them of the programs, told them how to qualify, and then helped them apply. As they went about their work, they never "compromised the principle of legitimacy" and were "proud of the progress for Spanish-speaking Americans that we created."

For the rest of Marumoto's time before the committee, his examiners pushed him on how legitimate the Brown Mafia's activities actually were. Samuel Dash was the majority counsel for the Senate Watergate Committee. He asked the first round of questions. When Dash zeroed in on the Brown Mafia's involvement in the process of grant-making to Hispanic businesses and organizations, Marumoto acknowledged that the Brown Mafia helped them obtain federal funds. But the Brown Mafia also coordinated special events, such as a visit to the Southwest by the president of Mexico, and were responsible for communications and public relations between the government and Hispanic communities. They advised the president on high-level government appointments, and arranged speeches by cabinet officers and other surrogates at campaign events in Hispanic communities.

Under pressure, Marumoto also stated that Brown Mafia meetings related to federal grants were primarily political in nature, relating to their fund-raising efforts and winning Hispanic votes. But he insisted that they hadn't done anything illegal. Even their political motivations stemmed from a desire to be the best possible advocates for Spanish speakers. Their mission, Marumoto said, was also historically significant. They were members of the first administration ever, he claimed, to work for the full inclusion of Hispanics in the social, political, and economic life of the United States. Nevertheless, the grants demonstrated how the Brown Mafia tried to leverage its financial and political power for the benefit of the president. As the Brown Mafia itself put it, it was through the grant-making process that they sought to maximize the "political impact" of their work.

Marumoto told the committee about the Nixon administration's so-called Responsiveness Program, established in late 1971 or early 1972 to give contracts to Hispanic businesses and organizations, appoint Hispanics to government posts, and publicize Nixon's accomplishments with respect to Hispanics. Specifically, the Responsiveness Program awarded Hispanics and other "traditionally Democratic" voters federal grants and contracts, hoping to receive financial support or an endorsement for the president's reelection campaign.

Even though an overwhelming majority of Hispanics historically had voted for Democrats, they believed that Hispanic votes were winnable. They decided to target Mexican Americans in key swing states such as Texas, New Mexico, Arizona, Colorado, and California, as well Cubans in Florida and Puerto Ricans in New York and New Jersey. In a memo to then attorney general John Mitchell, the CRP made the case that Hispanic voters would be key to victory. Even better, they could be convinced to vote for Nixon, the CRP argued, if the president responded to their "social and economic problems."

The Brown Mafia therefore began soliciting applications from Hispanic-run businesses and organizations, and then, according to a local paper in California, "sifted" them "through a political screen" to gauge whether the applicant was a supporter of the president or could be convinced to support the president. If they couldn't win their support, they hoped to at least "neutralize" it to prevent the grant recipient from openly supporting Nixon's opponents.

The reelection campaign strategy for dealing with Hispanics was laid out in a document titled "Capitalizing on the Incumbency." The executive branch of government could "control" government assistance to Hispanics, and, in doing so, could "fill in any gaps in the President's record and generate favorable publicity for the campaign persuasion effort." Among other initiatives, they would offer grants, keep campaign staff updated about the administration's Hispanic programs, use spokespersons from various government agencies to publicize Nixon's support for Hispanics, and, especially in the months leading up to the election, supervise Hispanic organizations that received federal subsidies, to make sure they were "solving the problems of the Spanish speaking poor," for which they had received government support. This was the blueprint, elements of which future Republicans sought to duplicate.

Marumoto told the Senate Watergate Committee how they carried it out. The members of the Brown Mafia moved between federal government agencies and campaign fund-raisers. If a Hispanic-run business or organization received a grant, Fernandez, as chairman of the National Hispanic Finance Committee, would liaise between the Brown Mafia and the CRP, and shortly thereafter would approach the grant recipient to ask them for a donation to the president's re-election campaign. If a grant application was pending, members of the Brown Mafia might make a recommendation to award the grant or not based on the recipient's past or present support for the president, or their openness to supporting him in the future.

The Brown Mafia, in other words, coordinated with granting agencies and the CRP to make sure that Hispanics receiving support were also giving it. If the recipients of federal funds supported Nixon, money was more likely to keep flowing. If the Brown Mafia and the CRP learned that Hispanic grant recipients supported Nixon's opponent, or were on the fence, the well might run dry.

In total, the Brown Mafia facilitated more than $60 million in federal grants and contracts to almost fifty entities, from government agencies including the Department of Health, Education, and Welfare; the Department of Labor; the Department of Transportation; the Office of Minority Business Enterprise; the Department of Housing and Urban Development; the Small Business Administration; and the Office of Economic Opportunity.

They gave to Hispanic advocacy groups such as the League of United Latin American Citizens, the American GI Forum, and the National Council of La Raza; Hispanic social, vocational, and educational organizations including the Hispanic Baseball Association, the Truck Drivers School in Texas, and a group that received a Yellow School Bus grant to pay for student field trips; and Hispanic-run businesses and business groups including the Chicano Builders Consortium, Ultrasystems, and J. A. Reyes and Associates. The federal government spread money around with an eye toward recruiting Hispanics to support Nixon.

Some of the grants went to companies whose executives were closely connected with Nixon's reelection campaign. Ultrasystems, a Newport Beach consulting firm run by World War II veteran Fernando Oaxaca, received a $300,000 grant from the Urban Mass Transit Administration. J. A. Reyes and Associates, whose president, Joe Reyes, headed the National Hispanic Finance Committee's mid-Atlantic fund-raising efforts, was contracted to review some of the projects that received Brown Mafia–sponsored grants. The Urban Research Group of Austin, Texas, headed by Claudio Arenas, received a grant and then donated a small amount to the campaign that was used to pay for receptions and "souvenirs to be distributed to Nixon Supporters," including matches, pens, photographs, and a gold medallion with the president's face on it. In addition to federal grants, the Brown Mafia offered judgeships and a car dealership to a group of Mexican Americans in San Antonio.

Meanwhile, those who were cut off from funding were known to be closely aligned with leading Democrats, or for other reasons were resistant to helping Nixon win reelection. The members of the Senate Watergate Committee harped, in particular, on the experience of a grant recipient named Leveo Sanchez, whose Washington, DC, consulting firm, Development Associates, received $2 million in federal grants in Nixon's first term. During the campaign, Development Associates applied for another $100,000 grant, but they were denied.

Sanchez and Marumoto met on July 17, 1972, one month to the day after the break-in at the Watergate. During their encounter, Sanchez said he didn't plan to support Nixon. With the election only a few months away, Marumoto wrote a letter to a colleague at the

White House saying that Development Associates was a "classic example of a firm not necessarily being on our team which is making a comfortable living off us." One week after Marumoto sent his memo, his Brown Mafia colleague Alex Armendariz wrote a concurring letter. He added that Sanchez had "close ties" with the McGovern campaign, the DNC, and César Chávez, the farm worker organizer who was an enemy to many Hispanic Republicans.

Two months later, in September, Sanchez received a letter that congratulated him for having "graduated" from the federal grant program. He had been cut off. Sanchez wrote a letter of complaint, but to no effect. Marumoto testified that the memo only expressed his opinion, one that the agency did not have to follow. But the sequence of events made clear that Marumoto's wishes had been influential if not decisive.

The case of Leveo Sanchez and Development Associates was a clear example of the Nixon administration's abuse of power. A federal grant recipient was denied further funding because he did not support the president. Dash explained that his inquiry had nothing to do with whether the administration sought to give grants to minority groups, which would have been a perfectly reasonable, even admirable goal. Instead, it had to do with whether "certain members of that minority group were made enemies," excluded for not supporting the president and dismissed as potential grantees. Politics should not determine how agencies spent their money. Nor did politics justify that grant monies be given only to grantees who supported the president. But in this case, grantees who "refused to pledge themselves" to Nixon were "dropped brutally," Dash said.

If the committee focused on the case of Development Associates, it didn't overlook smaller offenses, such as the time the National Economic Development Agency (NEDA) refused to make its employees picket the *Los Angeles Times*, which had criticized Nixon's bombing of Cambodia. Similarly, the Southwest Council of La Raza was promised a $30,000 grant from the Department of Labor, but when the

nonpartisan group refused to endorse Nixon, the Brown Mafia recommended that the White House not fund their grant proposals.
Finally, a powerful Mexican American in Texas alleged that when he
refused to cut ties with the McGovern campaign, the Brown Mafia
told him that several charitable groups he worked with would stop
receiving federal funding. All of this, according to one investigator,
amounted to "political extortion" and an "effort to politicize the federal bureaucracy."

WITH MARUMOTO'S TESTIMONY, THE SENATE WATERGATE COMMITTEE HAD THE ACTIVITIES
of the Brown Mafia laid out before them. Still to come was Fernandez's denial of his involvement in a particular pay-to-play allegation.

Fernandez had become the most prominent Hispanic Republican
in the nation. In addition to his position as chairman of the National
Hispanic Finance Committee, he served as chairman of NEDA and
was a member of Marumoto's Brown Mafia. He had a reputation to
defend when he appeared before the Senate Watergate Committee
to answer questions about his dealings with Florida contractor John
Priestes.

Wearing a three-piece suit and red-and-blue-striped tie, Fernandez looked like a consummate professional as he flatly denied Priestes's
allegation that he offered the contractor a quid pro quo: Fernandez
would make Priestes's troubles with the US Department of Housing
and Urban Development (HUD) go away in exchange for a $100,000
donation to the National Hispanic Finance Committee. Fernandez
insisted that he "never promised [Priestes] any favors, directly or indi-
rectly." Defiantly, he stated that Priestes had done "everything in his
power to stain my good name" and he was "appalled, shocked, and
disgusted."

Fernandez explained to the committee how the affair came about.
The National Hispanic Finance Committee had been formed in early
1972 to raise money for Nixon. Fernandez held a press conference in

Miami to inaugurate the committee's efforts. It was well known that many Cuban Americans were well-to-do and were increasingly politically active. It made sense for Fernandez to begin his fund-raising efforts there, with them.

Fernandez addressed the guests at a cocktail party hosted by Cuban American doctor Manolo Giberga, another member of the National Hispanic Finance Committee. The gathering was an early example of how Mexican Americans and Cuban Americans were trying to come together to build a national Hispanic Republican movement. As chairman of the committee, Fernandez tried to meet everybody. He said he must have shaken a hundred hands that evening, including that of the Cuban American businessman who first told him about Priestes.

The unnamed Cuban American said he knew a "dynamic, aggressive, young multimillionaire" who wanted to contribute to the Nixon campaign via the National Hispanic Finance Committee. His name was John Priestes, and he had several Hispanic business associates. Fernandez testified that he was "pleased and delighted" to hear the news. It would be his first large donation. He claimed that several Cuban Americans at the party asked him to meet with Priestes in his hotel room, so he finally relented.

The next day, before Fernandez left town, Priestes visited his room to talk both about the donation and his troubles with HUD. One of Priestes's "associates" arranged the details of the meeting. Priestes showed up with a stack of *Miami Herald* clippings full of negative articles about him, but he did not have a check. Before he gave Fernandez money, he wanted assurances that Fernandez would do something for him in return. Fernandez said he could guarantee no such thing, but if Priestes had done nothing illegal, then he had nothing to worry about. Fernandez and Priestes parted ways but agreed to meet in Washington, DC, where they would discuss the matter in person with CRP head Maurice Stans.

Fernandez and Priestes next saw each other in March 1972, in

Fernandez's hotel room in the capital, where he was having a drink with Oaxaca. José Manolo Casanova was also there. He was the state chairman of the Florida Hispanic Finance Committee and was an increasingly important player in Cuban American and Republican Party politics. These were Hispanic Republican power brokers.

In his testimony before the Senate Watergate Committee, Fernandez recounted how the meeting with Stans didn't go as anyone hoped. Fernandez didn't get his contribution, and Priestes didn't get any guarantees that Stans would help him with HUD. Rather than taking his money, Stans concluded that Priestes was someone from whom they should distance themselves, quite the judgment given how the CRP was spending some of the money it raised.

Even though Fernandez was called before the committee to answer to Priestes's accusations, he used almost every second of his prepared remarks, at the beginning and end of his testimony, to advertise and defend his role in the building of the national Hispanic Republican movement. Fernandez recounted how he had spent the better part of two decades "fostering the free-enterprise system among Spanish-speaking people throughout the United States," bringing "Spanish-speaking voters into the two-party system of government," and doing everything he could to support Richard Nixon, who, he said, was the first president to embrace Hispanics "as first-class citizens."

Fernandez emphasized his success and commitment in each of these pursuits. Yet again, he said Nixon had introduced Hispanics to capitalism and gave them a taste of economic success. As the founder, chairman, and president of NEDA, Fernandez said he worked to "lead our people out of the economic wilderness." He claimed responsibility for increasing the number of banks that were "organized, managed, and controlled" by Hispanics. He also said he traveled some 250,000 miles in 1972 and raised $400,000 for the president's reelection. This was less than $2 per mile traveled, but still a strong sign of his commitment. Even considering all of Nixon's troubles, Fernandez, speaking for Hispanic Republicans generally, said "our dedication has not wavered."

Fernandez's testimony before the Senate Watergate Committee was supposed to be a somber affair. But he used his appearance before the cameras as his national coming out as the leader of the Hispanic Republican movement.

MARUMOTO AND FERNANDEZ TOGETHER REVEALED KEY DYNAMICS OF HISPANIC POLITICS in the early 1970s. Hispanic leaders represented a group of voters that was just as susceptible to the influence peddling and patronage politics that plagued other groups and that many Americans found so revolting about the Watergate scandal. Their testimony also showed that it wasn't only Hispanic Republicans who were vulnerable. The Watergate investigation implicated and exposed Hispanic leaders on all points of the political spectrum, conservatives and liberals alike. The members of the Brown Mafia had tried to cut deals with some of the most prominent leaders of the Chicano Movement for civil rights, and these leaders proved willing to play ball.

Linda Chavez, a Judiciary Committee employee who was in her midtwenties at the time of the Watergate scandal, later wrote in her memoir *An Unlikely Conservative: The Transformation of an Ex-Liberal (or How I Became the Most Hated Hispanic in America)* that she was one of the first people to look through the papers of the CRP. When she realized the extent of the Brown Mafia's dealings with Chicano Movement leaders, she forwarded some of the documents she looked at to Senator Montoya. He was from New Mexico like she was, and was also a member of the Senate Watergate Committee. Chavez knew that if she sent the documents to him he was likely to show them to the other members of the committee. She also leaked them to a reporter at the *Dallas Morning News*. This was how the Brown Mafia's actions became a subject of inquiry, she claimed.

The papers Chavez made public revealed how, in the summer of 1972, the Office of Economic Opportunity and the Department of Health, Education, and Welfare (HEW) awarded $2.8 million to the Zavala County Health Corp., which planned to establish a family health clinic in Crystal City, Texas, a border town in the heart of the Rio Grande Valley. Crystal City was the hometown of José Angel Gutiérrez, founder and chairman of La Raza Unida, the third-party spoiler that, for a brief period in the late 1960s and early 1970s, ran Mexican American candidates for local and state offices in Texas and then across the Southwest, against both Democrats and Republicans.

The proposed family health clinic was one of Gutiérrez's pet projects, and the Brown Mafia saw him as someone who could influence Mexican American voters in the valley. They didn't expect him to turn out Mexican Americans for Nixon, but they hoped he could help neutralize their support for McGovern. Traditionally, some 90 percent of valley residents, almost all of them Mexican Americans, had voted for Democrats, which was why the McGovern campaign sent teams of volunteers to canvass the area, including the recent Yale Law School graduate Hillary Rodham (later Clinton). The Brown Mafia and the CRP hoped to narrow such a lopsided margin. They thought that if they helped Gutiérrez with his health clinic, he, as the leader of La Raza Unida, could help them by suppressing the Mexican American vote for McGovern and limiting remarks that were critical of Nixon.

The Brown Mafia's dealings with Gutiérrez came to light during the Senate Watergate Committee hearings. When they did, Gutiérrez vehemently denied that he traded political favors for the grant. But reporters cited other evidence that made his denials seem false. Leading up to and during his organization's annual conference in El Paso, held in September 1972, he made decisions that could only have harmed McGovern and, conversely, benefited Nixon. Gutiérrez canceled an invitation to the UFW leader César Chávez, an important surrogate for McGovern, to speak at the conference. He also rejected the organization's "dump Nixon" resolution. Meanwhile, Gutiérrez's wife received an $18,000-per-year salary as the "medical director" of the proposed clinic, even though the HEW official working on the project considered her qualifications "marginal." It was unclear why Gutiérrez would have disinvited Chavez and abandoned the dump Nixon resolution, and why his wife got a job that she was seemingly unqualified for, if he hadn't made an agreement with the Brown Mafia.

Evidence presented to the committee also suggested that the Brown Mafia tried to cut a deal with the "King Tiger," Reies Lopez Tijerina, who was serving a federal sentence for a raid on the Rio Arriba Courthouse in Tierra Amarilla, on the edges of Kit Carson National Forest in New Mexico. A former Pentecostal preacher, Tijerina was the charismatic leader of La Alianza Federal de Mercedes, a movement that, at the height of its influence in the late sixties, claimed twenty thousand members who fought for the return of land stolen by the US government after the US-Mexican War in the middle of the nineteenth century.

A few years before the Watergate break-in, Tijerina and his followers had forced a confrontation when they entered the courthouse to demand the release of their compatriots, who had been arrested at a gathering of La Alianza. His men shot a sheriff's deputy and abducted two others. Tijerina hid in the forested mountains nearby, ultimately surrendered, and spent much of the next few years in and out of prison. He kept his following and served as a Chicano delegate to the Poor People's Campaign in Washington, DC, counting as friends activists such as Bobby Seale of the Black Panthers. Hoping to take advantage of, or to mitigate, Tijerina's influence with fellow Chicanos, the Brown Mafia allegedly promised Tijerina a presidential pardon for his actions in Tierra Amarilla in return for his support of Nixon.

THE LONE HISPANIC ON THE SENATE WATERGATE COMMITTEE, JOSEPH MONTOYA, WAS from Tijerina's home state of New Mexico. Montoya served in the

House of Representatives until 1964, when he was elected to fill New Mexico's open Senate seat, as part of the same Democratic wave that elected Lyndon Johnson and rejected Barry Goldwater. By the early 1970s, Montoya faced questions about his own efforts to launder $100,000 in campaign contributions, and even though the charges were never proven, his political reputation was damaged. During the Watergate hearings, Montoya grew increasingly frustrated as he listened to the testimony by Marumoto and Fernandez. He challenged them when he could, suggesting that the leaders of the CCOSSP, many of whom were closely allied with the Brown Mafia, had spent a great deal of time campaigning for Nixon but very little time actually working for Hispanics. He reminded Marumoto that he himself had introduced the legislation to renew the CCOSSP, not Nixon, as Hispanic Republicans claimed. As Fernandez proudly recited his accomplishments as chairman of NEDA, Montoya drew the audience's attention to criticism of Fernandez's leadership. But above all, he called the Brown Mafia's pay-to-play schemes an "incredible insult" to the Hispanic people.

Perhaps the greatest tragedy of the Watergate scandal for Hispanics was the revelation that Hispanic votes could be bought just like anyone else's. Hispanic politicians such as Montoya had dedicated their careers to the idea that Hispanics were independent political thinkers and equal participants in the political process. They weren't voters on the fringe who were valuable only insofar as they might swing elections, which made campaign teams draw up schemes to manipulate them. Fernandez had the same goal as Montoya, even though he was a Republican. Hispanics, he believed, had been manipulated by Democrats who sought their votes but then took them for granted. That's why he thought Hispanics should participate in the two-party system. In working to recruit Hispanics as Republicans, though, Fernandez and other members of the Brown Mafia ended up participating in the same corruption they criticized. The "grave consequence" of their "shady activities," one Hispanic writer opined, was the "substantial number of Latinos who have become disillusioned and bitter towards the two-party system."

Yet this wasn't what all Hispanics took away from Watergate. The legacy of the Nixon administration among Hispanics inspired not only disillusionment but also celebration. When it looked like the break-in was the work of anti-Castro freedom fighters, Cubans in Miami saw them as heroes. For some, the break-in should have been lauded, not condemned. For others, such as Ben Fernandez or Henry Ramirez of the CCOSSP, the Watergate investigation was a witch hunt, part of the effort by Democrats to destroy Nixon. When the Watergate scandal broke, Nixon's Brown Mafia stood behind their president. Nixon was the first president to take Hispanics seriously and include them in the Republican Party. After showing Nixon tepid support in 1968, their support for him increased dramatically. Because of the role Hispanics had played in his reelection, he rewarded them with several high-level posts.

While they were sad to see Nixon go, Hispanic Republicans were initially encouraged by Vice President Gerald Ford, who told them in a meeting at the White House—as impeachment proceedings were under way, but Nixon's fate not yet sealed—that he intended to "move ahead" with the administration's "domestic programs and politics." Nixon's Hispanic supporters took Ford to mean that Hispanics would continue to receive their "fair share from government." Nixon had guaranteed them a "piece of the action," but he wouldn't be president much longer.

Just three months after Nixon resigned, the nation watched as the so-called Watergate Babies swept the November 1974 midterm elections. Democrats picked up forty-nine seats in the House of Representatives and five in the Senate. They rode a wave of anti-Nixon, anti-Republican sentiment. Yet despite Nixon's and the Republican Party's damaged reputations, Hispanics continued to admire the disgraced president because he had created opportunities for them to climb the party ladder. If only for this reason, Nixon's Hispanic supporters remained loyal. He had been critical to their success. They forgave him, if they blamed him at all. They wouldn't turn their backs on him or on the Republican Party.

CHAPTER 5

POLITICAL POWER

The new president, Gerald Ford, appointed Fernando De Baca as the special assistant to the president for Spanish-speaking affairs. It was part of his plan to extend Nixon's efforts of inclusion. The announcement came on September 16, 1974—Mexican Independence Day, and the beginning of Hispanic Heritage Week—only eight days after Ford controversially granted a "full, free, and absolute pardon unto Richard Nixon for all offenses against the United States."

Only thirty-six years old at the time, De Baca already had more than a decade of government experience, both in the Nixon administration and New Mexico's state government, as the state manpower coordinator, regional tax director, and commissioner of the Department of Motor Vehicles. Right before his move to the White House, he was the western regional director of housing, education, and welfare. He had been offered the job as special assistant back in June, but said he didn't want it "until the Watergate scandal was resolved through a change of administration." He accepted when Ford extended the offer again.

De Baca's appointment was historic, he said, because there had "never been a person of Hispanic origin . . . on any president's staff" in the two-hundred-year-year history of the United States. Earlier in the year, the recently formed Spanish-Speaking Advisory Committee of the Republican National Committee, chaired by Manuel Luján Jr., had recommended the appointment of a special assistant. When he made the recommendation, he already had De Baca in mind because of his "outstanding Republican credentials" and deep familiarity with the needs of the Hispanic community, and because he believed De Baca would be an effective bridge between Ford and Hispanics across the United States. De Baca was eager to get to work.

Three months into the job, De Baca talked to the *Los Angeles Times* reporter Frank del Olmo about his ambitions and predictions for how Ford would relate to Hispanics. The president was from Michigan, and while he had some Hispanic acquaintances, he didn't have the natural connection with them that Kennedy, Goldwater, Johnson, Nixon, or California governor Ronald Reagan had. De Baca traveled across the country to learn about the "special concerns" of Cubans in Miami, Puerto Ricans in New York, and Mexican Americans in Los Angeles, and to convey the Ford administration's agenda.

Along the way, De Baca talked with Hispanics about the "common concerns" they shared with each other. All Hispanics wanted better health care and housing. They supported bilingual education and job training programs. There were more than ten million Hispanics in the United States, and half of them spoke only Spanish. The government needed to be able to teach them in their own language. He added his own priorities as well. An accurate counting of the Hispanic population was critical, he thought, and the 1970 census had undercounted them. He also wanted to encourage more Hispanics to go into business and for Hispanics to be more visible in general. So many of the problems Hispanics experienced, De Baca believed, stemmed from the fact that they had been invisible.

Finally, De Baca noted that unauthorized immigration was an increasingly thorny topic, one that neither Democrats nor Republicans had tackled "head on." Any letters the White House received about immigration landed on his desk. They expressed a wide range of opinions. There were screeds that said Mexicans stole jobs, didn't pay taxes, and brought diseases, and letters from police officers who claimed that Mexican immigrants "get all the welfare rights and benefits" that should be reserved for citizens. On the other hand, there were letters from Catholics who argued that "amnesty" must be part of any immigration reform, and from the United Farm Workers union, which opposed undocumented migration because it constituted a "black market" for "slave labor."

De Baca personally favored amnesty for immigrants who had lived in the United States for five years, as well as increased border enforcement. He said Ford felt the same way, but he didn't know whether the president would propose new immigration legislation. Congress had begun debating visa issues and employer sanctions, and the commissioner of the Immigration and Naturalization Service (INS), Leonard Chapman, characterized illegal immigration as a national emergency and asked for more resources as the number of undocumented immigrants went up. But because of the impasse, including within the Republican Party, between different sides of the immigration issue—especially over employer sanctions and amnesty—Ford opted for the status quo. De Baca acknowledged how tricky immigration had become when he told a reporter, "There will be no simple solutions."

Yet De Baca was optimistic. The year 1974, he said, "will be remembered as a year which marked a resurgence in the political process by Spanish-surnamed citizens." Two Hispanic governors had been elected that year, in Arizona and New Mexico. Hispanics also scored successes at the local level, on school boards and city councils. Taking over local politics was a strategy of the new right and the new left. "We know now that we can do it," that Hispanics could have "fuller political representation than we have had" in the past, De Baca said. During his travels around the country, he heard "a great deal of excitement" about the growing political engagement and activism by Hispanics.

He may have been one of the few Americans at the time to express such optimism about the years ahead. Many felt that the walls of the nation were crumbling around them. While Nixon responded to the Paris Peace Accord and the end of the Vietnam War with the slogan "Peace with Honor," most Americans knew deep down that the United States had lost. The Watergate scandal, rising crime rates, high unemployment, and shocks to the economy caused by OPEC's decision to stop exporting oil to the United States, resulting in the quadrupling of oil prices in late 1973 and early 1974, contributed to the malaise as the United States approached its bicentennial. Within the Republican Party, a conservative insurgency akin to the one waged by Goldwater's supporters was again threatening to unseat the moderates. Historians have used different metaphors to describe the 1970s—it was an age of fracture, an end to the Cold War liberal consensus, a period of deepening fault lines—but De Baca's optimism about the moment was rooted in the opportunities for political newcomers, including Hispanic Republicans. As divisions grew wider, Hispanics and others jumped at the chance to fill the void.

DE BACA WASN'T ALONE IN THINKING THAT HISPANIC POLITICAL PARTICIPATION WOULD surge as both Democrats and Republicans fought for their votes. The number of Hispanics in the United States increased by more than 50 percent during the sixties, from 6.4 million to 9.6 million, and would again increase by more than 50 percent in the seventies, from 9.1 million to 14.6 million. Thanks to a revision to the 1965 Voting Rights Act, the Hispanic electorate would also grow. Slowly but surely, the number of Hispanics in Congress was increasing. At the local and state levels, more Hispanics were running for and winning elected office. Their greater involvement was in many ways the fruit of the civil rights movements of the 1960s. Hispanics, African Americans, Asian Americans, Native Americans, and others fought for political representation, including by members from their communities, at the same time that they waged campaigns for racial, educational, and economic equality. Both Democrats and Republicans, to remain the purveyors of progress on these fronts, borrowed from the language of access, inclusion, and power that was critical to the success of civil rights activists.

Ford nodded to the growing political power of Hispanics by describing how they made the United States a more diverse and therefore stronger country. As the number of Hispanics increased, so did their influence on American life. When corporations recognized the potential of the Hispanic market, they blitzed Hispanics with advertisements. When many Hispanics became more upwardly mobile, they opened businesses, attended college, and consumed and produced mass media. As they represented a greater share of the electorate in important states such as California, Texas, Florida, and New York, politicians sought their support. The upshot was that encouraging Hispanic participation in all areas of American life—in economics, education, and politics—seemed critical to the ability of both parties to win future elections.

Even if the Republican Party was reeling after Nixon's resignation in 1974, Hispanic Republicans felt more optimistic about their place in the GOP than ever before. In 1973 RNC chairman George H. W. Bush inaugurated the Spanish-Speaking Advisory Committee. The year after that, the RNC made the Republican National Hispanic Assembly (RNHA) an official auxiliary. The purpose of the new organization would be to "consolidate and expand Republican gains" among Hispanics.

The first national gathering of the RNHA was held just two months later, in July 1974, as the Watergate scandal neared its denouement. This "organizational conference" at the Stouffers Hotel on Jefferson Davis Way in Virginia mixed business with pleasure. On the Friday night of the gathering, they had a "Fiesta Latina" and handed out "Grassroots '74" kits that taught Hispanics how to run an effective campaign. Several of them tried, in California, New York, and Florida, for example.

The main purpose of the conference was to stress the RNHA's two main reasons for being: bringing Hispanic Republicans together at the same time would correct the impression that all Hispanics were Democrats, and the speeches at the conference would convince Hispanic Democrats that the Democratic Party took their votes for granted, so it was important for Hispanics to participate in the two-party system of government to get both parties to fight for their support.

In a press release on the new union between the RNC and the RNHA, Luján announced Fernandez as the group's first chairman, praising his "many years of experience in organization, administration, and Party activism." The position would give him an opportunity to fulfill his longtime dream of uniting Hispanics across the United States, to connect them locally, regionally, and nationally to bridge their differences. His presence at the conference was one of his first official acts as head of the RNHA. He called the event the "most important Hispano meeting in the history of our country." It was a first step, he said, toward giving Hispanics a "permanent home in the GOP."

WHILE THEY WERE INITIALLY ENCOURAGED, HISPANIC REPUBLICANS QUICKLY DEVELOPED mixed feelings about Ford's presidency and the state of the Republican Party generally. Nixon had done a lot for them, so they stood by him leading up to and after his resignation. Yet he also brought negative attention to the party, ruining its chances for electoral success in 1974 and maybe 1976 as well. Hispanic Republicans were pleased when Ford, after becoming president, acknowledged the important role Hispanics had played in US history. In September 1974, to mark the beginning of National Hispanic Heritage Week, he said, "History shows that before Plymouth Rock there were Spanish-speaking individuals" settled in Florida, New Mexico, and Puerto Rico; they were "in the very vanguard of the settlement of the new world," and "contributed significantly with their deep religious convictions" and "their dedication to an outstanding family life."

But some of Ford's positions gave Hispanic Republicans pause, especially those of the foreign policy hard-liners. Experts expected relations with Cuba to soften under the new president. Since the Cuban Missile Crisis in 1962, there had been an embargo on trade with Cuba and no diplomatic mission to the island. Many suspected that Nixon continued the embargo in part because of the influence of his best friend, Bebe Rebozo, who was deeply rooted in the Cuban exile community in Miami. By contrast, Ford was expected to usher in a new era of détente with Cuba. He was seen as a pragmatist and a moderate, not an ideologue, even though he aided and abetted the long history of US support for right-wing dictators. In Chile, for example, Nixon had helped oust the socialist Salvador Allende and install Augusto Pinochet. Nixon then launched Operation Condor, which lent U.S. financial and technical assistance to South American countries seeking to defend neoliberalism against its opponents in the region. What hard-liners wanted was a continuation of Nixon's strident anticommunism, but they feared they'd get something more conciliatory instead.

Castro railed against continued US interventionism but also welcomed the détente. Messages from Cuba's foreign minister to Secretary of State Henry Kissinger, who promised a "new dialogue" and was seen as détente's main advocate within the administration, were relayed through Mexico's foreign minister. Cuba's better financial shape in the 1970s allowed it to purchase goods from the United States. US officials began talking about opening trade relations to gain access to the island's sugar, nickel, tobacco, and rum. In 1974, when the Organization of American States voted to allow Cuba's reentry as a member, the United States abstained instead of voting against it outright.

These all sounded like prudent measures from which the United States and Cuba would both benefit, but the United States was forced into its more flexible stance. After the loss in Vietnam, the country's loss of clout affected its ability to assert its will in the hemisphere.

The foreign policy establishment, including members of the Ford administration, felt that the United States would have to make concessions in countries from Cuba to Panama, where it began negotiating a treaty that would cede control of the Panama Canal.

None of this was good news in "right-wing circles," especially to Cuban exiles and Cuban Americans in Miami. To them, any degree of warming in the relationship between the United States and Cuba, or even the appearance of warming, was a misstep. As an article in the *Washington Post* explained, the opening of a dialogue with Cuba would mean "the curtailment of American hegemony in the hemisphere, violation of the Monroe Doctrine, and all manner of supposedly bad things." Ford soon found out how his support for détente with Cuba would affect him personally.

As Hispanic Republicans debated what Ford's presidency would mean for them, the RNHA came to symbolize the growing presence of Hispanics within the Republican Party. The Nixon administration had gotten the ball rolling with key appointments, economic programs, and a Hispanic outreach initiative that was more robust than ever before. But by the mid-1970s, Hispanic Republicans had become part of the official party apparatus.

As ever, the RNHA's goal was to bring Hispanic groups together in common cause. The leaders of the organization said that their first national conference had been a "historic" effort to unite "Mexican, Puerto Rican, Cuban and other Latin heritages." Their objectives were fourfold: first, "to promote and to encourage" the "participation of Spanish-speaking citizens in Republican Party activities and to work for the advancement of the Party's principles, policies, and candidates"; second, to "stimulate" the participation of Hispanics in the "regularly constituted structure of the Republican Party," as opposed to forming their own splinter organizations that the RNC would have little control over; third, to "endorse and support" Republican candidates for office; and fourth, to fund-raise for the Republican Party among Hispanics.

The RNHA also developed an impressive infrastructure and anticipated that the 1976 presidential election would for the first time "see them in action as an organization." Before the 1972 election, the mobilization of Hispanic Republicans happened more at the local and state levels instead of nationally. The RNHA had a plan to stitch together the local, state, and national levels by creating a national executive committee that would communicate with state and local chapters everywhere they were organized. The local chapters would do the on-the-ground work of courting voters and recruiting and supporting candidates.

As a Ford campaign pamphlet said, "While others have been talking about doing things for Hispanic Americans, President Ford is way ahead—he's already doing them." It was another jab at Democrats, who, Hispanic Republicans had long argued, took Hispanic votes for granted and were all talk but no action. A radio ad asked Hispanics to instead imagine a future in which Republicans like Ford, having earned Hispanic votes, continued to fight for them for fear of losing them. To Hispanics worried that Democrats would punish them for turning their backs on the party, the ad argued to the contrary, that Democrats would finally realize they had to compete for their votes. "The difference between political death and succeeding politically," it concluded, was remaining vital to both parties. This was the core of the Hispanic Republican argument. RNHA chairman Fernandez couldn't have said it better himself.

FORD HEARD APPEALS THAT THE REPUBLICAN PARTY HAD A REAL OPPORTUNITY TO RECRUIT Hispanic voters. When "Hispanic leaders" requested meetings with Ford, he got them on the calendar. Ford said he had a *puertas abiertas* (open door) policy for Hispanics, and encouraged them to call on him whenever they needed his help. It was part of his administration's posture of transparency, intended to restore faith in government after Watergate. Hispanic Republicans had formed the RNHA precisely to command this kind of attention. With "clout" came "power," Fernandez said. Their meetings with the president suggested that their plans were coming to fruition.

Ford met with Hispanics on three occasions in 1974 alone. They first met on April 11, when he was still vice president. Their second meeting came less than a month after his swearing-in ceremony, on September 4. Their third was only two months into his presidency, on October 17. These meetings, whose attendees were mostly Hispanic Republicans, were opportunities to discuss "the concerns of the Spanish-speaking community," including education, jobs, and political participation. Using the same kind of patronage politics as Nixon, Ford promised to appoint a White House liaison to the "Spanish-speaking community"—the position eventually filled by De Baca. "You will be proud of my administration's performance on behalf of Hispanic Americans," he said.

By early 1975, however, he had yet to meet with the leadership of the RNHA, and Fernandez, the leader of the organization, read this as a sign that the president was not taking Hispanics seriously. Fernandez sent RNC head Mary Louise Smith a letter that read, "I have never in my life seen an Administration dissipate a block of votes as swiftly as the Ford Administration seems to be doing." All the RNHA wanted to do was organize Hispanic Republicans nationally to help Republicans win elections. But so far, President Ford hadn't thought it a priority to sit down with them. Perhaps Ford thought he had covered his bases by meeting with other Hispanic groups, but Fernandez wanted to put his group front and center in the president's mind as the Hispanic group that could, or could choose not to, help him secure reelection.

It took Ford almost a year to respond to Fernandez's request for a meeting with the RNHA, which ultimately took place on December 11, 1975, just as Ford began his campaign for reelection. Until only recently, it looked like Reagan might not enter the race. Even after Reagan declared himself a candidate, members of the Ford for

President Committee disagreed about the threat he posed. Many saw him as a lightweight, but others recognized his appeal. On the eve of the primary elections, Ford didn't want to take any chances. He began to line up his supporters.

Ford had reason to be confident heading into the meeting with the RNHA. He had expanded the Voting Rights Act signed ten years earlier by Lyndon Johnson. Before 1975, many Hispanics couldn't vote because they spoke Spanish and couldn't read ballots. The expanded Voting Rights Act changed this by legislating the publication of ballots in Spanish and English in areas where the native language of more than 5 percent of the voters wasn't English. Ford also allotted $100 million to bilingual education programs, and continued the brown capitalism initiatives of the Nixon administration.

But Ford still approached the meeting with the RNHA cautiously. The RNHA needed his support, but he also needed theirs. Many of them would be inclined to support Reagan, who they believed would be successful among Hispanics in states such as California, Texas, and Florida, given his experience as the governor of a state with a large Hispanic population, his record of Hispanic appointments, and his strident anticommunism, which was sure to be popular among Cuban Americans. If Ford hoped to earn the RNHA's endorsement, he would have to give them something more. Hence the meeting at the White House.

The RNHA saw the meeting as an opportunity to brief Ford on the issues and to let him know that they were "presently uncommitted in the upcoming Republican Presidential primaries." Such bold expressions marked a turning point in the Hispanic Republican movement's experiment with hardball ethnic politics. Not only was the 1976 presidential election at stake, but also, according to Fernandez, all "future Presidential, Senatorial, and Congressional campaigns." Fernandez would remind Ford of how successful Nixon had been with Hispanics. They cast 2 percent of all votes for him, enough to tip a close election. All Hispanics were asking of Ford, Fernandez said, was that he "respond to the requests of a noble people, of a gracious people, of decent Americans who wish only the right to participate as Americans and as Republicans."

The leaders of the RNHA flew to Washington from all across the United States, to meet with President Ford in the Roosevelt Room. More than thirty of them were in attendance, including state chairpersons José Manolo Casanova of Florida, Nelson Carlo of Illinois, Martin Castillo of California, and Rita DiMartino of New York. This collection of representatives—Cuban Americans, Puerto Ricans, and Mexican Americans—reflected the shifting leadership of the Hispanic Republican movement, away from the primarily Mexican American leaders of Johnson's Inter-Agency Committee on Mexican American Affairs and Nixon's Cabinet Committee on Opportunities for Spanish-Speaking People, toward a more pan-Hispanic movement that was representative of different national groups from different areas.

The White House's task, as administration officials understood it, was to "give recognition and visibility" to the RNHA. They believed that the gathering would be a prime photo opportunity that could have a significant political upside. Photographers and reporters were on hand, anticipating that "favorable press coverage" might persuade the RNHA to officially back Ford.

THE TWO MOST IMPORTANT PEOPLE IN THE ROOM WERE FERNANDEZ AND DE BACA. FERnandez was there, of course, as RNHA chairman. He seemed to be everywhere. To the Ford administration, he represented Hispanics nationally. The state-by-state chairs of the RNHA were there as his lieutenants. De Baca's job, meanwhile, was to listen to Fernandez and the state-level leaders, then translate what they said to the president and his aides. Through Fernandez and De Baca, the RNHA and the White House hoped to advance their respective agendas.

By and large, Fernandez and De Baca agreed about what Ford's aims should be. In his first few months at the White House, De Baca spread the word that Ford recognized Hispanics as an indispensable segment of the electorate. Politicians had to harness their power as voters and better represent their needs. He promoted the Ford administration's support for bilingual education, the amendment to the Voting Rights Act, and Hispanics working for the federal government.

Like the RNHA, De Baca wanted to bring Hispanics together; to bridge their many differences, "somehow sorting out all those problems which are common to many of us and then developing programs that address the specific needs of the individual subgroups." He also believed that bringing Hispanics together would be their greatest shot at political power. Politicians who responded to their interests would be voted in, while those who did not would be voted out. The RNHA had supported De Baca's appointment because they believed that he would bring them "instant visibility" and establish "in the eyes of the community the eagerness of President Ford to give Hispanos a voice in his Administration." With his arrangement of the meeting, and his kindred desire to unite Hispanics nationally, De Baca met their expectations.

At the White House, Fernandez and De Baca guided the discussion of a wide range of issues, including Cuba, presidential appointments, and the role that Hispanic voters would play in the upcoming election. All agreed that Ford needed to appoint more Hispanics. Fernandez stressed the symbolic importance of high-level appointments. President Kennedy, he noted, had appointed three Hispanics to super-grade positions such as secretaries, under secretaries, or assistant secretaries. President Johnson had appointed six. President Nixon had appointed a whopping fifty-five. In the Ford administration, there were sixty-one, but almost all of them were Nixon administration holdovers. Ford himself had appointed only two—De Baca and Oaxaca, who worked for the Office of Minority Business Enterprise (OMBE)—a dismal number that Fernandez called "poor."

To show Hispanics that he was committed to them, Ford had to make a "concerted effort" to appoint many more "qualified Hispanic Republicans to higher positions in the Administration." The RNHA wanted to see more super-grade appointments and appointments as heads of commissions, boards, advisory councils, and independent agencies, such as the Federal Communications Commission or the Federal Reserve Bank. A few judicial appointments would be nice, too, including, perhaps, a nominee to the Supreme Court. They asked Ford to issue an executive order committing to his prior statements of support for Hispanics in the areas of jobs, education, and political inclusion.

The leaders of the RNHA also urged Ford to stand strong against Fidel Castro—something he had already shown himself to be con-flicted about—and do everything he could to support Cuban exiles living in the United States. Casanova advised Ford to maintain a "firm posture" against Castro. Despite reports that Castro had pulled back from his ambitions in the hemisphere, Casanova claimed that he continued to export revolution to other Latin American countries, "perhaps even more actively and overtly than before." The thing Cu-ban Americans feared most, Casanova stated, was an administration that advocated "coexistence" with Cuba.

Casanova also had a specific complaint: Cuban exiles waited too long to become legal permanent residents and then citizens. It took them two to four years, whereas the process used to take only a few months. According to Casanova, the Cubans in line to become citi-zens were already "living, working and paying taxes in the United States." Casanova asked Ford to expedite the process for them so they could "be absorbed at an accelerated rate into the mainstream of the American system." For months after the December meeting, the issue preoccupied Ford. The president had heard Casanova, and he wanted something done, in part because the Florida primary was approach-ing, and Reagan was making inroads among Cuban Americans there.

Ford directed his aides to talk to lawyers and the heads of the INS to find out the reasons for the delays. Some aides didn't understand the urgency of the matter, while others, including De Baca, didn't grasp the details of how immigration law applied to Cubans. Casanova suggested that Ford might issue an executive order to speed up the processing of applications for permanent residency. The INS said that wasn't possible, since the president was constrained by current immigration law.

The Immigration and Nationality Act of 1965 for the first time limited the number of immigrants from Western Hemisphere countries to 120,000 per year, and no more than 40,000 from any one country. Whether Cuban entrants counted toward the quota was a matter of some confusion. Many were humanitarian parolees who didn't count toward the quota. Others applied to become legal permanent residents, but immigration laws required them to enter from a third country—usually Canada or another Caribbean country—and to pay visa fees, so only Cubans with means could take this step. As immigrants, not refugees, those who became legal permanent residents counted toward the quota. Even though the 1966 Cuban Adjustment Act was to make it easier for Cuban exiles to become legal permanent residents, by the mid-1970s Congress still hadn't decided exactly what their status should be.

The law also established the Select Commission on Western Hemisphere Immigration, which had the task of reporting on its fairness, its application to Cuban refugees, and other matters. In the commission's final report, delivered in 1968, it noted that at least some Cubans were immigrants who were counted toward the overall quota. Their applications, like those submitted by all other immigrants, were processed on a first-come, first-served basis. Most of the members on the select commission felt this shouldn't be the case; Cubans were protected refugees, not immigrants, so they shouldn't be counted. But when the leaders of the RNHA met with Ford at the White House, Casanova complained both that Cubans were waiting too long to have their applications processed and that many were still counted toward the Western Hemisphere quota.

In 1975, the number of Cubans who had submitted applications and were waiting in line was seventy thousand, half of them in Miami alone. The INS reported that it was processing the applications of Cubans at a rate of about nineteen thousand per year; hence the long wait that Casanova complained about. Once their applications were processed, they still had to wait thirty months to be naturalized, which was half as long as the five years that other Western Hemisphere immigrants had to wait. It was a special dispensation that was "unique," as one administration official put it, because the United States encouraged their defection from Cuba. For Casanova, these nuances didn't matter. He just wanted something done about it.

According to lawyers and INS officials, the best solution would be further legislation to benefit Cubans. They discussed already proposed bills to eliminate the distinction between Eastern and Western Hemisphere quotas, which would increase the number of visas available to Eastern and Western Hemisphere immigrants to three hundred thousand. Such an approach might have the added benefit of alleviating pressure caused by undocumented immigration from Mexico, since a greater number of Mexicans would be able to receive visas, reducing the percentage who came illegally. Other solutions—the ones ultimately endorsed by the president himself—were to exempt Cubans from the Western Hemisphere quotas and to prioritize the processing of applications from Cubans in Miami above the applications from all other Latin American immigrants.

To demonstrate his responsiveness to the needs of Cuban Americans, Ford wanted to announce the new plan when he visited Miami on February 28, 1976, at the beginning of the primary season. Shortly before the Florida primary, De Baca also announced his resignation as Ford's special assistant. It was another move to placate Cuban Americans, who thought De Baca embodied the idea that Mexican Americans held too much power in Hispanic Republican politics.

Because Ford seemed amenable to the demands they had made at the White House, Fernandez expressed confidence that the meeting had achieved its desired ends. It "laid the foundation," he said, "for a positive long-term relationship" and expressed his organization's desire to be an "integral part of this man's present and second administration." Ford got what he wanted: the RNHA decided to back him.

Fernandez then sent a letter to all RNHA members that gushed, "Never in the history of the Republic has the President of the United States met with a permanent Republican organization consisting of a cross-section of Americans of Hispanic origin!" He underlined "cross-section" to emphasize his goal of bringing Hispanics together. The meeting was both historic and "superb," he said. Ford's "attitude toward the goals and objectives of Hispanic-Americans was positive and absolutely wholesome." He concluded with some of the most oft repeated, ultimately most damning, sentiments about the president: "He is a good man; a sincere man; one who is trying to be a good President." Fernandez's tone about Ford was entirely changed after the meeting at the White House, from the skepticism he expressed in early 1975 to the optimism he conveyed in the New Year. On the road, during a visit to Michigan, Fernandez told his audience that the president was "doing a sensational job."

THE RNHA HAD CONFIDENCE IN FORD, EVEN IN THE FACE OF THE SERIOUS CHALLENGE POSED by Reagan. The two men were a study in contrasts. Ford was part of the political establishment; Reagan postured as an insurgent railing against insiders. Ford represented traditional midwestern and eastern conservatism; Reagan blazed a trail of success in the Sunbelt states stretching from Georgia to California. Ford stabilized the Republican Party after the fallout from Watergate; Reagan denied that the party even had a Watergate problem. Ford was wooden; Reagan epitomized charisma. Reagan knew how to eat a tamal, a popular Mexican food made of ground corn. Ford did not.

On April 9, at a reception at the Alamo in San Antonio hosted
by the Daughters of the Republic of Texas to commemorate the
bicentennial, Ford suffered a highly symbolic setback. He had just
eked out a victory in the important Florida primary, even though he
lost to Reagan in Miami's Dade County, where most Cuban American voters lived, by a margin of 71 to 29 percent; Ford had the backing of Florida's RNHA chapter, but other Cuban American leaders
supported his opponent. Even if the concessions to Cuban Americans
after the White House meeting didn't go exactly as planned, Ford had
momentum after victories in Florida and other states. Many politicos
were saying that Reagan's campaign was on the verge of demise, but
Ford still needed a strong showing in Texas.

That's when the president visited San Antonio, where he awk-
wardly, now infamously, bit into a tamal before removing its tie and
cornhusk. The *San Antonio Express-News* printed a photo that caught
Ford "looking like he was in the middle of gulping something painful," one historian wrote. Some attendees said they didn't notice the
mishap. They learned about it only when they read the news the next
day. Nevertheless, it didn't take long for the episode to become known
as "The Great Tamale Incident."

The incident helped shift momentum toward Reagan, who ended
up winning the Texas primary by a margin of 66 percent to 34 percent.
Ford reflected on his loss when he returned to Washington. "There
were two things I learned in Texas," he said, "One is—never under-
estimate your opponent, and two, always shuck the tamale." Reagan
had other advantages in Texas besides Ford's missteps. He was winning all of the western states and was the chosen representative of the
new conservative movement, whereas Ford's areas of strength were
the staid Northeast and Midwest. He wouldn't clinch the nomination
until the convention itself.

Newspapers big and small had a field day at Ford's expense. One
in California claimed, "this Fordian performance at the Alamo is not
a minor breach of ordinary dinner table etiquette"; rather, it was a

"catastrophe of major dimension." Mexican and American cultures were thoroughly blended in San Antonio. Tamales had been "canonized" as an important part of American gastronomy. The made-up word *Fordian* was a play on *Freudian*, suggesting that the gaffe revealed something primal about the president. It told voters all they needed to know. Any American "who cannot handle a hot tamale," the author wrote, "had better forget about politics—in Texas anyway."

According to one historian, the tamal incident was significant because it showed Ford "choking on his own cultural ignorance." Another has written that the president's Hispanic supporters were "crestfallen" because of the gaffe at the Alamo, since it forced them to acknowledge that "the candidate from Michigan knew little about Spanish-Mexican culture of the Southwest." But the tamal incident also didn't strike them as a true measure of the man. It was an unfortunate incident that didn't look good, but it also couldn't erase the support Ford gave them. The leaders of the RNHA stuck with him, hanging on to their positive feelings after the meeting at the White House. They spent the rest of the spring primary season honing their talking points in support of Ford: that he supported statehood for Puerto Rico, passed the 1975 amendment to the Voting Rights Act, advocated immigration reforms including amnesty and speedier naturalization for Cubans, and appointed the first special assistant for Hispanic affairs. Ford had things moving in the right direction.

Such positive strides led Fernandez to write another letter to the national and state executive committees of the RNHA. He boasted of the hundreds of new members the organization had recently enlisted, in California, Colorado, Florida, Michigan, New Jersey, and elsewhere. He especially praised the Florida chapter for devising incentives to sign up as many members as possible. If a member recruited twenty-five others, they received recognition from the state chapter. If they recruited fifty members, they received recognition from the national organization. If they recruited a hundred members, they were recognized by the RNC. And if they recruited two hundred members, they received an all-expenses-paid trip to Washington, DC. His letter also included news about Hispanic appointments to the boards of directors of the Public Broadcasting System and the Panama Canal Corporation. These weren't the cabinet secretary or Supreme Court justice positions the RNHA hoped for, but they were steps forward.

EVEN AFTER FORD AND REAGAN CLOSED THE PRIMARY SEASON WITH A ROUGHLY EQUAL number of delegates, raising the possibility that the convention would be contested, the president and the RNHA made July a month of celebration. First, there was the birthday of the United States itself. Second, the RNHA was planning its first annual banquet. They sent out two thousand invitations and expected one thousand people to attend. Chefs at the elite Mayflower Hotel planned a French-inspired menu of caviar, smoked oysters, beef tartar, and pastries.

The evening would begin at six thirty, with cocktails and hors d'oeuvres with Vice President Rockefeller. Dinner would begin at seven thirty and would include speeches by Congressman Luján, President Ford, and Texas senator John Tower, who delivered the keynote address. Such distinguished speakers gave the banquet the air of import its organizers had hoped for, not only for themselves, but for Ford as well. The television personality "Smilin'" Jack Smith would be the master of ceremonies, and "Mariachi América" provided the musical entertainment. They were an all-Hispanic band whose members drove into the district from Virginia and Maryland.

Ford's critics insisted that with his attendance at the event, the president was only demonstrating how desperate he was for Hispanic votes. The RNHA, though, saw the event as a sign of how far they had come in just a few short years, all because of support from the Republican Party and the efforts of leading Hispanic Republicans such as themselves. It would an opportunity for them to display how well connected and influential they were. Significantly, the RNHA did not invite Reagan. They instead "steered all of the emphasis in the direction of the President." They pitched the event, held just a few weeks before the national convention in Kansas City, as "an excellent opportunity for the President to show to the Hispanic Voters of America that he is with them."

Ostensibly, the event was a "salute" to the Hispanics serving in the Ford administration, including Thomas Aranda of Arizona, who had replaced De Baca as the special assistant to the president for Hispanic Affairs; Alex Armendariz, director of the OMBE; Carlos Conde, public affairs officer at the Inter-American Bank; Fernando Oaxaca, associate director of operations at the OMBE; Philip Sanchez, who had moved on from the Office of Economic Opportunity to become the ambassador to Colombia; Albert Zapanta, assistant secretary for management at the Department of the Interior; and Fernando Penabaz, the former Goldwater campaign worker and current chairman of the National Advisory Council on Economic Opportunity. There were a few Democrats, too, including Raymond Telles, the former Mexican American mayor of El Paso and ambassador to Costa Rica, who at the time was the commissioner of the Equal Employment Opportunity Commission.

After Alicia Casanova, José Manolo Casanova's wife, recited the pledge of allegiance, Luján stepped up to the podium and began talking about the meaning of the bicentennial. He said it was an especially important anniversary because the United States had been riven by domestic and international turmoil, including tensions caused by the Vietnam War, Watergate, and high rates of unemployment. The country was "being chopped down on every front," he said. Bicentennial celebrations would help Americans "get ahold" of themselves and remember that the United States still had the "best system of government" and held "great promise for the individual." Luján said that for Hispanics, in particular, the celebration demonstrated how their history in the Americas was "intertwined" with the history of the United States itself.

By the time of his speech, Luján was the dean of Hispanic Republican politics. Others, such as Fernandez, served in presidential administrations as aides, in advisory roles, or as heads of government agencies, but Luján had been a congressman for eight years. He had helped found the RNHA, served as chairman of the RNC's Spanish-Speaking Advisory Committee, and climbed the ranks of Congress as a member of the House Interior and Insular Affairs Committee, which oversaw western lands and island territories including Puerto Rico. He came from a district that was 40 percent Hispanic, 10 percent Native American, and 50 percent non-Hispanic white and where Democrats outnumbered Republicans by more than two to one. His success in such a district signaled the potential crossover appeal of Hispanic Republicans. At the Mayflower Hotel, Hispanics listened to one of their most distinguished representatives.

Luján had said that his thinking wasn't all that different from the thinking of his Democratic colleagues in Congress. But once he spoke on issues from land conservation to social protest, his conservatism became clear. Luján supported the economic development of western lands, including mineral extraction, water use for commercial purposes, and the burning of fuel without regard for pollution. He recognized that the Spanish language enriched his life and opened doors of opportunity to him, yet he reduced funds for New Mexico's bilingual education programs. Discrimination was real, he acknowledged; many Americans held stubborn attitudes toward Hispanics, African Americans, Native Americans, and other minorities. Yet he rationalized a preference for lighter skin by suggesting it was a matter of personal comfort. "You can identify better with that guy who's not jet black," he said. Other Hispanics called that racism.

Lastly, like other Hispanic Republicans, Luján held hawkish views of US–Latin American relations. He spoke out against the "leftist" Mexican president Luis Echeverría for his alleged support of Castro and Allende. Unlike Ford, he did not want to return the Panama Canal. The president had begun negotiating with Panama to do just that, but ultimately decided it was too risky during an election year. Luján also dismissed charges of US imperialism in the hemisphere as akin to the tinge of envy—*poca envidia*, he called it—that everyone felt toward the "big rich guy who lives on a hill." Other Hispanics thought imperialism was more insidious.

Luján and other Hispanic Republicans did agree with liberals on one issue: illegal immigration. The Spanish-Speaking Advisory Committee he chaired passed a resolution in 1974 that supported granting permanent resident status to undocumented immigrants who had lived in the United States for fifteen years. They "significantly contributed" to the American economy, worked at difficult jobs for little pay, paid taxes without receiving benefits, had lived in the United States for a long time, and had children who were American citizens.

Hispanic Republicans and Democrats almost uniformly opposed employer sanctions, but for distinct reasons. Democrats opposed them because the threat of penalties could lead companies to discriminate against Hispanics in the hiring process. Many Republicans believed this as well, but others, including Luján, believed sanctions were an unfair punishment of businesses. It was the government's responsibility to prevent undocumented immigration, not the responsibility of business owners. Employer sanctions seemed to have gained momentum after the raids on Ramona's Mexican Food Products, but the issue fell by the wayside for the rest of the 1970s; or, rather, it sparked sufficient controversy that it didn't find support in Congress.

Ford's prepared remarks echoed Luján's. The bicentennial, he said, recognized "our rich national heritage—including our Hispanic heritage." Ford told them that the Republican Party needed their "abilities" and "skills." They swooned when he said that Hispanic "genius," "energy," and "pride" had strengthened the United States. He concluded by telling them that the GOP "provides the channel for a new surge and flow of Hispanic involvement from California and the Southwest, from Florida, from New York, and from all over this great land as we unite to preserve the principle of unity from diversity that makes America great." At the end of the president's speech, Fernandez handed him a plaque that praised him for his "untiring dedication and devotion to the promotion of Republicanism among Hispanic Americans in the United States. *Presentado con todo respeto.*" Presented with respect.

The banquet gave Ford and the RNHA momentum heading into the Republican national convention, where Fernandez was given a speaking slot. As it turned out, the Mexican American restaurant owner Hector Barreto had established the Hispanic Chamber of Commerce in Kansas City, Missouri, earlier in the year, so it was a fortuitous coincidence for him to have Hispanic Republicans visit his hometown to discuss their shared interests. With Nixon's support, his Hispanic appointees—especially Sanchez and Fernandez—had helped create opportunities for Hispanic-run businesses through the Small Business Administration and the National Economic Development Agency. The establishment of the Hispanic Chamber of Commerce was a testament to how much Hispanic businesses had grown during Republican administrations.

As for his speech at the convention, Fernandez argued that Hispanics could no longer be taken for granted by either party. This was the message he and other Hispanic Republicans had delivered for some time, and he would state it again in Kansas City. Even though he had praised Nixon during his testimony before the Senate Watergate Committee, he changed his tune. After Fernandez spent so much time and energy raising money for Nixon, Hispanics were "dropped like hot potatoes" after the election. But that wouldn't happen again, he said, because now Hispanics had power within the Republican Party. (At least he had power within the party, as the first Hispanic given a seat on the Republican National Committee.) "We're saying *basta*!" (enough), Fernandez exclaimed. "We're tired of Anglo politicians wearing big hats and serapes during their campaigns." He said it was insulting.

Because of the RNHA, he said, Hispanics now had a "power base within the GOP." He challenged the delegates at the "overwhelmingly white convention," as one newspaper article described it, to "demonstrate by bold, direct moves that you care; that you really want to expand the political base of the party." Pulling no punches, he warned Republicans about "not keeping in step with the times—and that step has a Spanish accent." His closing words were particularly forceful: "We're coming in whether the conservatives like it or not. They can protest all they want, but we're coming in, with dynamism. We're hoping to be invited but we're not waiting to be. We're moving in." Such language was representative of the demands for inclusion made by civil rights activists, and an example of how they shaped the rhetoric of Democrats and Republicans alike.

And so Fernandez and Hispanic Republicans charged out of Kansas City ready to hit the campaign trail. Emboldened, Fernandez sent a letter to Stuart Spencer, now of the Ford campaign, to suggest a budget for Hispanic outreach that included salaries for campaign workers, travel expenses, office supplies, and postage for mailings. He wanted to target Washington, DC, California, Texas, New York, New Jersey, Florida, and Illinois in particular. He said he would work "hand in glove" with the campaign and was confident that his efforts would lead a million and a half Hispanics to turn out for Ford, if he were given the support he needed.

FERNANDEZ AND THE RNHA SOON HAD THEIR HOPES DEFLATED. AS AUGUST TURNED INTO September, and the election neared, concerns again emerged about Ford's concern for Hispanics. For Fernandez, it was a return to the grave doubts he expressed back in early 1975. Hispanics had considered before whether they were shaping their own political destiny or whether they were pawns. They asked this question again after the convention in Kansas City.

If they had gotten their hands on one particular confidential memorandum, they would have concluded that they were pawns. "Mexican

Americans in California and Texas are different from one another (and jealous) and can be played" against one another, the memo said. "But better still they can be played off against the negroes to break the monolithic low class man that goes Democrat," and "we should do what we can" to "show the Chicanos we'll do more for them and not prefer the negroes (as the Democrats do)." The author concluded, the "terrain is ripe for this since the Chicanos are acquiring greater cognizance of their strength and heritage and the fact they figure last on the 'minority' totem pole." This was the old divide-and-conquer strategy, not the approach of a campaign that took Hispanics seriously and wanted to work with them, hoping they might be convinced that Republicans represented their interests.

The prospect that the Republican Party didn't really care about them was haunting, but they also had specific grievances against the Ford campaign that gave them pause and led them to anticipate a dire outcome. There were the ongoing critiques from liberals that the Ford administration had never really committed to addressing their problems and was insensitive to their needs. Some again questioned the very idea that political engagement would pave the way to success. One professor, Richard Santillan, wrote an essay called "Chicanos and the American Bi-Centennial," which criticized the assumption that politics was the key to "social and economic mobility." What, he wondered, had Hispanics gained from participation in politics? Little to nothing was his answer.

As Americans planned their celebrations, he argued, Hispanics were being denied "basic human rights." They suffered from political and economic inequality, discrimination, high rates of unemployment, poor housing conditions, impoverished schools, mistreatment by police, and hunger. He said it was "tokenism" to assume that the number of appointments and elected offices held by Hispanics signaled progress in the United States. July Fourth and election day, he concluded—echoing Frederick Douglass's famous speech "What, to the Slave, Is the Fourth of July?"—would only be painful reminders to the nation's fourteen million Hispanics that they remained second-class citizens.

These criticisms stung, but even more painful were the doubts of fellow Republicans. In September 1976, a couple of months after Santillan wrote his essay, the chairman of the Illinois chapter of the RNHA complained about Ford's lack of effort among Hispanics. At the convention, he wrote, he had been told that the president was "very interested in the Hispanic vote." But then the campaign didn't give them the resources they thought they needed. "It's quite obvious," he concluded, "that you don't have any regard for the Hispanic vote." Hispanic Republicans also continued to chide Ford about his unimpressive record of appointments; three high-level and two lower-level appointments wouldn't cut it.

Finally, Ford compared his accomplishments to those of Kennedy and Johnson, rather than to what Nixon had done. It seemed to many Hispanic Republicans that the momentum of the Nixon years had stalled. No matter how much Ford's glossy campaign pamphlets said he did for Hispanics, it felt to them like he made promises he hadn't kept, and this was the same criticism they had of the Democratic Party. Even within one abridged presidential term, Hispanic Republicans felt their support from the Republican Party waning and waxing and waning again. It was a precarious position for them to be in.

When Ford lost the election to Jimmy Carter in early November, some analysts pointed to the tamal mishap as the incident that sealed his fate. Others pointed to his missteps in US-Cuban affairs. The truth is that there were many reasons Ford lost the election, including the ongoing fallout from Watergate, fears of American decline generally, how uncompelling some found him as a candidate, and the intriguing Democratic candidate. Whatever the cause or causes, the president won only 19 percent of the Hispanic vote, reversing the gains Nixon had made. If Ford had won over more Hispanics in Texas or Florida or New York, he would have needed only a few more electoral votes to win the election, which he might have picked up somewhere else.

That he wasn't able to duplicate Nixon's success among Hispanics was one among several failures for a president who had expressed high hopes and had accomplished some of the things that were important to them.

The legacy of the mid-1970s for the growing Hispanic Republican movement would not be the continuation of Nixon's success through Ford, but rather the establishment of the RNHA, which connected Hispanic Republicans nationally and brought them into the Republican Party. Right after Ford's election, there were eight Hispanics in Congress, all of them Mexican American and Puerto Rican. Seven of them were Democrats, while Luján was the only Republican. They came together in late 1976 to form the Congressional Hispanic Caucus, to pursue "issues of common concern to the Hispanic Community." The group's members became the leading advocates and strategists in Congress for Hispanic political affairs, including presidential appointments, bilingual education, and voter registration. These were the same issues that had animated the RNHA's activism for years.

BEFORE FORD LEFT THE WHITE HOUSE IN JANUARY 1977, HE HAD ONE MORE TRICK UP HIS sleeve. In December, the lame duck president spent two weeks in Vail, Colorado. On New Year's Eve, reporters waited for him at the bottom of the mountain slope as he pivoted back and forth toward the end of his run. They readied their microphones and video cameras to record his response to the surprise announcement made earlier in the day, that he would "propose Puerto Rican statehood legislation to the new 95th Congress before he leaves office." The timing of his proposal was right, Ford said, since Puerto Ricans, on the same day as the presidential election, had elected the mayor of San Juan, Carlos Romero Barceló, as the fifth governor of the commonwealth of Puerto Rico. According to John Chancellor, the *NBC Nightly News* anchor delivering the report, Romero Barceló said that his main job as governor would be to "tell the people of Puerto Rico of the advantages of statehood."

Before Romero Barceló was elected governor, he had served for eight years as the mayor of San Juan. A graduate of Philips Exeter Academy and Yale University, he was the city's first mayor from the Partido Nuevo Progresista (PNP, or New Progressive Party). When Ford assumed office, Romero Barceló had recently published his pro-statehood manifesto *Statehood Is for the Poor.* In it, he argued that statehood was the best way for Puerto Ricans to achieve equality with other American citizens, and that statehood was not, in fact, the design of rich people like the former governor Luis Ferré. Instead, it was commonwealth status that benefited the rich, who profited from special provisions that allowed American companies doing business on the island to avoid taxes. Yes, it was true that Puerto Ricans would have to pay taxes if Puerto Rico were to become a state, but poor Puerto Ricans wouldn't pay taxes, just as poor Americans on the mainland didn't pay taxes. The island's wealthiest citizens would shoulder much of the burden. Ferré, he wrote, supported statehood despite the fact that it would harm him financially, because he believed it was best for the island. Support for statehood was still the official position of the Republican Party, and even though Ford wouldn't be in office long enough to see his call come to fruition, he wanted to signal his support for Romero Barceló's and Ferré's desire.

Ford claimed that Romero Barceló's position on statehood was a "good reflection of the attitude of the Puerto Rican people," so it seemed to him that "we ought to take an initiative here in Washington"— never mind that he was in Colorado—"to indicate our full support for statehood." Why hadn't he left it to Carter to propose statehood after he took office? one reporter asked. Ford, slightly winded but smiling, answered, "Because I'm president until January twentieth." Moreover, during the transition period, after he had already lost the election, he couldn't be accused of proposing statehood for political reasons, to woo Hispanic voters.

At the end of the broadcast, Chancellor wondered, "What would happen to the stars on the American flag?" Would the "flag be redesigned to put them in a circle, or a star shape, or would Puerto Rico have a column all to itself?" These were "serious questions," he said.

When Ford said that "the people of Puerto Rico had spoken," he could only have meant the 48 percent of Puerto Ricans who voted for Romero Barceló and the New Progressive Party. More than 46 percent of Puerto Rican voters had supported Romero Barceló's opponent from the Popular Democratic Party (Partido Popular Democrático, or PPD), which favored maintaining commonwealth status. Another 6 percent had supported the candidate of the Puerto Rican Independence Party (Partido Independentista Puertorriqueño, or PIP), which, as its name suggested, favored independence. Ford knew that the opinions of Puerto Ricans were far more divided than he let on. During trips to Puerto Rico, he didn't voice his support for statehood because he "did not want to spur demonstrations." His parting words on his way out of office foreshadowed how Puerto Rico would loom large in the minds of future Republican candidates.

In a letter Fernandez received in the summer of 1977, a few months into the Carter presidency, the House minority leader, John Rhodes of Arizona, stroked his ego by telling him how important and powerful the RNHA had become. The summer before, Rhodes had attended the RNHA's first annual banquet. He wrote, "In just three years, your organization has moved from an idea to a force within the Republican Party." It exemplified "the diversity and outreach that our Party must embrace." Republicans had already come a long way. Hispanics had made significant progress during eight years of Republican leadership. "GOP," Rhodes wrote wishfully, "means the Grand Opportunity Party." Party leaders had men such as Fernandez to thank for it. He and the RNHA had made "Republicanism the choice of many Hispanic Americans." Hispanics, Rhodes was saying, would help revive the Republican Party after its years of disgrace following Watergate.

PART III

DOUBT

CHAPTER 6

FERNANDEZ *IS* AMERICA

On the cover of its October 16, 1978, issue, *Time* magazine displayed a collage of Hispanic faces. They belonged to men and women, young and old, light-skinned and dark-skinned. The words scrolling across the cover, rising diagonally from left to right, like a graphic representation of the growing population itself, read, "Hispanic Americans: Soon—the Biggest Minority." Surveys conducted that year showed that the Hispanic population stood at about twelve million, which was almost a 15 percent increase over the previous five years. This group of immigrants and Americans—the fastest-growing, according to the magazine—faced many challenges, including lower incomes, higher rates of unemployment, and an alarming percentage of high school dropouts compared with other Americans. But they were also on the brink of transforming the nation.

The article caricatured Miami as home to upwardly mobile Cuban entrepreneurs; Los Angeles as the capital of the Mexican American Southwest, with a strong middle class despite many who suffered from the high incidence of gang killings and educational discrimination; and New York as home to Puerto Ricans, who crowded into the tenements of the Lower East Side and Spanish Harlem, yet displayed their cultural vibrancy during the Puerto Rican Day celebration every June. Demonstrating how immigration was growing as a national issue, *Time* categorized the undocumented Mexican immigrant community as an altogether separate group of Hispanics. The question of what to do about them consumed a good bit of Carter's first year in office. Hispanics who were American citizens had "mixed feelings" about them, the article said. It quoted a Mexican American police officer in Los Angeles who thought they were a problem. The "only way we're going to stop them is to build a Berlin Wall," he said.

The differences among diverse groups of Hispanics could be diminished, and the whole lot could be improved, the article suggested, if Hispanics were to become more politically engaged. This was the main argument of the individuals quoted in *Time*, Hispanic and non-Hispanic alike. To become more politically active, more of them had to become citizens, and more of them had to register to vote. This was just what they were about to do, the article implied. The potential was there; things were about to change. All of this would have sounded like a conservative argument to Hispanics who were skeptical of participation in the American political system. But *Time* focused its attention on the faithful, quoting leaders such as Raul Yzaguirre, the director of the National Council of La Raza, one of the most influential Hispanic advocacy organizations of the era. He said, "The 1980s will be the decade of the Hispanic."

If Hispanics were about to have their "turn in the sun," as California governor Jerry Brown, who had been one of Carter's main rivals for the Democratic nomination, predicted, the article in *Time* claimed that it would come at the expense of African Americans. It furthered the old idea that gains for one group necessarily meant losses for another. "Black power" would give way to "Latino power," the article said. African Americans and Hispanics competed for jobs and federal aid. The publisher of an African American newspaper in Miami warned that "black hostility" toward the "Cuban influx" would be the logical result of their competition. According to *Time*, the Cubans were winning.

The story elsewhere was different. In New York, the article said,

Puerto Ricans were the "most beleaguered" group of Americans, earning less money, drawing more welfare, and receiving less job training than African Americans. In California, meanwhile, even though Hispanics had surpassed African Americans as the state's largest minority population in 1970, they still held fewer elected offices in state government. Whereas more than a hundred predominantly black colleges and universities received "multi-million-dollar subsidies from Congress," according to a member of the Los Angeles School Board, there wasn't a "single goddamned Mexican-American institution of higher education." The reporter noted that there were, in fact, five such institutions, but all of them were "struggling" and received "little or no federal aid." The point stood, though, that if Hispanics were going to play a greater role in national affairs, they needed more attention and support than they'd been given in the past—and the more attention they received could, and maybe should, mean less for African Americans.

Over the course of the 1970s, Hispanics had indeed gained more political clout, but some of the qualifications in the *Time* article suggested that they still hadn't been fully integrated into national politics. They still lagged behind in several areas, and Hispanic Democrats and Republicans were working to change this. They were optimistic that their moment was about to come. On the Republican side, Benjamin Fernandez placed himself at the center of the conversation. This was his purpose at the 1976 convention in Kansas City, when he said that the Republican Party was building a Hispanic power base. He imagined himself as its leading representative. His life and career were models for the whole country.

EVEN BEFORE CARTER'S INAUGURATION ON JANUARY 20, 1977, THE OUTGOING COMMISsioner of the Immigration and Naturalization Service (INS), Leonard Chapman, called the influx of undocumented immigrants a "silent invasion." Their number was estimated at six million to eight million, which meant that there were half to two thirds as many Hispanics who weren't citizens as Hispanics who were. More than seven hundred thousand had been apprehended in 1976 alone, but beyond the numbers, the issue became increasingly divisive because it pitted different interest groups against one another. Newspapers reported the growing incidence of violence against undocumented immigrants, and Carter's campaign was based in part on a promise that he would pay greater attention to human rights abroad and at home. But there was also the growing din of voices such as Chapman's who saw undocumented immigration as one of the biggest problems the nation faced. It represented the complete breakdown of law and order, a loss of control over national borders, and even the erosion of national sovereignty.

Carter named a cabinet committee to study the problem and make recommendations. He also nominated the first Hispanic commissioner of the INS to replace Chapman and study the issue with him. He was a Mexican American from Texas named Leonel Castillo, and his nomination drew mixed reviews. Labor unions opposed him because they thought he'd support the hiring of undocumented workers, while many Hispanic advocacy groups supported him. Castillo and Carter's new secretary of labor made several announcements in the spring of 1977 to let Americans know what the new administration's immigration priorities would be. "In the short run," Castillo said, "better enforcement will have to be done, but you simply cannot seal the border." In the long run, he continued, "the United States cannot afford to wall itself off from the poor people of the world." Instead, he thought it would be best to support economic development programs in sending countries. For his part, the labor secretary said, "I cannot emphasize strongly enough that we have never considered closing off the border with Mexico or beginning mass roundups of suspected undocumented aliens." Their words were reassuring to Americans who favored greater leniency but unsettling to those who favored greater restrictions.

After months of planning, the White House released a formal proposal, Carter's Alien Adjustment and Employment Act of 1977. It was the first major immigration bill debated since the 1972 Rodino bill. Administration officials agreed that the newly proposed law would revive Rodino's employer sanctions and visa provisions. There were no objections to the visas, but employer sanctions proved to be as controversial this time as they were before. Most of the opposition still came from Hispanic organizations, which argued that they would have the effect of employer discrimination against Hispanic job applicants.

Even more controversial about Carter's bill, though, was its proposed amnesty for undocumented immigrants who had lived in the United States for five years. This was the "Alien Adjustment" in the bill's title, and it hearkened back to the phone call that Ramirez had with Robert Finch, when Nixon apparently told him that he wanted to grant amnesty to undocumented Mexicans to thank Hispanics for the boost they had given him. The amnesty in Carter's bill was its most controversial aspect, leading to an extended debate about its merits and drawbacks. In the end, immigration proved too divisive for the moment. As one Carter aide put it at the very beginning of the process, immigration involved an "unbelievable thicket" of sensitive issues. Reform was postponed for another day.

Undocumented immigration may have been the trickiest and most nationally audible issue challenging US–Latin American relations during Carter's first year, but there were certainly others as well. They included the ongoing negotiations over the future of the Panama Canal; Carter favored some kind of timeline for returning it to Panama. He also favored reestablishing relations with Cuba, though he may have gone farther than Ford by supporting the lifting of the trade embargo and allowing Cuban exiles to visit Cuba and Cubans to visit the United States. Finally, Nixon and Ford's general approach to Latin America—doing whatever was necessary to ensure stability and the protection of US interests in the region, including turning a blind eye toward or even supporting ruthless dictators—contributed to the outbreak of violence in Central America, which became another challenge for the Carter administration.

BENJAMIN FERNANDEZ BELIEVED HE COULD BE THE ONE TO DEFEAT CARTER AND ADDRESS the national and international problems the Democratic president had created. He fancied himself as the person who was best prepared to take advantage of the promise of growing Hispanic political power, and to solve the many challenges the United States faced in the area of US–Latin American relations. In December 1977, he assembled a group of twelve Hispanics to travel to Latin American countries as a "fact-finding commission," to "review the status of relations between the United States and their governments." At the end of the tour, the members of the commission promised to send a report to President Carter, members of Congress, and the "American public." The fact-finding commission was described as a public service, but in fact it was designed to establish Fernandez's credentials, to prepare him to run for president.

Several of the Hispanics Fernandez took with him were leaders of the Republican National Hispanic Assembly (RNHA), including Philip Sanchez, who had just ended a stint as the US ambassador to Colombia, and Esteban Taracido, the president of the New York chapter of the RNHA. To document the tour—it was a grand performance, after all—Fernandez also paid for a young photographer, Jeff Widener, to travel with them. Widener, for whom it was all a big adventure, recalled having dinner in Nicaragua at Anastasio Somoza's house. Somoza wasn't home, but he had arranged a lavish spread for them. At the time, the US-backed dictator, after years of violently repressing his opponents, was faced with a cross-class opposition movement that would, in short order, remove him from power. The tour, according to Widener, was staged to demonstrate Fernandez's experience in Latin America. A decade later, Widener would snap his iconic Tank Man photograph in Tiananmen Square.

In the New Year, not even a month after he returned from Latin America, Fernandez formed the Hispanic Presidential Candidate Committee, which registered with the Federal Election Commission "to find a qualified, electable Hispanic Republican to run for President in 1980." For the next several months, he traveled around the country raising the $20 million he anticipated it would cost for their selected candidate—not necessarily him—to run a presidential campaign. Surely he imagined it as a way to begin raising funds for himself.

Fernandez interrupted his work for the committee only once, in August 1978, when he reconvened the fact-finding commission for a final visit to Mexico. He had taken to calling it simply the "Fernandez Commission," and reporters began to call him a "candidate for the Republican nomination for president," even though he wouldn't declare his candidacy for another few months. In Mexico, he managed to arrange a meeting with President José López Portillo to discuss undocumented immigration and Mexico's energy resources—quite astounding considering Fernandez didn't hold an official government position. They also met with the popular comedian Cantinflas and other artists and entertainers about the possibility of them visiting the United States to campaign for Fernandez.

When he returned to the United States, he hit the road again for the Hispanic Presidential Candidate Committee, which needed help because it was falling far short of its fund-raising goals. By the end of April, when Fernandez visited the Cuban American Republican Club of Palm Beach, Florida, they had raised only $500,000 of the $20 million they hoped for. Fernandez wanted to solidify support among all Hispanics, and just as he had in 1972, when he was the head of Nixon's Hispanic Finance Committee, he began his fund-raising efforts among Cuban Republicans, who had money and increasing clout. If they were going to contribute to his committee, they wanted to know who the candidate might be. "Right now," Fernandez told them, "only five Hispanic Americans fully meet the requirements to become a candidate, and I am one of them." He didn't name the other four, or what their qualifications were, but he said he was "an idealist, a pragmatic man," and if he entered the race "it would be to win."

FERNANDEZ DECLARED HIS CANDIDACY IN THE FINAL DAYS OF NOVEMBER 1978, ABOUT A month after the publication of the *Time* issue featuring Hispanics, as if he were trying to capitalize on the momentum created by a national publication's focus on his community. It would be the wave that would lift him up. He certainly believed in the article's message, that Hispanics were on the brink of unprecedented political influence.

A few weeks earlier, Hispanics in Texas had just helped Senator John Tower get reelected to his fourth term. Tower had mastered the patronage politics characteristic of Nixon's outreach efforts among Hispanics. He wrote letters criticizing the Carter administration's cuts or changes to federal programs that benefited Hispanics, such as bilingual education, health care, job training, and housing initiatives. Then he made sure that Hispanic organizations such as the American GI Forum and the League of United Latin American Citizens (LULAC) knew about his complaints. He also forwarded press releases to reporters so they would write about them in their newspapers.

Hispanics in Texas recognized his support and stood by him. "*Tower con nosotros*" and "*Nosotros con Tower*," they said (Tower's with us, and we're with Tower). Nosotros con Tower was the name of the group that formed to support Tower's reelection campaign. It was made up of members of the Mexican American Republicans of Texas, which had merged with the RNHA in the mid-1970s. The GI Forum gave him an award for his efforts to make Hispanics part of the "mainstream of the business community." At a LULAC convention, he received the loudest cheers of any Republican. His Democratic opponent practically got booed off the stage when he criticized Tower's record, especially his votes against the Civil Rights Act and the Voting Rights Act.

Warm receptions at Hispanic conferences had become par for the course for Tower, but in 1978 he added a new weapon to his arsenal of Hispanic outreach. He hired the San Antonio firm Ed Yardang & Associates to run his ad campaign targeting Hispanics. A young employee named Lionel Sosa directed the effort, and in addition to the usual radio spots, phone calls, and flyers, they also wrote and produced a *corrido*—a popular Mexican ballad—called "El Corrido de John Tower." Its lyrics were professions of love and appreciation for all Tower had done for Hispanics. He was "beloved by many," he "helps the worker," there were "few like him," he always "lends a hand" to the small-business owner, he "keeps his promises," and he stood "with all the Mexicans." It was a hit. Tower won 37 percent of the Hispanic vote that year.

Because of how effective their campaign was, Sosa's firm took off. They picked up new clients such as Bacardi rum, Dr Pepper, and Coors beer, who wanted Sosa to direct their Hispanic advertising efforts. The firm "balked" at becoming a "Hispanic agency," so Sosa struck out on his own and started Sosa and Associates.

Tower had demonstrated again that Republicans could find success among Hispanic voters locally and statewide. Fernandez sought to mobilize Tower's supporters in Texas and reproduce Tower's success on a bigger stage.

When Fernandez announced his candidacy, he said, "I fully expect to be the next president of the United States." He was fifty-three years old then, and his presidential campaign would be the first time he ran for elective office. He had spent the previous decade building the Hispanic Republican movement. He headed the National Economic Development Agency (NEDA) as well as the National Hispanic Finance Committee to reelect Nixon. Then he founded the RNHA and served as its first chairman. Still, he aimed higher. "I think big," he said.

That year, Fernandez was one of many Americans who thought they could become president. Months before the first primary, more than a hundred candidates had entered the race. In 1972 Shirley Chisholm, an African American congresswoman from New York, became the first person of color to run for president, so Fernandez couldn't claim that mantle. But he hoped the fact that he was a fresh face in national politics would be a boon. Carter's own success made this seem like a realistic expectation. Outsider status had its downsides, though. As criticisms of Carter mounted, many attributed his shortcomings to his relative lack of experience. Americans the next time around might not have had the same appetite for an outsider, and at least Carter had served as the governor of Georgia.

Understandably, Americans reacting to Fernandez's announcement didn't have the optimistic outlook he did. A poll of more than fifteen hundred adult Americans taken almost a year after he declared himself a candidate found that only 9 percent of them had heard of him, making him the least well known of any candidate. In light of this finding, Fernandez's confident tone aimed to mitigate the improbability that he would succeed. Still, while some members of the RNHA were peeved because they saw his campaign as a sign that his individual ambitions—always strong—had become more important to him than the success of the organization, his decision to run, most Hispanics agreed, didn't make him "crazy" or a "kook" who would ultimately "be an embarrassment to the Latino people." They agreed that he was "intensely serious, a man of poise and assurance who has already achieved a great deal both in politics and business." As the first Hispanic to run for president, he could "make it just a little easier" for the next one.

As soon as Fernandez entered the race, reporters became curious about him. It was one of the advantages of being one of the first to declare his candidacy that he could command attention before the field got too crowded. But they were also curious about who the Hispanic was who thought he could become president of the United States. He used the exposure to criticize Carter, which didn't distinguish him from other Republicans. But he also used it to talk about the need for a Hispanic president, someone with the qualifications that only he possessed. Reporters wrote stories about his background and his rise from poverty. They wrote about the Fernandez Commission, which, he told them, demonstrated his chops in the area of US–Latin American relations. It was as though he had anticipated their stories for years.

Like other Republicans, Fernandez made his campaign, in part, a referendum on Carter's presidency. Carter was a weak leader who didn't solve national problems of unemployment and inflation, he argued, and with Carter as president, the United States lost prestige around the world. Regarding Hispanics in particular, Fernandez repeated the old line that Carter asked for their votes but then ignored them after the election. Hispanic Democrats had made the same complaint in the early months of Carter's presidency. Edward Roybal and the other members of the Democrat-controlled Congressional Hispanic Caucus argued that Carter wasn't repaying the support Hispanic voters had given him. They believed that Carter had won thanks to Hispanics in Texas, Florida, and New York, who gave him the votes necessary to defeat Ford. Slight shifts in any combination of these states might have returned Ford to the White House.

But the article in *Time*—published a year and a half into Carter's first term—said that Carter had recognized his earlier missteps and "reciprocated" by "appointing more Hispanics to federal positions than any of his predecessors." In his first months in office, he appointed the first ever Hispanic commissioner of the INS in addition to more than a dozen other appointments to posts in the Departments of Commerce, Agriculture, Housing and Urban Development, and Health, Education, and Welfare. He also appointed several Hispanics to work in the White House. In addition to appointments, Carter increased spending on bilingual education, created new schools in Hispanic communities, and while he didn't explicitly come out in favor of statehood, he said he would abide by any decision Puerto Ricans made about the status of the island.

Fernandez himself supported some of these initiatives, but he didn't credit Carter for implementing them.

Pressure on Carter came not only from Fernandez and other Hispanics, but also from within his own party. He was already worried in late 1978, a full two years before the election, about a potential Democratic challenger: Senator Edward Kennedy of Massachusetts. Hispanics liked Kennedy, in part because of their support for his deceased older brothers. John F. Kennedy was the young and attractive Catholic who had courted their support through the Viva Kennedy Clubs, and Robert F. Kennedy had been one of César Chávez's greatest champions. LULAC president Ruben Bonilla predicted that Kennedy could win 85 to 90 percent of the Hispanic vote if he were the Democrats' nominee instead of Carter.

REGARDLESS OF WHETHER FERNANDEZ'S OPPONENT WOULD BE ANOTHER REPUBLICAN OR Carter or Kennedy, he believed that the Hispanic vote was his to lose. In many ways his candidacy was based on this belief. His premise was that not only was America ready for its first Hispanic president, but also that America needed a Hispanic president, someone who was "bilingual in Spanish and English" and "fully understanding of both cultures." Hispanics represented a growing percentage of the population and an increasing share of the electorate. They were more politically active than ever before. More worked as employees or appointees of the federal government, and the civil rights era had prompted more to run for, and win, elective office than at any point since the nineteenth century. Fernandez would represent them and all Americans. "Ben Fernandez is America," he said. *"Fernandez para presidente"* (Fernandez for president).

In front of Hispanic audiences, he rehearsed the idea that Democrats took their votes for granted. Exaggerating considerably, he said that ninety-five of a hundred Hispanics were Democrats. If they weren't in the "hip pocket" of the Democratic Party, Democrats usually won a convincing majority of their votes, which meant that Democrats didn't have to "do a thing for them." Fernandez thought he could finally change the long-held belief that Hispanics were Democrats, and if that happened, he said, "American politics will never be the same." He tried to rally them by denouncing police violence against Mexican Americans in Texas, promising public sector jobs, making the Republican Party more inclusive, and pushing through immigration reforms that were tough but which included some form of amnesty.

At a November 1978 gathering of students and faculty at the University of Texas at El Paso, just weeks before he declared his candidacy, Fernandez described how an acquaintance of George H. W. Bush had told him, "Ben, my God, if you can rally the Mexicano, the Hispano, call him what you will, in California, Arizona, Texas, Florida, New York, and New Jersey, you'll win." His response: "the thought has occurred to me." He wanted to be the one to prove *Time* correct, that Hispanics would join forces and shake up American politics.

Fernandez knew as well as anyone how difficult it would be to unite Hispanic Republicans behind him. In his testimony before the Senate Watergate Committee a few years earlier, he claimed that trying to "work the Spanish-speaking people into a cohesive unit" was the "toughest thing I've ever done in my life." Hispanics, he said, had a "tradition of not working together." They came from different places and had different interests.

But Fernandez saw himself as a unifying force. He believed that his presidential campaign could bring Hispanics together because he had brought them together before. Mobilizing Hispanic Republicans and bringing them together under the same umbrella, as Hispanics, had been his mission for almost two decades, for as long as he had been involved in politics. He believed that all Hispanics, not only Republicans, would rally around him because of his humble origins and meteoric rise. By the time he declared his candidacy, he claimed to be a millionaire, although a friend later recalled that he was nowhere close. Regardless, where he stood in 1980 was a long way from where he began life—in poverty, like many Hispanics. His story was the fulfillment of the American Dream that all Hispanics aspired to. This was the story that reporters latched on to.

From the moment Fernandez said he was running, reporters obsessed over his biography. They asked him about his involvement in the Watergate scandal, but he insisted that he had been completely exonerated. In fact, he claimed to be "one of only two witnesses"—he didn't say who the other was—"who testified before the Watergate committee and improved their reputation." Not a single reporter said he could win, but they all found him to be a compelling character due to his story of accomplishment despite being born into difficult circumstances.

Fernandez had entered the world in a railroad boxcar in Kansas City, which earned him the nickname "Boxcar Ben." At campaign events, he relished the idea that Republicans might compare his birthplace to the log-cabin origins of another eventual Republican president.

His parents immigrated from the quaint town of Tangancícuaro, in the Mexican state of Michoacán. He had seven siblings: three brothers and four sisters. From age five, he worked alongside them in midwestern agricultural fields, picking tomatoes and sugar beets in Illinois, Michigan, and Indiana. They worked, he said, so that they wouldn't have to go on relief. It was a sentiment designed to appeal to antiwelfare Republicans.

Like thousands of Mexicans and Mexican Americans, his family moved from the fields to the industrial centers of the Midwest, eventually settling in the steel town of East Chicago, Indiana, where his dad had taken a job at the Inland Steel Company, and where Fernandez attended Washington High School. His yearbook said he wrestled and participated in intramural sports, but one of his best friends from childhood, Frank Casillas, said that they mainly chased girls. Together with another friend, they formed a clique called "the wolves."

Casillas later credited Fernandez with introducing him to the Republican Party. Casillas's family had crossed the border illegally when he was just a boy and then opened a restaurant they named La Michoacána, after the Mexican state from which many Mexicans, including the Fernandez family, had arrived in the United States. Fernandez ended up recruiting Casillas to serve as one of the original directors of NEDA and as the Illinois head of his presidential campaign.

Fernandez eventually committed to his education because he hoped it would help lift him out of poverty. He spent a year at Purdue University near home, then dropped out and left for California, where he paid his way through Redlands University by working as a waiter and dishwasher. After receiving his bachelor's degree, he earned a graduate degree in business from New York University. His schooling landed him a job at General Electric, the same firm Reagan had worked for, and from there he started his own business consulting company.

Fernandez believed about himself that he represented everything that was best about America. "Fernandez is America." He said that his experience would "restore faith" in the idea that "anyone born in this country who prepares himself, who works hard, and is willing to shoot for the moon, can one day become president of the United States." He wanted every supporter to believe with him that the United States was great. "Whatever I am," he said, "I owe to the opportunity given to me by America."

This, he said, was America's promise. It was the most powerful message he delivered as a candidate, that the United States was still a land of economic opportunity and uplift. It was an old story, and one that came in for close scrutiny in a decade when the gap between rich and poor grew, but it was one that had been meaningful to him personally. It's how he had come to understand his own experience in the country. When a friend in college told him that the Republican Party was for rich people, he responded, "Sign me up! I've had enough of poverty." In an interview with one newspaper, he cited his "own rise from poverty as an example of what the free enterprise system can do for any hardworking American."

Everything he had learned from his experience with NEDA, Nixon's National Hispanic Finance Committee, the RNHA, and now as a presidential candidate, had taught him that Hispanics were hungry for political representation. He was convinced that he "must continue to work with the Spanish-speaking people, that this must be a lifetime commitment" for him. Hammering this message over and over, first with Hispanics and then with all Americans, would punch his ticket to the presidency.

Fernandez made a compelling case for why he was the right man for the job. In addition to his ethnicity, he also happened to be a "trained economist" at a time when the nation needed an economist at the helm. He wouldn't only seek the uplift of his own people. Rather, he would "concern himself with the problems of the country as a whole." The crucial phrases in his statement were "trained economist" and "country as a whole." With the former, Fernandez foreshadowed how President Carter became a punching bag for all Republicans and even some Democrats. The nation needed an economist to lift the country. With the latter, he signaled his intention to bring Hispanics into a party that had marginalized them in the past. He would continue his life work as president.

Seeking to unite Hispanics nationally, he would reach out to Puerto Ricans in New York, even though the common wisdom about them was that they were steady Democrats. The Puerto Ricans living in Florida or the Washington, DC, area were more conservative, in part because they were more affluent. They were some of the lawyers and entrepreneurs conducting governmental relations and business between Puerto Rico and the fifty states. Yet New York had a handful of dedicated Puerto Rican Republican organizers, including Angel Rivera, Rita DiMartino, Taracido, and their protégé Hannie Santana, who later became chairperson of the eastern region's RNHA. They had been affiliated with New York's Republican Party for a long time, as Rockefeller Republicans. DiMartino started a group called the Spanish American Republican Club on Staten Island, where she lived, and it merged with the New York chapter of the RNHA once it was up and running.

Republicans and Democrats alike worked toward the economic uplift of Puerto Ricans, who earned less and suffered higher rates of unemployment than other ethnic groups. The city had lost six hundred thousand jobs between 1969 and 1977, mostly in the manufacturing, apparel and textile, and food and beverage industries, which employed mostly Puerto Ricans and African Americans. The job loss was compounded by white flight from the city to the suburbs, which hit New York's tax base hard and led to the slashing of social services. Puerto Rican Republicans in New York were largely Hispanic professionals who thought that the solution to the city's economic problems would come through educational opportunities, business development, and political integration instead of more federal welfare programs. They sought uplift by distancing themselves from the African Americans who had earlier been their partners in the city's civil rights struggles. They linked up with Hispanics across the United States, and claimed that they were culturally distinct white ethnics, rather than members of a racial group, as African Americans were.

Fernandez thought he would stand on more solid ground among Cuban Americans in Florida, where he told the group in Palm Beach that he wouldn't begin a dialogue with the Marxist leader until he "stops exporting revolutions and frees all political prisoners." This message may have satisfied some in the audience, but hard-liners would have questioned whether he was better qualified to handle Castro than Carter was. They believed that the United States shouldn't reestablish relations with Cuba under any circumstances as long as Castro was still in power. They wanted to punish nations that did business with Cuba. They didn't think that Cuban exiles or anyone from the United States should be allowed to travel to Cuba—even to visit family—or that Cubans should be permitted to enter the United States.

Cuban exiles and Cuban Americans looked for something different from Fernandez, not a replay of Carter's wrong move on the issue that was of paramount importance to them. Reacting both to Ford's and Carter's support for détente, the most extreme hard-liners used violence to make their point, bombing banks, travel agencies, and airplanes. In the bicentennial year alone, bombs exploded at the Cuban mission to the United Nations; travel agencies in Little Havana that booked trips to Cuba; and, notoriously, Cubana de Aviación's Flight 455, which was downed by a bomb as it headed from Barbados to Jamaica on October 6, killing all seventy-three passengers on board. But these terrorists were unrepresentative. As the chairman of the Florida chapter of the RNHA, José Manolo Casanova, put it, "The whole drift of the thing has been to move closer to relations with Cuba." This general shift motivated Fernandez's position on the matter.

In November 1978, the same month that Fernandez became a candidate, the White House supported a trip to Cuba by a Cuban American banker from Miami accompanied by seventy-five scholars, journalists, and businessmen from the United States, Puerto Rico, Spain, Venezuela, and Mexico. With members of Castro's government, they discussed political prisoners, family reunifications, and other issues that plagued US-Cuban relations. Castro promised that if those initial discussions went well, he would begin to release prisoners. They did, and he did. Over the first several months of 1979, Castro released thousands of Cubans, many of whom had been in prison for more than a decade. He stopped calling exiles *gusanos* (worms) and instead began referring to them more politely as the Cuban community abroad.

Carter then commuted the sentence of the Puerto Rican nationalist Lolita Lebrón, who along with three others had shot up Congress in 1954, which led Castro to release more prisoners. Exiles could visit their families in Cuba, and there were flights between Miami and Havana for the first time in more than fifteen years. By the end of 1979, more than a hundred thousand Cuban exiles had traveled to Cuba.

Most were loyal Americans who weren't trying to repatriate. Some were influenced by the civil rights movement, and they favored the rapprochement between Carter and Castro. But others—hard-liners, Bay of Pigs veterans—called those who visited Cuba "traitors," "cowards," and "opportunists." Instead of fighting to oust Castro, they propped him up by spending millions of dollars in Cuba. More than anything else Carter did, this opening between the United States and Cuba led to their sharp right turn.

The "drift of the thing," as Casanova described the state of Cuban exile politics in the late 1970s, was also toward political participation in the United States and the often painful realization that the United States, not Cuba, was their home. A greater number of Cubans applied to become citizens in the bicentennial year than ever before. The uptick had begun earlier in the decade, but the citizenship spike in the late seventies was part of a concerted effort. The Cubans for American Citizenship campaign hoped to initiate the naturalization process for ten thousand Cubans that year. They registered sixty-five hundred people on Independence Day alone, and by the end of the year they had registered twenty-five thousand. The initiative's success was just one sign of how Cubans were adopting American identities. As one historian has explained, over the course of the decade, Cuban exiles became Cuban immigrants who became Cuban Americans. Another was their success in the United States. Cubans had the highest income and education levels of all Hispanic groups, and high rates of home and business ownership.

Many of these new, successful citizens would become cornerstones of the Hispanic Republican movement nationally, but in the late seventies most of their influence was exerted at the local level, and not always as Republicans. Individuals such as Fernando Penabaz, Manolo Giberga, and Casanova had developed inroads with the Republican Party nationally. But as the *Time* cover story noted, there still weren't any Cuban American representatives in the Florida state legislature or US Congress. By contrast, in the Miami area, where more than half of the eight hundred thousand Cubans in the United States resided, there was a Cuban American Democrat on the school board and a Puerto Rican Democratic mayor named Maurice Ferré, who was Luis Ferré's nephew.

It wasn't a surprise, therefore, that Cuban Americans supported Carter over Ford by a margin of 75 to 25 percent, but Fernandez and other Republicans recognized that the community was in flux. He decided to make Florida one of his top priorities after returning from Latin America. In 1970, only 25 percent of Cubans were American citizens, but by the end of the decade the proportion was greater than half. Every month a thousand Cubans became citizens and eight hundred registered to vote. Fernandez sought to benefit from these changes and to continue connecting Hispanics in Florida with the national Republican Party.

Hispanic Republicans also worried that Carter's opening of relations with Cuba would embolden Castro to spread his influence throughout Latin America, including hot spots in Central America such as Nicaragua. In the later years of Carter's term in office, the most radical opponents of Somoza, whose family had waged violent campaigns against their opponents in order to control Nicaragua for much of the previous half century, seized power. Named after the radical leader Augusto Sandino, who was overthrown by Somoza's father when he was president, the Frente Sandinista de Liberación Nacional—FSLN, or Sandinistas—used armed struggle to wage their political revolution.

By the late 1970s, it was clear that Somoza had lost whatever support he had and increasingly resorted to violence, using weapons that had been supplied by the United States to suppress his opponents. Carter no longer wanted to support Somoza, but he also couldn't support the radical socialist Sandinistas, led by the Ortega brothers, Daniel and Humberto. Carter feebly called for a cease-fire and the formation of a government of national reconciliation in an effort to bring all sides to the table. He believed that this might prevent an outright Sandinista victory, but he was wrong. The Sandinistas entered Managua and Somoza fled to Miami. One would assume that Fernandez was welcomed in Somoza's home because they regarded each other as allies in the struggle against radicalism.

Fernandez alone had gone on a fact-finding mission in Latin America and therefore saw himself as uniquely qualified to clean up the mess made by Carter, who had allowed communism to flourish. Fernandez considered himself to be a Cold Warrior who was even more conservative than most Republicans. He saw communism spreading throughout the hemisphere, from the "Communist triangle" of Cuba, Jamaica, and Guyana, to the rest of Latin America and the Caribbean, especially Guatemala, El Salvador, and Colombia. It had become apparent to the "whole world," Fernandez said, that the United States was "surrounded by a positive communist movement across our soft underbelly." He said he was the candidate who was best prepared to make positive changes in Latin America and that he would respond to the region's challenges aggressively.

But he also believed that Republicans had erred in thinking that military intervention was the only answer. Sometimes it would be necessary, but Fernandez aimed to forge a middle road that would actually accomplish the peace and stability earlier Republicans had only dreamed of. He looked back to the last years of the Eisenhower administration, after Nixon's tough trip there in 1958, when the government sought solutions to poverty in addition to eradicating communism. For Fernandez, the two goals were connected, because poverty had opened the door to Communist influence. The United States needed a president who could provide leadership in the Americas, respond to the needs of "the poor," and "assume the leadership of development" in economics, housing, and education.

As Carter had found out, challenges in the hemisphere included the rising number of undocumented immigrants crossing into the United States, and Fernandez had ideas about this matter as well. Carter had proposed his bill, but it didn't go anywhere. Failing comprehensive reform, the INS supported more modest (but still controversial) measures, including the construction of a fence—a "tortilla curtain," many called it—in high-traffic urban areas. Many still advocated employer sanctions, the hiring of more Border Patrol officers, and some form of amnesty. Undocumented workers themselves protested their mistreatment. Three filed suits against ranchers in Arizona who detained, abused, and shot at them, and employees at Goldwater's citrus business, Goldmar Corp., went on strike because they said they lived in unsanitary housing and were paid below the minimum wage. Republicans didn't like that Goldwater had hired undocumented workers instead of citizens, but they also didn't think that immigrants without papers should have been in the United States in the first place.

Preventing the flow of drugs into the United States had become another focal point of US-Mexico relations. Republicans had waged a war on drugs for most of the seventies, supporting increased funding for drug-fighting government agencies, the ability to extradite Latin American kingpins, and mandatory sentences for drug traffickers and users. By the end of the decade they turned their attention to Mexico. Between 80 and 90 percent of all heroin in the United States came from Mexico, and Republicans didn't think that the Mexican government was doing enough on its side of the border to crack down on trafficking.

Still, Fernandez didn't believe that confronting these problems and maintaining amicable inter-American relationships were mutually exclusive goals. Solving problems required US and Latin American cooperation in the fight against communism, undocumented immigration, and drug trafficking. It also required cooperation to pursue the rising-tide-lifts-all-boats dream that increased trade and investment in Latin America would also be good for the United States.

Despite rising anti-immigrant and nationalist sentiment on the far right, Fernandez's views were in line with those held by other leading Republicans in the late 1970s, including Goldwater, who argued that the Americas needed fence mending, not fence construction.

With Fernandez as president, the United States would firmly enforce the law, but it wouldn't turn its back on Mexico. "Whatever we do must be done in conjunction with Mexico," he said, and it must be done with dignity, integrity, and compassion. Again, Fernandez was the voice of reason. What better way to deal with the challenges in the hemisphere, he asked, than to have a "Fernandez in the White House"?

FERNANDEZ HAD ENTERED THE RACE EARLY, BUT BY THE END OF 1979, ON THE EVE OF THE primary elections, the field had grown crowded. Senator Bob Dole from Kansas entered the ring, along with Congressman John Anderson from Illinois; Senator Howard Baker from Tennessee; and former Texas governor John Connally. The two front-runners were the former congressman, RNC chairman, and CIA director Bush, and Reagan, the California governor who went on to win the general election in this, his second bid for the presidency. It was a good thing that Fernandez had garnered a significant amount of attention before the first primary, because it became increasingly difficult for him to stand out once the elections began.

Fernandez nevertheless remained optimistic that he would capture the Hispanic vote and that his success among Hispanics would propel his success among other Americans. His longtime friend Fernando Oaxaca, who had taken over as chairman of the RNHA, delivered a strong message about why Hispanics would support Fernandez. Republicans weren't prepared to "concede the Hispanic vote to the Democrats," he said, citing a recent survey conducted by a university in Texas that found that 48 percent of Hispanics considered themselves Democrats, while 38 percent considered themselves Republicans. To Oaxaca, those numbers signaled that Republicans should fight for every single Hispanic vote.

Oaxaca reiterated that Republicans embraced Hispanics, while Democrats ignored them. The RNC was spending more than the

DNC on Hispanic outreach, he noted, and was hiring more workers and spending more on radio, television, and newspaper advertisements. He also repeated the notion that Hispanics were naturally
conservative. Hispanics had momentum, and Republicans stood to
benefit. Galvanized by Fernandez's campaign, they would shock the
nation.

Fernandez got on the ballots in New Hampshire, Michigan, Pennsylvania, Kansas, Connecticut, Massachusetts, and a handful of other
states, and even if he didn't ultimately earn votes in these places, he
wowed crowds wherever he went. He was extremely charismatic on
the stump. He knew how to work a crowd and was a dynamic speaker.
His off-the-cuff speaking skills distinguished him from some of the
other candidates according to Casillas, his friend from childhood and
campaign chair in Illinois.

Casillas recalled Fernandez's performance at a gathering of women
Republicans in the Land of Lincoln, who hosted a meeting with all of
the candidates. These meetings were all on the same day, but not all
candidates would take the stage at the same time. Casillas observed
how Connally of Texas had planted supporters in the audience with
prepared questions. After he had his turn, Casillas went backstage and
urged Fernandez to do the same. Fernandez wouldn't do it. "I take
'em as they come," he said. That's just the way Fernandez was, Casillas
recalled, and he could pull it off. They set up several other events in
Illinois, and, according to Casillas, Fernandez "came out way ahead"
in all of them. Women, in particular, clapped and screamed like Fernandez was some kind of rock star. "Women loved him," Casillas
said. "He was a good-lookin' guy."

Sometimes Fernandez let his talent get to his head. Casillas said it
was his ego that killed him. He remembered one episode when they
were together in a hotel room. Fernandez got a call from Reagan,
who, Casillas said, was calling to congratulate him on the campaign
he was running. The ex-governor told him he was doing a good job,
and that he wanted to meet with him when they were both back

A dollar bill signed by the treasurer of the United States Romana Acosta Bañuelos.

The head of Puerto Rico's pro-statehood party Luis A. Ferré (at left, smiling) and Governor Nelson A. Rockefeller (center, shaking boy's hand), at the Puerto Rican Day Parade in New York City, April 29, 1959. Others in the photo include Senator Jacob Javits (behind Rockefeller), parade chairman Jose Caballero (wearing ribbon), and Governor and Mrs. Robert B. Meyner of New Jersey. (*Courtesy of Bettman/Getty Images*)

Vice President Richard Nixon with sugarcane workers in Havana, Cuba, January 1, 1958. (*Courtesy of Haynes Archive/Popperfoto/Getty Images*)

Robert Benitez Robles and Arizona senator Barry Goldwater. (*Robert Benitez Robles Papers, 1907–1988, Box 2, Scrapbook of Robles's political career, undated, courtesy of the University of Arizona Libraries, Special Collections*)

Romana Acosta Bañuelos and President Richard Nixon in the Oval Office, October 21, 1971, a month and a half before her confirmation as the first Hispanic treasurer of the United States. (*Courtesy of the Richard Nixon Presidential Library and Museum, NARA*)

Watergate burglar Virgilio Gonzalez posing for a photograph while opening a safe in his driveway, May 13, 1992, to mark the twentieth anniversary of the break-in. (© *Acey Harper/The LIFE Images Collection/Getty Images*)

President Gerald Ford biting into an unshucked tamal in San Antonio, April 9, 1976. (© *Pat Hamilton/San Antonio Express-News/ZUMA Press*)

Benjamin Fernandez and Jacqueline Fernandez ordering ice cream in Santiago, Chile, December 1977, during Fernandez's "fact-finding" mission to Latin America, to prepare him to run for president of the United States. (© *Jeff Widener*)

"Benjamin Fernandez for President" campaign photo, 1980. (*Courtesy of Eugene Acosta Marin Papers, Chicano/a Research Collection, Arizona State University Library*)

Columba Bush speaking in Puerto Rico, with her husband, Jeb Bush, and her father-in-law, presidential candidate George H. W. Bush, by her side, February 1, 1980. Seated beside Bush is Puerto Rico's secretary of labor Julia Rivera de Vincenti. (© *Dirck Halstead/The LIFE Images Collection/Getty Images*)

Katherine Davalos Ortega speaking after President Ronald Reagan nominated her to become treasurer of the United States, September 9, 1983. (*Courtesy of Bettman/Getty Images*)

President Ronald Reagan lifting an eight-year-old girl at a gathering hosted by the Nicaraguan Refugee Fund, April 1, 1985. (© *Diana Walker/The LIFE Images Collection/Getty Images*)

Hispanic supporters of Texas governor George W. Bush in Ames, Iowa, August 14, 1999, after Bush won the local Republican Party's unofficial "straw poll." (© *Luke Frazza/Agence France-Presse (AFP) Photo/Getty Images*)

Hispanic supporters of President Donald Trump, at a rally with Trump and Senator Ted Cruz in Houston, Texas, October 22, 2018, right before the midterm election in which Cruz narrowly defeated his Democratic opponent Beto O'Rourke. (© *Sergio Flores/Bloomberg/Getty Images*)

in California. Casillas wanted him to take the meeting because he thought Reagan was considering Fernandez as a potential running mate, especially because he could have helped him with the Hispanic vote. Fernandez claimed that three Republican candidates had in fact asked him to serve as their candidate for vice president. But Fernandez declined because he thought he was going to win the race, and he didn't want to play second fiddle. Casillas said Fernandez genuinely thought he was headed to the White House.

Casillas may not have known that federal law prohibited the candidates for president and vice president from coming from the same state, but for him it was a telling example of Fernandez's hubris to imagine that he was going to win the race, especially considering that Fernandez wasn't raising any money. We "didn't have a dime," Casillas said. The fact that Fernandez was unable to secure enough donations from enough individuals in different states is what kept him off the ballot in many places.

FERNANDEZ PLACED WHATEVER CHIPS HE HAD ON THE PRIMARY IN PUERTO RICO. IT WAS the first time ever that Puerto Ricans even participated in the selection of a nominee. They still wouldn't participate in the general election, but they were eager to at least make their voices heard at the convention. Along with "dominos and cockfighting," wrote a reporter for the *Boston Globe*, "Puerto Ricans regard politics" as their "national pastime." Some 85 percent of voters on the island regularly participated in elections, which was more than anywhere else in the United States.

Puerto Rico's *ley de primarias presidenciales compulsorias* (the compulsory presidential primaries law) had passed in September 1979. Traditionally, the members of Puerto Rico's Popular Democratic Party (PPD) were associated with the national Democratic Party, while the members of the prostatehood New Progressive Party (PNP) were associated with the Republicans. But two leading statehooders in the late 1970s, Franklin Delano López and Governor Carlos Romero Barceló, sought to cut into what they saw as the PPD's monopoly on sending delegates to the Democratic national convention. To their mind, the prostatehood position had to have supporters in the fifty states who were both Democrats and Republicans. They encouraged fellow members of the PNP to "invade every democratic caucus" on the island and ended up winning nine of the twenty-two delegate positions.

From that moment on, the PPD and the PNP wouldn't align neatly with the national Democratic and Republican Parties; there would be both Democrats and Republicans in the prostatehood PNP, including Romero Barceló, the leader of the PNP himself. He would support Carter's reelection, while other members of the PNP would rally around one of several Republican candidates. Fernandez hoped they would choose him.

Fernandez needed them to choose him. He based his campaign on a rather unorthodox electoral strategy. He would win the island's fourteen Republican delegates. They were Hispanic like him, so he believed they would be drawn to him naturally. He would win Puerto Rico in a "landslide," he said, because Puerto Rico is "Fernandez territory." His success in Puerto Rico, which came a month after the always-first Iowa caucus but before the important New Hampshire primary, would be infectious and feed enthusiasm for him elsewhere. Fernandez predicted that he would head into New Hampshire as the "bona fide front-runner of the Republican Party." He could win at least 10 percent of the vote there because of the state's "large bloc" of French Canadian and Italian voters who would "identify with another Catholic ethnic." Then all of the other candidates could "watch my smoke."

In many ways, the Fernandez campaign would be a trial balloon to test his theory that Hispanics were on the brink of revolutionizing American politics and that they would support their own. The *Washington Post* columnist George Will, who was emerging as a leading conservative voice, cast doubt on Fernandez's ethnic strategy. Fernandez said his base would be the 16 million Hispanics living in the United States. Will threw cold water on his plan, noting that the Census Bureau counted only 11.3 million, only 2 million had voted in the 1976 election, and only 1,500 lived in the important primary state of New Hampshire. He drew attention to the persistent challenges to Latino political influence: the Latino population was growing but still represented a small percentage of the electorate overall, and not enough Latinos turned out to vote in order to tip elections.

Regardless of the overall wisdom of Fernandez's strategy, he was at a disadvantage in Puerto Rico whether he was Hispanic or not. Other Republican candidates had been organizing on the island for years and had better connections there than he did. As one Puerto Rican Republican later recalled, Fernandez just didn't have name recognition on the island. Among the small group of Puerto Ricans who had heard of him, he became known as the *Latino aspirante presidencial* (the Latino hopeful for president), but that didn't garner him a following.

Bush, on the other hand, became known everywhere. When he was contemplating a run for the presidency, he and his wife, Barbara Bush, attended a cocktail party in Washington hosted by the only Puerto Rican lawyer in town, Luis Guinot Jr., who had worked in the Nixon administration as deputy counsel for the Department of Agriculture, and had helped the former governor of Puerto Rico Luis Ferré, whose daughter Rosario was a childhood friend, to organize the Office of the Commonwealth of Puerto Rico in Washington (OCPRW). Guinot held the party at his house for the mayor of San Juan, Dr. Hernán Padilla, who was visiting Washington, and by the time of the event, Guinot had left his position as the director of the OCPRW for a job in private practice. He and Bush got to talking. Their wives were getting along fabulously. At the end of the party, Bush asked, "Listen, if you're not committed to anyone, would you consider joining my efforts to run for the presidency?"

As it turned out, Guinot wasn't committed, so he committed to Bush. Guinot became one of the first few members of the Bush campaign, along with James Baker, Chase Untermeyer, and others in a small inner circle. Bush told Guinot that as a Texan, he could gain the votes of Mexican Americans, but he needed to "get a foothold" in the Puerto Rican community, and he knew that gaining support in Puerto Rico would be critical to gaining the support of Puerto Ricans in the fifty states. He said he wanted to make Puerto Rico a "cornerstone" of his campaign.

Like Fernandez, Bush believed that a victory in Puerto Rico would give him momentum heading into Iowa and New Hampshire. He called momentum the "big mo'," Guinot said. So Bush sought the big mo' in Puerto Rico. When Bush made his pitch to Guinot, he also told him that Americans wrongly believed that Hispanics were a monolith. The host of the cocktail party immediately agreed. "We Puerto Ricans," he told Bush, were not like the Cubans, and they were not like the Mexicans. The only thing they shared in common was their language and culture, but other than that, they tended to be "different." From the outset, then, Fernandez and Bush's strategies in Puerto Rico were polar opposites. Fernandez's strategy relied on convincing Puerto Ricans that Hispanics were the same, while Bush's relied on tapping into the beliefs of Puerto Ricans such as Guinot that Hispanics were different from one another.

Bush and Guinot clicked, but as a naval officer and lawyer, Guinot later said that he "didn't know beans" about political campaigns, so he asked his friends on the island to help him. He spoke first with the current and former governors of San Juan—Romero Barceló and Ferré—who said they couldn't help because they were already committed to Tennessee senator Howard Baker. But they wouldn't stand in his way. Next he turned to the mayor of San Juan, Hernán Padilla, who put him in touch with Julia Rivera de Vincenti, "Doña Julita," Puerto Rico's secretary of labor. She was someone Guinot considered to be a "very accomplished political organizer." She began helping Guinot with the Bush campaign, and things got rolling from there.

Bush traveled to Puerto Rico several times. Guinot advised him that if he wanted to win, he had to support statehood, and he had to "say it often, and loudly." That's exactly what Bush did. At a press conference in San Juan he said, "The one message I bring to my fellow American citizens in Puerto Rico is *estadidad ahora*, statehood now!" In a later visit, to attend Ferré's birthday celebration, he said he hoped to be the president who would put the fifty-first star on the flag. He also sent his twenty-six-year-old son Jeb to live in Puerto Rico for two months leading up to the primary election. Jeb and his Mexican-born wife, Columba, lived in the apartment that Guinot kept in San Juan. They attended the seventy-five campaign events held for their father.

The result, according to the *Washington Post*, was that Bush "swamped" the other Republican candidates, winning 60 percent of an estimated 200,000 votes cast. Baker won 37 percent, the former Texas governor Connally won 1 percent of the vote, and Fernandez placed just behind Connally, also with 1 percent, or 1,912 votes. He did better than Minnesota governor Harold Stassen and Kansas senator Bob Dole.

Jeb said that his dad's victory marked "the first step toward statehood" for Puerto Rico. Former governor Ferré also called Bush's victory a "turning point" because "the people" demonstrated in a "wonderful manner, peacefully, affirming that the great majority are against independence or continuing as a commonwealth."

Guinot didn't recall what the Fernandez campaign had done in Puerto Rico, but he knew it was going to be very hard to beat Bush. Even Bush's big mo' proved to be short-lived, though. When Reagan announced his candidacy on November 13, 1979, he made a statement in support of Puerto Rican statehood, but then his team decided his name wouldn't be on the ballot in Puerto Rico. They believed it was more important for him to focus on Florida, South Carolina, and

New Hampshire instead. He gave a legendary debate performance in New Hampshire that vaulted him over the top of the bar.

Unfortunately for Fernandez, things only got worse after Puerto Rico. He charged that voter fraud "riddled" the election, calling it the "Puerto Rican Watergate." He asked the Federal Elections Commission to investigate the matter. But nobody was listening.

FERNANDEZ'S HOPES WERE DEFLATED WHEN HE REALIZED THAT A MEXICAN AMERICAN candidate wouldn't necessarily find success among other Hispanics. His lack of support from Puerto Ricans upset his idea that Hispanics would vote for him because he was Hispanic. He actually won more votes in Kentucky and Kansas, which were less populous than Puerto Rico and were states that very few Hispanics called home. After his stinging defeat in Puerto Rico, everyone wrote him off. Yet he kept going despite the fact, as he put it, that everyone considered him a "goner." He ran out of campaign funding, but paid for the campaign on his own until the very end. He began talking less about his future as president and instead talked about being an effective surrogate for Reagan in the general election, helping him court the Hispanic vote.

The runner-up, Bush, conceded on May 26, 1980, but Fernandez wouldn't concede until early June. It was just a few weeks before the Republican national convention in Detroit, where Reagan accepted the nomination, and where he chose Bush as his running mate. Bowing out as late as he did, Fernandez seemingly wanted to make a point, even though he claimed to be in it for more than show.

In his hometown of Calabasas Park, Fernandez cast a ballot for himself in the California primary, held on June 3, 1980, the same day as a handful of other states that were the sites of the last primary elections. The *Los Angeles Times* quoted "political observers" who said that the Fernandez campaign was "hopeless" from the start. The paper called the vote he cast for himself a "quixotic gesture," but Fernandez said it was the high point of his campaign. By the time he bowed out of the race, some twenty-five thousand Americans had cast ballots for Boxcar Ben, representing 0.2 percent of all primary votes cast.

Fernandez dropped out of the race the day after the California primary. His campaign was $150,000 in debt. If he had to do it over again, he said, he would follow the more traditional path of seeking support in Iowa and New Hampshire. Journalists asked him if he would consider joining the Reagan ticket as vice president. He only said, a "Reagan-Fernandez ticket would be mighty hard to beat." He returned to his business and to his family, who, he said, hadn't seen much of him for more than a year.

By traditional measures, Fernandez's campaign was a failure. He ruminated, in particular, on his lack of support among Hispanics. It was "crushing," he said, that he was unable to "fire up" the nation's Mexican Americans, Cuban Americans, and Puerto Ricans. Even though they weren't fired up, his campaign gave the Hispanic Republican movement a national voice. Fernandez traveled across the country, making visible the interests of a growing minority within the Republican Party.

Maybe Fernandez wasn't the right Hispanic Republican to run for president, though it is hard to imagine anyone who was better prepared. Or maybe the Republican Party wasn't ready for its first Hispanic presidential candidate. It was being pulled in at least two different directions, by people such as Fernandez and groups such as the RNHA, which sought to make the party more inclusive, and far-right activists who were nationalists and xenophobes.

The *Time* magazine cover had said that Hispanics would soon be the nation's largest minority, but that didn't mean one of them would soon be leading the nation. Nevertheless, Fernandez embodied the Hispanic Republican movement of the 1970s. He helped blaze a trail for future Hispanic candidates for office, and also for Reagan's path to the presidency, with a significant minority of Hispanic support. Almost as though he himself hadn't been in the running, he traveled across the country for Reagan, telling Hispanic audiences that their votes could decide the outcome of the 1980 election.

CHAPTER 7

REAGAN'S *REVOLUCIÓN*

As a surrogate for the Reagan-Bush campaign, Fernandez accepted the invitation of the Republican national convention's program committee chair, Mike Curb, the lieutenant governor of California, to speak on the first night. Fernandez's speech would come right after Gerald Ford's, live before a television audience of millions of Americans. Curb had already found an African American to serve as the convention's secretary, a state senator from Maryland named Aris Allen. "Every time there was a vote on the floor," one reporter wrote, "Allen's face would appear on the screen." But Curb wanted to feature a Hispanic as well. He asked a friend whether he thought Ben Fernandez would be good. His friend asked back, "The one born in a boxcar?" To which Curb responded, "That's the one . . . Boxcar Ben."

Fernandez got the invitation the week before the convention began, and arrived in Detroit on Saturday, July 12. He headed straight to the historic Leland Hotel, just a mile away from the new Joe Louis Arena—named after the African American boxer—where the convention would be held. The other Hispanic delegates stayed there as well. Cuban Americans from Miami huddled in the hotel room of one of their delegates, sipping Cuban coffee and discussing a party platform that criticized Castro and Carter in equal measure. It contained "everything that Cubans hope for," said Carlos Salmán, head of the Miami-based Reagan Para Presidente Committee. They were readying to hand out one thousand anti-Castro pamphlets to attendees entering the convention center. Fernandez complained about the smell of cigarette butts and wondered why the hotel didn't have air conditioning in the middle of summer. It reminded him more of the railroad boxcar he was born in than a hotel fit for a former presidential candidate. But the show had to go on. On Monday morning, the day of his speech, he had breakfast at the Plaza Hotel with members of the Republican National Hispanic Assembly (RNHA), bought himself a new shirt for prime time, and headed to the convention center.

The convention in Detroit was just the beginning of the sprint by Fernandez and other Hispanic Republicans to help Reagan and Bush win the election. Two Mexican Americans from California, Alex Armendaris and Fernando Oaxaca, were national cochairs of the Viva Reagan-Bush campaign. They were marketing and media executives, respectively, and both had worked for Nixon and Ford. As was customary, the Hispanic campaign picked up in early September, at the beginning of Hispanic Heritage Week. And again, for many Hispanic Republicans this was too late. After fielding enough phone calls from Hispanics who were confused about campaign strategy, Armendariz and Oaxaca sounded alarm bells, telling the higher-ups what they were hearing from New York, Florida, California, New Mexico, and Texas. Hispanics attended the convention and then wondered, "What now?" They had assembled the "troops, wagons, horses," but were "waiting for a goddam [*sic*] general!" Please, "let's don't blow it," Oaxaca pleaded in a letter to the higher-ups in the Reagan campaign.

Hispanic Republicans, like many other Republicans, believed that their party's prospects had changed for the better since 1976, when Ford and Reagan fought a divisive primary campaign and Ford ultimately lost to Carter. They were united behind the conservative candidate Reagan, who won the primary elections in all but a handful of states. Carter had the lowest average approval rating of any president since Harry Truman, and many Americans felt like the country was headed in the wrong direction. Hispanic Republicans had a candidate they could get behind, and they wanted to rally others to vote for him. They thought the election was theirs to lose.

The campaign finally responded to prodding by Armendariz and Oaxaca. Reagan, Bush, and their surrogates fanned out across the country to meet and talk with Hispanic audiences. Cuban Americans paid $1,000 per plate to have breakfast with Bush, and they swarmed Little Havana, gathering around the Bay of Pigs Monument, for the opportunity to meet Reagan. "Busloads of Hispanics" traveled to Liberty Park, New Jersey, and hundreds of members of the League of United Latin American Citizens crammed into a Washington, DC, ballroom to hear Reagan speak. But more often it was their surrogates who visited with the Hispanics they hoped to win over. Advocates such as Jeb Bush; Oaxaca; and, of course, Fernandez made dozens of stops all across the country.

Reagan again focused on Texas, a state Carter narrowly won in 1976—the "tragedy of 1976," Fernandez called it—and that the Republican had to win in 1980 if he had any hope of winning the election. The campaign camped out in the Lone Star State for much of September and October.

Reagan spent Mexican Independence Day, September 16, in San Antonio, with Fernandez and Texas senator John Tower. Two weeks later, Fernandez went to Odessa, a couple of hundred miles from El Paso. He told a group there that Mexican Americans in Texas would "put Ronald Reagan in the White House." Whichever candidate won the Mexican American vote would win Texas, and whichever candidate won Texas would win the presidency. He said that Hispanics liked Reagan's "macho image" and that they were natural Republicans because they were fiscally conservative and individualistic and feared government. He spent most of his time bashing Carter, who had cost the country jobs and defended the nation with a "second-rate military." It was a "track record" that spoke "eloquently of his incompetence," Fernandez said. He and Oaxaca repeated the message a month later in Seguin, on Columbus Day. Three weeks after that, right before the election, they returned to Seguin, this time with Jeb and Columba Bush and Roger Staubach, the popular quarterback for the Dallas Cowboys.

With their help, Reagan won Texas with a sizable share of the Hispanic vote. Nationally, he won the popular vote narrowly but the electoral vote overwhelmingly. His tally included an estimated 35 percent of Hispanics, which was even better than Nixon had done in 1972. Meanwhile, Republican support among African Americans continued to slide. Reagan won 11 percent of the African American vote, which was less than the share Nixon or Ford had won. It didn't help that Reagan had kicked off his general campaign with a speech in Mississippi in support of states' rights, not far from where civil rights workers had been murdered in the sixties.

Reagan's election marked the conservative movement's entrance into the political mainstream. It thrived in suburbs, claimed William F. Buckley Jr. as its voice, and lashed back against provocateurs who incited social unrest. Its economists advanced the free-market philosophies of Milton Friedman; its men and women preferred traditional gender roles; and its religious followers were evangelicals. Reagan had become a hero to conservatives, and about half of them were disappointed that he wasn't the nominee in 1976.

Historians have described the 1980 election as a revolution for the conservative movement, the Reagan Revolution. It was no less of a *revolución* for Hispanic Republicans that it was characterized by the shift away from courtship through patronage toward courtship through more ideological arguments about shared values and policy positions, as well as the growing prominence of Cuban Americans and Cuban exile politics within the Hispanic Republican movement. These transformations had a turbulent takeoff, but within a few short years they were complete, and resulted in Reagan doing even better among Hispanics in 1984 than he had done in 1980.

Many Hispanics continued to believe that Reagan was indifferent to them, or worse. But a significant minority now identified as loyal Republicans. They agreed with his strong statements against communism; his support for free-market capitalism; his anti-affirmative action and antiwelfare rhetoric; his statements recognizing the contributions of Hispanics; and his opposition to the construction of a border wall. But they still had lingering doubts about how much he would include them in government and whether he would include them as equals.

AFTER FERNANDEZ PUT IN HIS WORK FOR THE REPUBLICAN TICKET, HE FELL OFF THE MAP. His close friend from childhood Frank Casillas said he didn't hear from Fernandez after 1980, which was strange considering their friendship and the spotlight Fernandez stood under for more than a decade. It wasn't that Casillas had gone underground, either. He went on to be Reagan's assistant secretary of labor, in charge of two thousand people and an annual budget of $26 billion. When he worked at the White House, someone who wanted Fernandez to speak at an engagement called to inquire about Boxcar Ben, but nobody could find him. Casillas even called Fernandez's brother and sister, who said they hadn't heard from him either. He lamented Fernandez's exit from the stage. He may have had an overinflated sense of himself and had misread the political landscape for a Hispanic candidate, but Hispanics respected him. He had been an effective leader and advocate.

The Hispanic Republican movement forged ahead, as Fernandez's friends in the RNHA, Armendariz and Oaxaca, signed on to work for the Reagan-Bush campaign. They formulated the strategy to "attract Hispanic votes into the Republican column in November 1980." Their main objective was to highlight for Hispanics the "common traits" they all shared, and how their values aligned with the core values of the Republican Party. These included their individualism, patriotism, and belief in the "American Dream." Other Hispanic

Republicans felt that different Hispanic communities had to be targeted individually—as Mexican Americans, Cubans, and Puerto Ricans—thus pointing to the enduring tension in Hispanic culture and politics, liberal or conservative, between whether Hispanics were fundamentally similar or different.

To those who thought it best to appeal to different Hispanic communities with distinct messages, Oaxaca responded—much as Fernandez himself would have—that it was "divisive" and "flies in the face of what the RNC has worked on for years." The RNC had established the RNHA as an official auxiliary to represent the interests of all Hispanics, rather than a particular subset. Hispanics spoke the same language and had a "common heritage in Spain." They were proud, and they loved their "family, church, and work." This was the message they hoped regional and national leaders of the RNHA would deliver to Hispanic communities across the United States.

The 1980 presidential election was the first time that the RNHA exerted its full force in support of the Republican ticket. National leaders organized regional leaders who organized state leaders who organized local leaders, who encouraged the members of their community to vote for Reagan and Bush. It helped that the leaders of the Viva Reagan-Bush campaign were also national leaders of the RNHA, because they could coordinate their efforts as RNHA executives with their oversight of thousands of campaign volunteers who marched through neighborhoods, knocked on doors, and delivered pamphlets.

For the 1980 election, Reagan based his appeal to Hispanics on ideological issues such as religion, patriotism, and entrepreneurship. He appealed to their religious devotion more than earlier Republicans. He also opposed quotas and set-asides, so patronage didn't play the same role in his Hispanic strategy that it had in Nixon's. Instead, before an audience in East Los Angeles, Reagan emphasized how Hispanics preserved the "fundamental values of the family"; demonstrated a "deep and active concern for their parishes and neighborhood"; and worked "long and hard" to own homes, farms, and businesses. During wartime, no group of Americans was "more willing to risk their lives" to defend the United States.

Finally, it was the first time that a Republican presidential campaign launched a sophisticated media research strategy that included radio and television ads that targeted Hispanics specifically. The campaign brought the San Antonio ad man Lionel Sosa on board. After Sosa helped John Tower get reelected in 1978, Tower wanted to help him in return. He told Sosa that he was going to introduce him to Reagan, the next president of the United States. Sosa recalled that he and Reagan spent a few minutes together, and Reagan told Sosa that if he were to work for him, he would have the easiest job in the world. Just like Hispanics, Reagan said, he believed in country, family, faith, and opportunity. By the time of the 1980 election, these were the values Hispanics and other Republicans were presumed to share, and reciting them had become a refrain that had a ring of truth to it. But naming them also homogenized Hispanic Republicans in a way that minimized the varied reasons Hispanics found their way into the Republican Party, and downplayed distinctions between Hispanic Republicans and other conservatives.

It was in a conversation with Sosa that Reagan, working to convince Sosa to work for him, delivered his famous line that Hispanics were Republicans who just didn't know it yet. Sosa worked alongside Armendariz and Oaxaca on the Spanish-language advertising campaign.

Even though Armendariz and Oaxaca supported sending volunteers into Hispanic communities, they believed that bilingual television and radio ads in English and Spanish would be more effective. Oaxaca said that Hispanic voters responded to the "tube and radio" and therefore suggested a half hour television show that targeted Hispanics and that would portray Reagan as a "family man, happy, content, and understanding of family problems, the budget, the price of meat and beans and clothes and getting a good education for the kids and making sure that the wage earners will not lose their jobs in a Carter depression." Hispanics, he said, would also respond positively to the messages that Reagan would "ensure peace for the world through strength"; he "doesn't want to keep getting pushed around by Russia and others, and will improve relations with Latin America." These were symbols of Hispanic culture and politics that sought to establish their conservative values as much as reflect them, part of the steady drumbeat the Republican Party had played in its effort to court Hispanics since the beginning of the Cold War.

Oaxaca further visualized the scene. Ronald and Nancy Reagan would sit in a living room with "two or three Hispanics," one of them a woman. Bush could sit by their side. The commercial would answer the question *¿Quién es Ronald Reagan?* (Who is Ronald Reagan?) While they talked, "flashbacks" of East Los Angeles, Miami, and the Bronx would pop onto the screen, appealing implicitly to the Mexican Americans, Cuban Americans, and Puerto Ricans who lived there. The "right kind of music"—presumably salsa—would play in the background. The living room setting would transition to a Hispanic triptych: "happy smiling Hispanic kids on the school grounds, people on the way to work, coming out of Mass on Sunday." The narrator would speak with a "slight accent" reminiscent of Ricardo Montalban, the Mexican actor who appeared in dozens of popular films, including *Escape from the Planet of the Apes*, *Conquest of the Planet of the Apes*, and *The Mark of Zorro*.

The main aim of the commercial was to change the negative perception that many Hispanics had of Reagan. The folk wisdom about his relationship with Mexican Americans in California, as Oaxaca described it, was that he could be cold and callous. Importantly, his opponents included influential groups such as the generally moderate Mexican American Political Association. Its conservative members complained that Reagan rarely if ever came to their conventions, sometimes addressed them by phone but wouldn't answer their questions, and wouldn't schedule meetings with them in Sacramento. He had supported California growers in their struggles against César Chávez's United Farm Workers union, and vetoed bills that would have made it easier for Spanish speakers to vote. Reagan's opponents in California called his positions on housing, integration, agricultural labor, and other issues some of the most racist they'd ever seen.

Oaxaca argued that Hispanics outside of California barely knew who Reagan was. Cuban exiles in Florida remembered him from the strong anti-Castro statements he made there during the 1976 primary campaign, but his Hispanic circle didn't get much wider than that. Considering the grievances of Mexican Americans in his home state, perhaps it was better that way. If Hispanics did know him, Oaxaca wrote, they were likely to believe that Reagan "doesn't like minorities," did a "lousy job for poor people in California," was a "warmonger," and was a "tool of the big companies and rich people." The goal of the Hispanic campaign was to change that impression in the key states where it might make a difference.

The leaders of the Viva Reagan-Bush campaign felt they had plenty that was good to say about their candidate, and it was only a matter of targeting the right Hispanics to teach them what Reagan's conservative supporters already knew. As their governor, he had appointed more Hispanics to government positions than any governor before him. He amassed considerable Hispanic support through his endorsement of bilingual education and antibusing campaigns, which many Hispanic parents opposed because they preferred to send their children to the "neighborhood schools" where they were more comfortable. And throughout his time as governor, he had always been on the side of business.

Reagan was also a westerner, like Nixon, and fancied himself a cowboy, like Goldwater and the rancheros of the US-Mexico borderlands. He ceremonially rode on horseback alongside *Californios*—the elite Mexicans who settled in California before the US-Mexico War—from the border near San Diego up to Sacramento. Hispanic Republicans in California saw Reagan as a candidate who cared about them and was familiar with their interests. For fellow Hispanics who were still wary of the Republican Party, the Californians who headed the Viva Reagan-Bush campaign hoped to convince them to vote for "Reagan the man" if not for the Republican Party.

OAXACA ALSO WANTED CONSERVATIVE HISPANICS TO KNOW ABOUT REAGAN'S POSITIONS on US–Latin American relations, because they were highly sensitive to whether the Republican candidate held views that were similar to their own. Cuban Americans and Cuban exiles wanted Reagan to "pledge" not to enter arms control negotiations with the Soviet Union, to reduce all "non-defense" federal spending to balance the budget, and to keep the embargoes against commerce with Cuba in place.

They also wanted to know what he would do about the new wave of Cuban exiles arriving on Florida's shores. Between April and October of 1980—in the middle of the primaries, as the parties crowned their nominees, as Reagan and Carter sparred—more than 120,000 Cubans sought refuge in the United States. The new Refugee Act of 1980, which took effect on April 1, said that a refugee was a person with a "well-founded fear of persecution" at home. Even though many of the Cubans interviewed upon their arrival said they left Cuba for economic reasons, in part because the recent visits by Cuban exiles to Cuba had given them a glimpse of the wealth they had accumulated in the United States, most were granted political asylum because they were fleeing the Castro regime. When they were still primary candidates, Reagan and Bush preempted Carter by stating that the United States should welcome the Cubans. Carter followed suit, but only after the statements by the Republicans gained favor in South Florida.

At first the refugees were considered to be participants in a new "freedom flotilla," like those who had arrived in the 1960s, but that impression changed as their numbers grew and news reports circulated that they were less desirable than this earlier cohort—a more criminal element that was poorer, darker, gay. They were placed in

detention camps and became subjects of negative depictions in popular films such as *Scarface* (1983), starring Al Pacino.

Puerto Ricans, meanwhile, as one of Oaxaca's strategy documents
put it, believed that "what's good for the island is good for all Puerto
Ricans." For the Puerto Rican Republicans in Puerto Rico, New York,
Florida, Washington, DC, and elsewhere, that meant statehood. Their
participation in their first primary election in 1980 meant to many
prostatehooders that the time was right for another plebiscite on the
island's status, which Governor Romero Barceló planned for 1981.
And even though former governor Luis Ferré had stepped down as
the president of the Puerto Rican Senate in January 1981, right before
Reagan's inauguration, he would continue to advocate for statehood.

Finally, Mexican Americans wanted to hear what Reagan would
do about immigration and border control. A small number may have
wanted tighter enforcement, maybe even a wall, but the vast majority
of Mexican Americans, conservative and liberal alike, asked for leniency. As Oaxaca put it, Mexican Americans felt a "decided sympathy"
for undocumented immigrants, and they believed that "barbed wire,
armed guards and sentry dogs are not the answer."

In each case, Reagan gave Hispanic Republicans what they asked
for instead of bending to the will of far-right nativists.

Reagan's anti-Communist rhetoric matched Goldwater's almost
twenty years earlier. The Cuban Revolution had become increasingly
antagonistic toward the United States, Reagan argued, and Castro
was still trying to export the Revolution to other Latin American
countries. If the Caribbean Basin—from Mexico to Nicaragua, Panama to Venezuela, and the Dominican Republic to Jamaica—was a
smoldering political volcano about to erupt, Cuba was the white-hot
center that would cause it to spill over.

Reagan seemed almost incapable of talking about communism
in the Americas without referring to Cuba or Castro. Puerto Ricans
who called for independence were Cuba-inspired Marxists, he said.
The "Dominican Republic of Cuba" was his name for the Dominican

Republic. The situation in Nicaragua—where the leftist Frente Sandinista de Liberación Nacional, or Sandinistas, had overthrown Somoza in 1979—bore a "Cuban label." The Sandinista rebels had been
"Cuban-trained, Cuban-armed and dedicated to creating another
Communist country in this hemisphere." As a result, the entire Caribbean, Reagan said, was "a Communist lake in what should be an
American pond."

Even in the United States, Reagan saw leftists including the Young
Lords, Brown Berets, and Black Panthers—made up primarily of Puerto
Ricans, Mexican Americans, and African Americans—as sharing a
common cause with Latin American and African revolutionaries, inspiring fears of upheaval and social unrest. Reagan had made a career
railing against leftist activism domestically, especially against African
Americans during the Watts riots of 1965 and student protesters at
the University of California at Berkeley, San Francisco State University, and other institutions. His anticommunism and antiradicalism
merged in his critiques of the domestic left, and Hispanic Republicans were receptive to his arguments. They were part of the silent
majority, not troublemakers whose anger erupted in the streets.

On the campaign trail, Reagan's proposed solutions to these international and domestic troubles included a combination of new initiatives: greater military and policing powers at home and abroad;
support for the enemies of his enemies; the backing of capitalist democracies; and, if necessary, regime change. It was a direct rejection
of Carter's policies, and a return to the more conservative aspects of
Eisenhower's, Nixon's, and Ford's approaches. Even if the funneling
of resources to unsavory dictators began in the last years of Carter's
term, Reagan used harsher rhetoric than all of them. This was part of
his appeal to Hispanic Republicans.

Reagan's policies toward Mexico, meanwhile, perfectly combined
his anticommunism and support for friendly cross-border social, cultural, and economic relationships. Like Fernandez, Reagan believed
that cultivating capitalism, democracy, and international cooperation

would help reduce threats to the United States. He argued that Mexico would be a buffer between the United States and radical leftists in Central and South America. As the fringe right wing of his party increasingly harped on the dangers posed by immigrants—they stole jobs, were criminals, and were changing the face of the nation—Reagan struck a different chord. He supported immigration policies seen as sympathetic to immigrants, and rejected the construction of a border wall that would divide the United States and Mexico.

During a campaign stop in Texas, Reagan had proclaimed that the United States should "document the undocumented workers" and give them visas "for whatever length of time they want to stay." Governor William Clements walked back Reagan's enthusiasm, denying that the candidate said what reporters heard. But when they played a tape of the speech for him, Clements insisted that Reagan did not mean to suggest that undocumented immigrants could stay indefinitely. Farther down the campaign trail, Reagan told Mexican president José López Portillo that they "could make the border something other than a locale for a nine-foot fence." Just before the convention, he had sat for an interview with *El Sonorense*, a newspaper published in the Mexican state of Sonora, where he was getting a little rest before the final push of his campaign. He said that Mexico was America's most important ally.

Because of his statements about US–Latin American relations, especially in the Caribbean, Central America, and Mexico, Reagan began to develop a following in Latin America as well. Even though Latin Americans—including Puerto Ricans in Puerto Rico, despite their citizenship—were unable to vote in US elections, they offered to do whatever they could to help him get elected. The Mexican lawyer Agustín Navarro Vázquez wrote a weekly column for the Mexico City magazine *Impacto*, where he praised Reagan for saying he would support free enterprise, shrink government, lower taxes, and fight communism. A member of the secretive Mont Pelerin Society, which promoted free market economics, Navarro wished Mexico's leaders would do the same. He asked Reagan's Hispanic campaign chief, Alex Armendariz, to circulate his articles among the "26 million Spanish speakers in the United States, especially in Texas and California," hoping that he might convince them to vote for Reagan.

Reagan received more support from El Salvador, another Central American country where leftist guerrilla movements—which came together as the Farabundo Martí National Liberation Front (Frente Farabundo Martí para la Liberación Nacional, or FMLN)—were on the rise and a US-backed military dictatorship was struggling to contain it. Writing from his home there in the month before the election, Mario Moya Roldán lamented the economic, diplomatic, and military decline of the United States under Carter. Countries throughout the Americas still had all the advantages of their natural resources, human capital, and technological superiority, he wrote. With Reagan as president, the Americas would rise again. A letter sent from Argentina conveyed the same sentiment. Before Carter, its author wrote, the United States stood for "freedom and security." Since Carter's election, the country had come to represent "weakness and stupidity." Carter opened the door to communism; Reagan would shut it. "You will be the winner in November," the author wrote. "You must be the winner!"

Meanwhile, a Cuban woman from Maryland wrote Reagan on behalf of her ninety-two-year-old grandmother, who had been born in Cuba but had lived in the United States since the 1940s. Her grandmother had just become a citizen for the "sole purpose" of casting her vote for Ronald Reagan. She had "seen America deteriorate in the last 35 years—and especially the last four," since Jimmy Carter took office. She was confident that Reagan would turn things around. Another letter came just a week before the election, from a Cuban American woman in Miami. She wanted to wish "Mr. Ronald Reagan" and "Mr. Jorge Busch" the best of luck. Her birthday was on November 5, and all she hoped for was a bouquet of flowers and the chance to celebrate *el Triunfo Republicano* (the Republican victory).

She got her wish. Reagan and Bush flipped Texas, Florida, and New York. They won a third of the Hispanic vote nationally, including 18 percent of the Puerto Rican vote, 30 percent of the Mexican American vote, and a whopping 80 percent of the Cuban American vote. They found success in the Texas districts they targeted, winning 20 to 30 percent of the Hispanic vote where Ford had won only 13 percent. They did well in the heavily Mexican American communities surrounding Los Angeles, where, the *Los Angeles Times* reporter Frank del Olmo wrote, "upwardly mobile Chicanos" had moved after "leaving the inner-city barrios." Reagan and Bush had courted the Hispanic vote, and Hispanics responded. Hundreds of them spent thousands to travel to Washington, DC, for the inauguration, then waited to see what happened next.

The revolution was coming, but it was a revolution on the right. The fanatical support for Reagan united the Latin American right and Hispanic Republicans across borders.

BY THE TIME HE WAS SWORN IN, REAGAN HAD ALREADY BEEN CONFRONTED BY SEVERAL issues of importance to Hispanics. Just a week after the election, the Department of Education announced a proposal that would have required "bilingual classes in the public schools for students not knowing English." Reagan had supported bilingual education when he was the governor of California, but would not support it as a federal initiative. Advocates for bilingual education said it protected the rights of non-English speakers under Title VI of the Civil Rights Act of 1964, which prohibited discrimination on the bases of "race, color, or national origin" in programs that received federal funding. Opponents, channeling critiques of the Civil Rights Act itself, said it represented a government overreach that went beyond what the Department of Education was authorized to do. Whatever Reagan decided, his decision would affect Hispanics disproportionately. They represented 80 percent of the 3.5 million students in the United States with "limited

English proficiency." Among Hispanics, there was almost universal bipartisan support for bilingual education, although some opposed it on the grounds that native Spanish speakers could speak Spanish at home but should be required to speak English in school.

The increasing number of undocumented immigrants in the United States also became a focal point for the administration-in-waiting during the transition period. Presidents Nixon, Ford, and Carter had proposed reforms of Johnson's Immigration and Nationality Act of 1965 but were unsuccessful. Reagan hoped to triumph where they had failed. He wanted to be the one to pass a sweeping immigration law that would be lenient toward undocumented immigrants already in the United States while also halting their flow in the future.

In December 1980, the Select Commission on Immigration and Refugee Policy proposed a "limited amnesty" for undocumented immigrants who had resided continuously in the United States since January 1 of that year, a provision that migrants who moved back and forth seasonally wouldn't be able to take advantage of. The commission also proposed, once more, an increased number of visas for temporary workers, employer sanctions, and a national identification card. Hispanics gave amnesty and the visas their bipartisan support, but they opposed the sanctions and identification card. Reagan had spent considerable time on the campaign trail talking about how undocumented immigrants should be allowed to work in the United States, and his belief that there should be no border wall, so Hispanic Republicans were optimistic that he would forge a compromise they would approve of.

Last, Reagan's cabinet selections drew close scrutiny from Hispanic Republicans who felt that Reagan owed them something in return for their support. Because the Viva Reagan-Bush campaign had successfully turned out votes, leading Hispanic Republicans expected him to make some high-level appointments. Causing them great consternation, however, was the fact that Reagan's first ten cabinet selections were "all white men" from the East and West Coasts, chosen from among his longtime supporters and political associates. Hispanics and others applied pressure, and newspaper reports began to suggest that with Reagan's last five appointments, he was likely to pick a woman, a southerner, an African American, or a Hispanic. There was much speculation that Philip Sanchez, the highest-ranking Hispanic in the Nixon administration and former ambassador to Colombia, would become the secretary of housing and urban development. That didn't happen, and Hispanics continued to wait for their first cabinet secretary.

The month after the election, Reagan discussed bilingual education, undocumented immigration, political appointments, and other Hispanic issues with a group of his most prominent Hispanic supporters. They convened at Blair House in Washington, DC, where Puerto Rican nationalists a quarter century earlier had tried to assassinate President Truman. Those at the meeting included Manuel Luján Jr.; former campaign leaders Oaxaca and Armendariz; the former chairman of Nixon's Cabinet Committee on Opportunities for Spanish-Speaking People, Henry Ramirez; the former special assistant to the president for Spanish-speaking affairs under Ford, Fernando de Baca; the head of the Hispanic Chamber of Commerce, Héctor Barreto; leading Cuban Americans Carlos Salmán and Al Cardenas from Florida and Tirso del Junco from California; leading Puerto Ricans from New York Rita DiMartino and Ray Maduro; and the former governor of Puerto Rico Luis Ferré. It was a war room full of the leading Hispanic Republicans over the previous decade, minus Fernandez.

Reagan told them he planned to visit Latin America and to appoint Hispanics to government posts. He said he wouldn't cut Social Security programs or public assistance but that he would work to prevent their abuse. Luján told reporters afterward that Reagan also expressed his "preoccupation" with Fidel Castro's influence over all of Latin America.

Winning the Cold War was of paramount importance to all in

attendance, but Cubans, in particular, had finally found a kindred spirit in Reagan. After the opening symbolized by the dialogue between the Castro regime and Cuban exiles, the reestablishment of commercial flights between Miami and Havana, and the many thousands of Cuban exiles who visited Cuba, Reagan and Castro settled back into the more familiar dynamic of shared animosity. Castro returned to calling exiles worms, and to him the more recent exodus of *marielitos*—refugees who departed Cuba by boat from Mariel harbor—was both an embarrassment and a kind of revenge. The new refugees embodied the failures of the Cuban Revolution and enduring poverty on the island, which had caused them to want to flee en masse.

As for the hard-liners, they saw an opening in this renewed period of tension. Their war against Castro was back on. They sought to take their message to Washington.

A former freedom fighter turned businessman, Jorge Mas Canosa became the most influential voice against Castro in the capital. Reagan's national security adviser thought Cuban exiles should form a group that would focus on Cuba policy, and that they should adopt as a model the American Israel Political Action Committee (AIPAC), which had effectively shaped government policy toward Israel and the Middle East. In response, Mas Canosa and others established the Cuban American National Foundation (CANF) in July 1981, just months after Reagan's inauguration, and quickly got to work pressuring legislators and the president himself to aid their mission, as the group's articles of incorporation put it, to develop a "stable, democratic, free-market society in Cuba that can be a prosperous home for all Cubans." The CANF held its first meetings at the DC offices of the National Security Council.

One of Mas Canosa's early ideas was to establish a radio station akin to Radio Free Europe that would send anti-Castro, pro-US news and entertainment over the airwaves in Spanish, from Florida to Cuba. Reagan soon after proposed legislation to create the government-funded Radio Martí, which took a couple of years to get up and running due to opposition from Democrats, progressive Cuban exiles, and radio executives who feared that Castro would retaliate by trying to jam the frequencies not only of Radio Martí but other stations as well. Mas Canosa also believed that it was critical to curb the influence of the Cuban Revolution in other Latin American countries, especially hot spots in Central America such as Nicaragua. As he liked to say, "The way to Havana begins in Managua."

Reagan agreed. As early as 1981, he began planning covert military operations against the Sandinistas. He embraced Mas Canosa's idea that it wasn't wise to attack Havana directly, but rather to isolate Castro by thwarting his allies in the region. Through the CIA, he began to supply weapons to anti-Sandinista "Contras" who were trained in Honduras by Argentinian officers. By the early 1980s, Castro was also increasingly isolated from his longtime backers in the Soviet Union, which had already begun a slow but steady decline. He therefore clung to his alliances with other international socialist movements in the Americas. The Sandinistas were especially provocative in this respect. They pronounced their alliance with Castro, and also supported leftist revolutionary movements in El Salvador and Guatemala. The Reagan administration had tried to get the Sandinistas to stop their support for revolutions elsewhere, in exchange for a "less antagonistic policy" from the United States. But Daniel Ortega replied that the Sandinistas were "interested in seeing the guerrillas in El Salvador and Guatemala triumph." All Reagan could do was contain them. His anti-Sandinista army grew to fifteen thousand men. The arming, training, supplying, and military direction of the Contras came from the United States.

Reagan's Hispanic Republican supporters backed his policies toward Central America, including his arming of anti-Sandinista forces, even though the policy was controversial in the United States, not to mention illegal. Growing numbers of Central American refugees sought asylum in the United States. In 1982, the first churches in the United States offered them sanctuary, beginning in Tucson and Chicago, and then across the country. Their faith-based activism became known as the Sanctuary Movement. The Reagan administration claimed that sanctuary workers were breaking the law by harboring Central American refugees. But participants in the sanctuary network, which was compared to an underground railroad, responded that in providing safe haven, they obeyed the laws of God instead of the laws of man. Many of the sanctuary workers had been motivated by learning about the Reagan administration's involvement in the conflicts in Central America. Their activism in solidarity with Central American refugees was an explicit rebuke to Reagan's policies.

Hispanic Republicans and sanctuary workers both welcomed Central American refugees, but they saw different groups of refugees as distinct. After Somoza's overthrow, some fifteen thousand Nicaraguans arrived in Florida, many of them former members of the Nicaraguan National Guard, commanded by Somoza. They began to train alongside Cuban exiles at camps "in the brush and swamplands in Miami, just beyond new housing developments and a trash dump," preparing to attack the Sandinistas in Nicaragua. Hispanic Republicans, especially Cubans, saw the Nicaraguans as exiles akin to the Cubans who fled Havana after Castro took power. Sanctuary workers, meanwhile, saw Salvadoran and Guatemalan refugees as victims of right-wing dictators whose military henchmen, with US backing, slaughtered the innocents they saw as enemies. Those who made it out alive should be allowed to remain in the United States.

The arrival of many thousands of refugees from Central America only complicated the debate over immigration, which focused primarily on undocumented immigration from Mexico. The recently formed Federation for American Immigration Reform (FAIR), cofounded by a eugenicist ophthalmologist from Michigan, John Tanton, ramped up its attacks against undocumented immigrants. The rise in immigration to the United States, legal or illegal, "hurts American workers," FAIR argued, including African Americans, Asian Americans, and Hispanics. They also claimed that the population growth of the United States was unsustainable and that new immigrants were largely to blame. The United States was no longer a boundless frontier. Resources were limited. The border had to be closed, immigration had to be controlled.

The Reagan administration's approach was different. FAIR's ideas wouldn't move into the mainstream for another decade. The president sought compromise. Following up on the Select Commission on Immigration and Refugee Policy's proposal during his transition, two congressmen, Republican senator Alan Simpson of Wyoming and Democratic representative Romano Mazzoli of Kentucky, drafted a bill that contained the measures Reagan wanted: employer sanctions and border enforcement, yes, but also more visas for temporary guest workers and amnesty.

A draft of the Simpson-Mazzoli bill, named after the two congressmen, was made public in March 1982. The Hispanic reaction against it was swift and included opposition from leading Hispanic Republicans. Oaxaca, the cochair of Reagan's Hispanic campaign in 1980, said Simpson-Mazzoli demonstrated a "lack of sensitivity" to the "deep-rooted feelings of the Hispanic community." Fernandez himself, in one of his rare public statements after the 1980 election, said, "Hispanic America is going to be pretty disgusted with both parties if this bill passes." He promised to do everything in his power to discourage Reagan from signing it if it landed on his desk. Hispanic Republicans joined immigrant rights organizations, labor unions, and increasingly powerful Hispanic lobbyists and elected officials in arguing that the proposed law would enable the continued exploitation of Mexican farm workers, discriminate against Hispanic employees, and do nothing to address the root cause of migration: poverty in Mexico.

THE CHALLENGES OF IMMIGRATION AND US-LATIN AMERICAN RELATIONS WENT UNRE-solved in Reagan's first term. FAIR continued to advocate for immigration restriction, and in 1983 formed the spin-off US English to try to make English the official language of the United States. Tanton led US English with a Republican senator from California, Samuel Ichiye Hayakawa. Their goal was to secure the passage of an amendment that Hayakawa had first proposed in 1981. It was only two sentences long: "The English language shall be the official language of the United States. The Congress shall have the power to enforce this article by appropriate legislation."

FAIR and US English were complementary efforts. FAIR sought to restrict the number of immigrants, while US English aimed to ameliorate what its members saw as the detrimental impact of the official use of languages other than English, both on immigrant assimilation and American culture more broadly. The Reagan administration, meanwhile, continued to seek compromise. Immigration and US–Latin American relations were increasingly divisive subjects, and they became key campaign issues as soon as Reagan announced that he would run for reelection in the summer of 1983.

The official announcement was a mere formality; of course he would run again. In fact, he had been doing campaign-style events since spring, including a visit to Miami for Cuban Independence Day on May 20, 1983, which commemorated the day in 1902 when the US military turned over control of the island to Cubans themselves, and Tomás Estrada Palma became the first president of the independent Republic of Cuba. (Notably, the Cuban exile community continued to observe May 20 as Cuban Independence Day, in defiance of Castro's changing the date to July 26 to mark the day in 1953 when he and other guerrillas launched their first attack against Fulgencio Batista.)

Reagan had lunch at the Little Havana restaurant La Esquina de Tejas, where he sat next to Carlos Salmán, the head of the Dade County Republican Party. When he saw his helping of chicken, black beans, and rice, he asked, "Is there any place in Cuba, outside of maybe the presidential palace, where this menu could still be served?"

To him, the meal was a symbol of the "difference between freedom and what they now have there." Reagan's remarks at lunch were off the cuff, for a small group of leaders of the exile community and the Dade County Republican Party. But he delivered a blistering speech later that day at the Dade County Auditorium. More than twenty-five hundred Cubans packed into the auditorium to hear him, and another sixty thousand heard him through the loudspeaker that projected his speech on the streets outside.

Mas Canosa introduced Reagan and said that the Cuban exile community and the United States in general were engaged in the same fight for human freedom. He quoted Abraham Lincoln's line about how the United States could not survive as a society that was half slave and half free, to make his point that the Americas in the late twentieth century were half free, like the United States, and half slave, living under the boot of Communist governments. He thanked the United States for giving exiles a taste of freedom, and he thanked Reagan for the resolve he showed in confronting communism in Central America, where the "future of our free society" would be decided. He told Reagan directly that Cubans believed in the strength of his character, the strength of his ideals, and his unwavering fight for the cause of democracy. "We support you, Mr. President," he said.

Reagan walked to the podium as the popular song "Guantanamera" played over the sound system. The lyrics were written by the hero of Cuban independence José Martí. Reagan cast the leaders of the exile community as the ideal immigrants. They arrived in the United States with nothing but now were wealthy. The people of the Americas, he said, generally shared the same values, of "God, family, work, freedom, democracy, and justice." But Castro threatened this natural order of things. With help from the Soviet Union, he sought to "impose a philosophy that is alien to everything in which we believe and goes against our birthright." He assured his audience that the United States would put a stop to it. Someday, he promised them, Cuba "will be free."

The *Los Angeles Times* said his "hard-line foreign policy speech" was "easily the most successful of a series of campaign-style trips Reagan has made in recent weeks, possibly the most successful of his presidency." The Cubans and Cuban Americans in the audience didn't disagree. They clapped, laughed, and gave the president one standing ovation after another. Reagan had pardoned Eugenio Martinez, one of the Cubans who broke into the Watergate Hotel, just the week before his visit to Miami. His Democratic opponents called it a political stunt. Martinez sat in the front row to hear the president's speech, and when he mentioned the "Soviet-Cuban-Nicaraguan axis" that threatened the United States, the locksmith led the cheers.

Reagan's frequent return to the subject of Castro and Cuba suggested the growing influence of Cuban Americans and the Cuban exile community within the Hispanic Republican movement. As further evidence, the RNHA replaced the Mexican American Fernando de Baca with the Cuban American doctor Tirso del Junco as the chairman of the organization. Del Junco was the rare Cuban exile in California, and while his appointment as chairman rankled some Mexican Americans who saw it as a hostile takeover of the organization, the move also recognized that Cuban Americans would play an important role in the 1984 Hispanic campaign. Yet Reagan did not forget the Mexican Americans in California who had been his first Hispanic Republican supporters. He ran a national Hispanic campaign that combined his aggressive foreign policy with appeals to their religious traditionalism and belief in free-market capitalism.

ANOTHER EARLY STOP ON THE CAMPAIGN TRAIL CAME IN AUGUST 1983, WHEN REAGAN MET with almost a thousand Mexican American businessmen at a luncheon at the Biltmore Hotel in downtown Los Angeles. It was an opportunity for him to articulate his main arguments about the economy, which he believed would resonate with the "Latino business executives" gathered there. He wasn't interested in "welfare or handouts," he said, but rather "jobs and opportunity." He assured the crowd that he was deeply concerned about the poor and unemployed, but he warned that the solution to poverty and unemployment was not a return to tax-and-spend policies but rather to keep slashing taxes, lowering inflation, and cutting interest rates.

The Hispanic business community knew this better than any other group of Americans, he argued. "To every cynic who says the American Dream is dead, I say: Look at the Americans of Hispanic descent who are making it in the business world—with hard work and no one to rely on but themselves." He promised that his administration would continue to "reward" their efforts because "we believe in the dignity of work." Even the Democrats in the audience were persuaded. The president "said all the right things," one remarked, and "to a business person, the speech sounded real good."

Nevertheless, more than 250 demonstrators protested across the street from the Biltmore, in Pershing Square. They offered a referendum on Reagan's first term, repeating many of the ideas that Oaxaca and Armendariz had predicted would be the bases of Hispanic opposition. To the demonstrators, Reagan's performance only reaffirmed their long-standing critiques. Some Hispanics had made economic gains under Reagan, but high rates of unemployment demonstrated that Hispanics hadn't fared as well as he liked to believe. Reagan also claimed that his administration had accomplished much in the area of educational attainment, even though Hispanics suffered higher dropout rates than other groups. Many Hispanics also criticized him for his military support to counterrevolutionaries and his cuts—and proposed future cuts—to Social Security, veterans' benefits, bilingual education, and voting rights legislation.

The proposed immigration bill also continued to spark heated debates in Hispanic communities, even among Reagan's Hispanic supporters. They organized a conference in early 1984 where Democrats and Republicans united in opposition against its employer sanctions provisions, in particular. Oaxaca said he wouldn't support it even if it had the endorsement of a Republican president. "I am an American first, a Mexican American second, and a Republican third," he said. Meanwhile, Fernandez proclaimed, "As a lifetime Republican, I want the entire country to realize that I am united with our Democratic Party colleagues in the defeat of this bill." It may have been the first time he had ever expressed such bipartisan sentiment, but beginning with the debates over Simpson-Mazzoli, Hispanic Democrats and Republicans were often bedfellows in the immigration wars.

Because of Simpson-Mazzoli, Hispanics vowed to come together as a united front and "play hardball" with their "Latino voting power," according to the *Los Angeles Times*. Politicians supporting the bill could face "stiff opposition from Latinos during the next election." Hispanics might vote against them or stay home. Versions of Simpson-Mazzoli passed the Senate on two separate occasions, and squeaked through the House of Representatives in June 1984 by five votes, 216 to 211. Hispanic officials continued to lead the opposition against it. They weighed it down with three hundred proposed amendments. Edward Roybal, a Democrat from California, wrote one hundred of them himself. The House and Senate gave up trying to reconcile their different versions of the bill in September, and immigration wouldn't come up again until after the election.

In the face of criticisms of his economic, Central America, and immigration policies, Reagan hoped that his Hispanic surrogates would help him change the narrative. Most of them were his appointees, including several Cuban Americans and Mexican Americans. The Cuban American appointees included the Georgetown political science professor José Sorzano as ambassador to the Economic and Social Council of the United Nations; José Manolo Casanova as US executive director of the Inter-American Development Bank; and Carlos Benitez as a member of the Republican Finance Committee. The Mexican American appointees included Linda Chavez as staff director of the US Civil Rights Commission, where she fought against what she described as the "civil rights establishment" and its support for affirmative action and bilingual education; and Catalina Vasquez Villalpando, an active member of the RNHA in Texas and an employee with the regional branch of the Minority Business Development Association (MBDA), who accepted a post as the special assistant to the president for public liaison.

Other surrogates included the highest-ranking Hispanics in previous Republican administrations, or Republicans who were influential among Hispanics. They were Nixon's treasurer Romana Acosta Bañuelos; Congressman Manuel Luján Jr.; and Jeb Bush, who had spent the years between the 1980 and 1984 elections in Miami, building his own business and political career. They logged many thousands of miles for the Reagan-Bush reelection campaign, traveling to Hispanic communities across the United States to talk about Reagan's accomplishments on behalf of Hispanics, to explain away criticisms of the administration, and to stand before them as Hispanic leaders who embodied the kinds of success other Hispanics might have with Reagan as president.

Reagan provided his Hispanic surrogates with their talking points. The highest-ranking Hispanic to serve in the Reagan administration during his first term was Katherine Ortega, who became the second Hispanic treasurer of the United States, after Bañuelos. When the president nominated her for the position, during Hispanic Heritage Week in 1983, he gloated that 125 Hispanics already held posts in his administration and that he planned to appoint another 25. She and other surrogates then noted that Reagan had appointed more Hispanic women to his administration—Chavez, Villalpando, Ortega, etc.—than any president before him. They praised Reagan for supporting Hispanics throughout his political career, as governor of California and as president of the United States.

Hispanics again charged that Reagan nominated Ortega only to woo Hispanic voters, but the president scoffed at the idea. The *Washington Post* said it was "the prevailing political wisdom" that Reagan had to win 30 percent of the Hispanic vote in 1984 to "offset anticipated Democratic gains among black voters." He demonstrated his sincerity by insisting that Ortega be given the most coveted spot at the Republican national convention, as the keynote speaker. Del Junco, the newly appointed chairman of the RNHA, was given a slot, too, but not in prime time. Leaders of Mexico's Partido Acción Nacional (PAN) also flew to Dallas to observe the proceedings. PAN was the Republican Party of Mexico, according to Albert Zapanta, and Reagan was trying to solidify relationships with conservatives throughout the Americas.

ON AUGUST 21, 1984, LESS THAN A YEAR AFTER SHE BECAME THE THIRTY-EIGHTH TREA-surer of the United States, the *New York Times* ran an interview with Ortega. She told the reporter she spoke with that she was "born a Republican." Just the evening before, she had stepped confidently to the podium in her royal blue, long-sleeved, primly collared dress, rel-ishing the applause that filled the packed convention center for a min-ute and a half before she could even begin speaking. "Thank you." "Thank you." "Thank you very much." "Thank you."

Ortega had come a long way from her hometown along the east-ern border of White Sands National Monument, in Tularosa, New Mexico. She was the youngest of nine children. Her first language was Spanish, and she didn't learn English until elementary school. Ortega's father, Donaciano, worked as a blacksmith and then owned a café and furniture business. She said that her mother, Catarina, after whom she was named, was "very religious, a strong Catholic." She credited her background and upbringing as the source of her con-servatism. Her parents had taught her the virtues of hard work and perseverance. Her father said it was better to "do for ourselves rather than having somebody do for us." That's why she disliked big govern-ment, high taxes, and welfare. She was a surrogate par excellence.

The path to Ortega's appointment as treasurer was paved by her thirty-year career as an accountant and banker. Before college, she worked as a teller at the Otero County State Bank. Then she pursued a degree in business from Eastern New Mexico University, about three hours from home. After college, she moved to Los Angeles, where she eventually became vice president of Pan American National Bank in East Los Angeles; the bank had been founded by Bañuelos. Ortega then served as the president of Santa Ana Bank in Orange County before moving back to New Mexico, where she headed the family-owned Otero Savings and Loan Association. She still worked there when Reagan nominated her. Now she was onstage in Dallas—again, a long way from Tularosa.

When the applause died down, Ortega commenced by telling her audience how much it meant to her to be there. She was the first Hispanic keynote speaker at a national convention, Republican or Democrat. Hispanic Republicans had begun to organize only a couple of decades earlier, and now one of their own stood in the spotlight. Newspapers had called her honest and sincere, and those were the traits she displayed in Dallas. She wasn't known for her public-speaking skills, but she said her speech at the convention came "from the heart." She was there to share her "deep conviction that our country's future lies not in the empty rhetoric" at the Democratic national convention, held in San Francisco the month before, but in the "courage and vision of a president who in four short years has restored America's faith in itself."

As a keynote speaker, she said she didn't represent a particular interest group. She was there as an American, she said, not as a woman or a Hispanic. She therefore hit on the campaign's overarching theme, that the United States and all Americans were better off in 1984 than they had been when Reagan and Bush were elected in 1980. She passed judgment on the Carter years both to highlight the progress Reagan had made and to yoke Carter to Reagan's opponent, Walter Mondale, who had been Carter's vice president. Confidence in the United States had "eroded" during the Carter-Mondale administration. The United States couldn't go back to the Carter-Mondale years when weak leadership "left the door open" to Communist threats far and near. The nation had rebounded from double-digit inflation, high interest rates, and "economic misery," all hallmarks of the Carter-Mondale years.

Where Carter and Mondale had been weak, Reagan and Bush were strong. Republicans would run on Reagan's record of "peace, prosperity, and pride in America." He wouldn't cave to Soviet threats. The economy was growing, more Americans had jobs than ever before, and median income was up for all groups. Reagan and Bush should be given the opportunity to finish the job they started. The United States under their leadership, Ortega said, stood for freedom, opportunity, and the "right of every individual to fulfill his or her potential as members of the family of God, not creatures of an almighty government."

These campaign themes, spoken by one of their own, certainly resonated with Hispanic Republicans, but they also connected with Hispanic Democrats who were contemplating a switch that year. In passages that appealed directly to them, Ortega, as treasurer, offered a reading of the symbols on US currency. Bills had a picture of the Statue of Liberty, whose one hundredth anniversary, in 1986, was fast approaching. The Americans who entered the country at Ellis Island shared much in common with immigrants from Latin America and elsewhere. Coins had another message of diversity and unity: "E Pluribus Unum"—out of many, one. To make sure that the Spanish speakers who tuned in heard her, she said of the Republican Party, *"Nuestra casa es su casa.* Our home is your home." She ended with *"Dios bendiga America.* God bless America."

CHAPTER 8

IMMIGRATION DIVIDES

Katherine Ortega's performance at the 1984 Republican national convention failed to inspire. Ronald Reagan and George H. W. Bush called her on the night of her speech to say she had done a great job. The maids, waiters, and waitresses she ran into at her hotel thanked her. As she walked around the Dallas Convention Center the morning after, delegates congratulated her. But the reviews in the news were much less favorable.

Mostly they said she was boring. Instead of substance, she delivered banalities. "Apple pie was stacked high as Ortega brayed her way through a talk that had millions of viewers lunging for the channel selector," one columnist wrote. He joked that Ortega was a six-letter word for the sleep aid Sominex. Republicans in the convention hall apparently felt the same way. They were restless; during her speech, they got up from their seats, walked around, and weren't paying attention. David Brinkley wrote that Ortega was "not a hot, steaming, stamping, ranting orator." Another writer put it more plainly: Ortega's speech was "among the worst ever given at any convention, anywhere, any year."

Instead of criticizing Ortega's lack of oratorical skill, Hispanic liberals called her speech a betrayal that was full of omissions. Ortega had praised Reagan's use of the military in Central America and the

Caribbean, but ignored the cries of Latin Americans for justice and human rights, as well as Reagan's support for cruel dictators. In case Reagan had used Ortega "to make a good impression on the Latins," one writer warned, "we don't sell that easily." In the words of the journalist Juan Gonzalez, "Ortega lowered tokenism to profound new depths in an unabashed effort to convince conservative white Americans that Hispanics really aren't all that bad, and should be given more jobs as kitchen help." He called her "Katie" and said she was a "*Tia Tomasina* (a Latin Uncle Tom) to put right up there in the Pantheon of traitors to their race or ethnic group."

Ortega said she was unaware of these criticisms, but acknowledged that she was chosen for more than her skills as a speaker. Critics said that her chief attributes were that she was a female and a Hispanic in a party that was too male and too white. In some ways, those were the only things that mattered. After the convention, Ortega hit the campaign trail for Reagan not as the boring keynote speaker but as the first Hispanic to ever deliver a keynote speech at a Republican *or* Democratic convention. It mattered less what she said, and more that she said it. She may have been a token Hispanic Republican at the convention, but Reagan's support among Hispanics increased after her appearance. Hispanic Republicans believed it was largely because of how the president had acknowledged them in Dallas as important members of the Republican coalition.

In addition to Ortega's keynote on opening night, Reagan's nod to Hispanics included a luncheon on the last day of the convention hosted by the Republican National Hispanic Assembly (RNHA), where the president was the featured speaker. He opened by saying, "Having come from California, I wouldn't feel at home unless there was a strong Hispanic flavor to these festivities." He was just getting the crowd warmed up. "We Republicans see you as representatives of the mainstream of our party and our country," he said. "We are not a party of special interests that divides Americans into camps," like the Democrats did. "We are a party of people who share the same love of country and God, who have the same respect for family and hard work." Because Hispanics and the Republican Party had the same values, he predicted that Hispanics would "flock" to the GOP in "ever-increasing numbers." In partnership, Reagan and the RNHA raised $1 million for the Republican Party that afternoon.

After the convention, surrogates including Ortega picked up where Reagan left off. She went to California, Texas, and places not known to be Hispanic strongholds. She was the featured speaker at a gathering of the RNHA of Louisiana, an auxiliary of the Louisiana Republican Party. She praised Reagan for bringing down inflation, and said that the Hispanic business community had prospered under Reagan. The president had appointed more than two hundred Hispanics to posts with the Federal Communications Commission, the Federal Credit Union Administration, the Civil Aeronautics Board, and other agencies that had Hispanic representation for the first time ever. Again, these weren't cabinet appointments or Supreme Court nominations, but, she said, they made Hispanics proud.

Ortega's efforts appeared to pay off. The national chairman of the RNHA, Tirso del Junco, had predicted that Reagan would win 30 to 35 percent in the important states of Texas and California and about 40 percent nationally. He wasn't far off the mark. According to exit polls, Reagan won 37 percent of the Hispanic vote in 1984, which was slightly higher than the 35 percent he won in 1980. Moreover, one million more Hispanics voted in 1984 than in 1980, so the number of Hispanics who supported him increased as well. Hispanics in California and New York gave him the national average of about 40 percent, but the most impressive result was in Miami-Dade County, where Cuban Americans gave him 90 percent.

A doctor in Miami named Carlos Márquez-Sterling offered evidence of Cuban American enthusiasm for partisan politics, which increased throughout the 1980s. He wrote a column in *Diario las Américas* in December 1984, just one month after the election, that speculated about who the Republican nominee would be in 1988.

The "brilliance" of Reagan's victory hadn't even worn off, he said, but it was worth weighing the pros and cons of the four Republicans who were most likely to run next time: Vice President Bush, Howard Baker, Jack Kemp, and Bob Dole. There was no such thing as too early to begin plotting their next win.

For Márquez-Sterling, the choice was clear. He liked Bush, perhaps because their fathers shared a connection. Bush's father, Prescott Bush, a senator from Connecticut, and Márquez-Sterling's father had known each other. The sons themselves became acquainted over lunch at the Metropolitan Club in New York, after which Bush sought Márquez-Sterling's advice whenever he gave speeches on Cuba. Bush represented a type of public figure who was "threatened with extinction," Márquez-Sterling wrote, someone who was qualified to serve because he was knowledgeable, well-cultured, intelligent, hardworking, and had class.

Days after the column appeared, Jeb Bush wrote Márquez-Sterling a note, in Spanish, to thank him for his kind words about his father and grandfather. "Even though I can't claim to be objective," he wrote, "I also believe that George Bush is capable of leading this great country." They had to wait. Reagan still had four more years. His victory, an article in the *Miami Herald* stated, had "weaned an embarrassing chunk of the Hispanic vote from its traditional cocoon, the Democratic Party."

Hispanics had become "valuable plums" for Democrats and Republicans, and "unlike blacks," wrote a reporter for the *Fresno Bee*, they possessed "dramatic new influence" in electoral politics. But their influence didn't come easily, and it wasn't complete. Reagan's second term was in many ways a trying period. There were new power struggles within the Hispanic Republican movement. Tensions between Cuban Americans and Mexican Americans simmered as the relative influence of one group waxed while the other waned. The Mexican American Albert Zapanta recalled that Cubans weren't a significant part of the Hispanic Republican movement until Reagan brought them on board. Now one of their own—the Los Angeles–based Cuban American doctor Tirso del Junco—was the leader of the organization that Zapanta and other Mexican Americans had founded.

At the same time, the fate of immigration reform hung in the balance and would depend on several interest groups coming together to forge a compromise. Anti-immigrant activists moved from the fringes to the mainstream, even though their ideas were linked with debunked eugenicist thinking. Nativism and inclusion seemed to work at cross-purposes, and it wasn't clear which ideas would gain supremacy within the party. The two sides seemed to achieve an uncomfortable détente by focusing on illegality and citizenship as the key criteria for exclusion or belonging. Nativists increasingly argued that legal residency or citizenship were crucial to belonging in the United States, while Hispanic Republicans weren't willing to draw that line. Because of their personal and family histories of immigration to the United States, they believed that immigrants deserved a chance to remain in the country if they worked hard and didn't break the law after they entered, whether they had arrived legally or not.

Finally, one of Reagan's top international priorities—defeating the Sandinistas—came under close scrutiny when the corrupt and illegal shipment of arms to the Contra rebels in Nicaragua came to light. Because of the violence in their home countries, a growing number of Central American refugees sought sanctuary in the United States. These controversies forced Reagan's Hispanic supporters to defend him in the final years of his presidency, which highlighted growing divisions over immigration—by undocumented Mexicans, Central Americans, and a new wave of Cubans—that became wider over time. Hispanic Republicans came to his side like they came to Nixon's after Watergate, but they couldn't be certain that their support would be met with gratitude, or even acknowledged. Hispanic Republicans had hitched their wagon to the Republican Party, but had the Republican Party hitched its wagon to them?

AT 5:30 A.M. ON MAY 20, 1985—ANOTHER CUBAN INDEPENDENCE DAY, TWO YEARS AFTER Reagan's famous visit to Miami—Radio Martí made its first broadcast to Cuba from a station in the Florida Keys. The bill creating Radio Martí took a couple of years to wind its way through the legislative process. Opponents argued that it would further aggravate the already difficult relationship between the United States and Cuba. Supporters, especially the Cuban American National Foundation (CANF), argued that Radio Martí would finally present Cubans with a counternarrative of the Cuban Revolution. They learned this from the friends and family who traveled to the island in the late 1970s, but Radio Martí would disseminate the message even more widely.

Reagan created an Advisory Board for Broadcasting to Cuba, which acted like a lobbying group to drum up support for the Radio Martí bill (S. 602, the Radio Broadcasting to Cuba Act). The bill was first proposed in August 1982 but didn't become law until October 1983. Two of the board's members were Jorge Mas Canosa, the head of the CANF, and del Junco, the national chairman of the RNHA. Even though the first broadcast wouldn't happen for another year and a half, as the various stakeholders—the government and other Florida broadcasters, for example—sorted out the particulars of what would be broadcast and how, Cuba and the United States awaited it eagerly: Cuba with anger and fear, the US government and Cuban exiles with nervous excitement.

Named after José Martí, the father of Cuban independence who agitated against Spanish colonialism from New York City, where he ran several Spanish-language newspapers, Radio Martí broadcast for 17.5 hours a day, seven days a week. The broadcasts in the first week included messages to Cubans from congressional leaders of Florida and other states. Content in the first year included stories about kidnappings by the Cuban government, rebuffs of Castro by other Latin American leaders, Cold War conflicts around the world, and summit meetings between the leaders of the United States and the Soviet

Union. In addition to news stories, Radio Martí broadcast soap operas, comedy shows, and musical programming.

It was extremely effective. A year in, the *New York Times* editorial board reversed its earlier opposition to the station, saying they had been "unfair to Radio Martí." The station had already changed the minds of many Cubans and Cuban exiles. One listener in Cuba said that Radio Martí "fills the spiritual holes in the hearts of the Cuban people." An exile said he listened to the station before he left the island, where "Radio Martí's popularity spread and spread until everybody was listening to it." Cubans listened in the morning as they got ready for work and school. They listened in groups, at the homes of friends who had radios. They recorded shows and listened when they got home. They feared retribution for listening. They had to keep the volume down because it was "very dangerous to listen," one wrote.

Castro was enraged. He called Radio Martí an "act of aggression" and said that its programming was "subversive." He suspended the immigration agreement that Cuba and the United States entered into after the Mariel boatlift, and threatened to do more. He said he would broadcast signals to the United States on interfering frequencies, and would punish the Cuban Americans whom he believed to be responsible for Radio Martí by restricting letters, phone calls, and visits between friends and family members on the island and the mainland. To the supporters of Radio Martí, Castro's reaction was a demonstration of the station's effectiveness. Radio Martí's director, Ernesto Betancourt, who was also a member of the CANF's board of directors, said, "There will come a time, with respect to the Cuban situation, when we will have to talk about the time before and after Radio Martí went on the air."

AS RADIO MARTÍ MADE WAVES, CONGRESS REVISITED THE IMMIGRATION BILL THEY HAD put on ice before the election. They came back to the negotiating table because both Democrats and Republicans increasingly saw undocumented immigration as a problem to be solved, if for different reasons. Democrats believed that undocumented immigration led to the continued exploitation of workers, while Republicans believed that undocumented immigrants were invaders who threatened the rule of law.

In 1965, the year the last comprehensive immigration law was signed, the Immigration and Naturalization Service (INS) apprehended 40,000 undocumented immigrants. In 1975, that number grew to more than 500,000. In 1985, it more than doubled, to 1.2 million. Despite these mounting apprehensions, the undocumented population grew from a few hundred thousand in 1970 to more than 2 million by the early 1980s.

The members of Congress also came back to the negotiating table because they believed they were close to a deal. The battle lines hadn't changed much. Hispanics who were both Democrats and Republicans continued to oppose employer sanctions. Democratic congressman from California Edward Roybal wanted to introduce a new version without the provision. Roybal said that all Hispanics feared the "discrimination and barriers to employment that sanctions would create." Meanwhile, Republicans still argued that granting a blanket amnesty to undocumented immigrants rewarded their illegal activity and was unfair to immigrants who had waited their turn to become citizens through the usual application process.

Republicans claimed that undocumented immigrants took American jobs, drained state coffers by relying on public services, and failed to assimilate, as earlier immigrants from other parts of the world had done. They threatened American prosperity, security, and national identity. Just as Roybal wanted to scrap the employer sanctions, many Republicans supported the McCollum amendment, which proposed doing away with the amnesty provision. By the Reagan years, immigration restriction had become a Republican issue, whereas in the 1960s and 1970s it was a Democratic issue largely because labor unions opposed it.

In the two years between Simpson-Mazzoli's proposal in 1982 and the time that it died on the Congress floor before the 1984 election, Hispanics had adopted a strategy of "veto politics." They successfully blocked Simpson-Mazzoli, but in doing so were labeled obstructionists. In May 1985, New York representative Charles Schumer (before he became a member of the Senate) wrote an op-ed article for the *New York Times* in which he argued that special interests including Hispanics had killed the bill. Now the Hispanic minority needed to be silenced for the good of the American majority. Schumer maintained that the version of the bill that had passed through the House would solve many of the problems caused by illegal immigration, even if it wasn't perfect.

Hispanics in Congress began to adopt a more conciliatory tone. They faced pressure from constituents back home who wanted them to do something—anything—about immigration. The failure to pass legislation during Reagan's first term made it clear that an immigration bill wouldn't pass without some form of employer sanctions. It also made clear that a bill wouldn't pass without some form of amnesty. The goals for Democrats became strengthening the anti-discriminatory "safeguards" of employer sanctions and making sure that the law's amnesty provisions were as generous as possible. Meanwhile, the goal for most Republicans became insisting on employer sanctions while limiting the scope of the amnesty. Many Hispanic officials also felt that because anti-immigrant sentiment was on the rise nationally, the immigration bill was likely to become even more restrictive if they didn't act soon. The proposed bill might well be the best they would get.

In the year that it took Congress to reach a compromise on immigration, Radio Martí kept up its pressure on Castro, leaders of the RNHA continued to defend the White House's Central America policy, and Reagan appointed Linda Chavez as director of the Office of Public Liaison. She moved over from the US Civil Rights Commission and responded to the lobbying efforts of particular interest groups, including Jewish Americans; Catholics; evangelicals; labor unions; veterans; women; African Americans; and, of course, Hispanics. An avowed opponent of identity politics, she attempted to restructure the organization so that it wouldn't focus on particular interest groups but rather work toward goals they all shared. It mirrored Reagan's move away from patronage politics toward a more ideologically driven conservatism, but it failed, Chavez said, because the various interests had become so ingrained that it was impossible to shift course.

As Chavez continued her fight against the use of federal hiring, the immigration debate was coming to a head. By 1986, Hispanics in Congress felt even greater pressure to compromise. Between 1985 and 1986, the number of undocumented immigrants apprehended by the INS jumped again, from 1.2 million to 1.6 million. About half of them came from Mexico or other Latin American countries. Unemployment was on the rise, and elected officials were hearing from constituents who said that undocumented immigrants were responsible for their joblessness. Even Mexican Americans in their districts were saying things such as if immigration goes uncontrolled, "God knows how many people will come into this workforce." Hispanics therefore came to believe that "unbridled immigration" would "produce a backlash." They decided that their best option was either to support the bill or amend it, rather than oppose it outright.

By 1986, employer sanctions had come to seem like a commonsense solution to undocumented immigration. If employers feared fines or even jail terms, they would be less likely to hire undocumented immigrants, and if they were less likely to hire undocumented immigrants, fewer would try to enter the United States. The *New York Times* further described employer sanctions as a way of getting rid of the "sanctimonious stench" of the law as it stood, whereby it was illegal for undocumented immigrants to work in the United States but not illegal for employers to hire them. "The employer gets the cheap labor," the paper put it, while "the alien takes all the risks." The US treasurer, Romana Acosta Bañuelos, claimed innocence when her factory was raided back in 1971 precisely because of this loophole. Soon it would be closed.

About the only Republicans to oppose employer sanctions were Hispanics, the business owners who could be punished, and the growers who relied on immigrant workers to harvest their crops. They had the backing of their representatives and some in the Reagan White House. Members of the president's Council of Economic Advisers argued that sanctions would "reduce labor market efficiency" because they would require employers to work harder to secure low-wage labor and to more closely scrutinize the applications and credentials of prospective employees. Their complaint sounded a lot like what Congressman Manuel Luján Jr. had been saying since the late seventies, that employer sanctions would constitute a government overreach into the affairs of businessmen, an unfair imposition that went against the ethos of the free market.

But Luján's logic differed from that employed by most Hispanic Republicans, who tended to agree with Hispanic Democrats more than with non-Hispanics in their own party, that employer sanctions would likely lead to job discrimination. Luján's position on amnesty was another example of how Hispanic Republican views on immigration could differ. Aligned with the conservative wing of his party rather than the Hispanic Republicans who were generally immigration moderates, Luján opposed amnesty because, he argued, "it rewards those who break the law and penalizes those who wait in line to enter this country legally." Regardless of their differences on employer sanctions and amnesty, the alliance between Hispanic Republicans and employers made them important partners in the upcoming passage of immigration reform.

When an immigration bill came up for the third time in the Reagan era, it passed through the House with a vote of 230 in favor and 166 opposed. This time it was called the Simpson-Rodino bill, or the Immigration Reform and Control Act (IRCA) of 1986, and once again it was substantively similar to the Rodino bill proposed some fifteen years earlier, as well as the Carter and Simpson-Mazzoli bills considered more recently. Hispanics were more divided on immigration than they had been during Reagan's first term, and the divisions among them, one scholar observed, "contributed to the collapse of opposition against the bill." They took solace in the legalization provisions of the law, and waited to see whether discrimination from employer sanctions would materialize.

Reagan signed the bill into law on November 6, 1986, at a ceremony in the Roosevelt Room. He shook fifteen hands as he joked his way up to the podium to deliver remarks about IRCA. Standing behind him were white, middle-aged members of Congress who helped pass the bill, but not a single immigrant who would be affected by it, nor one of the Hispanic congressmen who had spent years in the trenches of the battle over immigration.

The president acknowledged how hard it had been to forge a compromise. He threaded a needle by trying to soften the blow to Mexicans and blunt the ire of anti-Mexican xenophobes. Immigrants had come to the United States from all around the globe. The "problem of illegal immigration," he said, "should not, therefore, be seen as a problem between the United States and its neighbors." He hoped the bill established a "reasonable, fair, orderly, and secure" immigration system to "humanely regain control of our borders."

As Reagan exited the room, a reporter threw him a curveball. Allegedly there to cover the signing of the immigration bill, the reporter asked, "Mr. President, do we have a deal going with Iran?" Caught off guard, Reagan responded, "No comment," then scolded the reporter for stirring a controversy that, he claimed, would make it harder to secure the release of American hostages who at the time were being held in Lebanon. He exited the room in haste. Three days earlier, on November 3, the Lebanese magazine *Ash-Shiraa* had published details of an agreement to sell arms to Iran in exchange for the return of American hostages. That article prompted the reporter's question.

He didn't know it at the time, but the topic of arms sales to Iran had implications for Hispanic politics, especially US–Central American relations and the growing stream of Central American refugees entering the United States.

THE 1986 MIDTERM ELECTIONS TOOK PLACE THE DAY AFTER THE PUBLICATION OF THE Lebanese article and two days before the immigration law's signing ceremony. Linda Chavez was on the ballot for one of Maryland's seats in the US Senate. She had left her post as director of the Office of Public Liaison on February 3. After sitting for a picture with Reagan in the Oval Office that morning, she got right to work campaigning. One of her opponents in the Republican primary was George Haley, an African American man whose brother, Alex Haley, had written *Roots*, a popular book and television series about the transatlantic slave trade and its legacies for his family. Chavez won that race, but lost in the general election by a margin of 69 percent to 31 percent despite her endorsement by the RNHA, and the fund-raising dinner that Reagan hosted for her at Fort McHenry, the site of the War of 1812 battle that had inspired Francis Scott Key to write "The Star-Spangled Banner."

Her Democratic opponent called her "a Hispanic who doesn't want to be identified with the Hispanic community." Chavez responded that she believed Hispanics would support her not because of her last name but because of her beliefs.

Another Hispanic Republican ran a successful campaign for a seat in the Florida Senate. Ileana Ros-Lehtinen, the daughter of the Cuban exile and author Enrique Ros, and her husband, Dexter Lehtinen, were elected at the same time. They took the usual anti-Castro positions but also said they would crack down on the multimillion-dollar drug trafficking business and would work to negate the "grave influence" it had on Florida youth. Fighting drugs became an even greater focus for Dexter Lehtinen when Reagan appointed him US attorney for the Southern District of Florida, during which time in that office he led the prosecution of Panama's Manuel Noriega. Ros-Lehtinen fast became a leading voice for the Cuban exile community whose career quickly jumped from statewide to national politics. But for a brief period she represented Florida's Thirty-Fourth District, including Fort Lauderdale and Hollywood, as its first Cuban American state senator. It was an important turning point in Florida politics.

President Reagan himself had brought national attention to the war on drugs just a couple of months before Ros-Lehtinen's election. On September 14, 1986, speaking from the White House residence to a national audience of television viewers, he and Nancy Reagan announced a "national crusade" against marijuana, heroin, and the new smokable cocaine called crack. He said Americans needed to wage an all-out campaign against drugs similar to the one the nation had waged during World War II. It would require that kind of effort. Many Hispanics believed their communities to be victimized by the war against drugs, because it led to the criminalization of Hispanic youth in particular. But Republicans such as Ros-Lehtinen and Reagan felt that keeping drugs out of American communities should be their priority; some amount of discrimination in pursuit of this goal was acceptable collateral damage.

A candidate in California named Mike Antonovich demonstrated how the war on drugs in California was part of the war on undocumented immigration. As the *Los Angeles Times* put it, immigration elicited a "gut-level reaction" from the state's Republican voters. Antonovich tapped into their anti-immigrant sentiments by claiming that illegal border crossers were also drug smugglers and terrorists. He argued that the US military should be sent to the border to back up Border Patrol officers and that the children of undocumented immigrants who were American citizens should be deported.

There were also other signs that his campaign was fueled by rising xenophobia. One was his call for a change to the Fourteenth Amendment, which granted birthright citizenship to "all persons born or naturalized in the United States." Welfare for the children of undocumented immigrants cost Los Angeles County more than $100 million a year, he claimed, even though his critics cited the much lower figure of $5 million a year. Another was his support for a ballot initiative—Proposition 63—to make English the state's official language. US English, the group that funded and organized the California campaign, called the proposed law an effort to defend the English language against its "erosion" in immigrant communities whose members continued to speak their native tongue. Antonovich lost in the Republican primary, yet he articulated sentiments that were spreading throughout California, and from there to the rest of the country.

By 1986, US English, founded only three years earlier, claimed two hundred thousand members across the United States. They had recruited some high-profile and well-respected members to their board of advisers, including Walter Cronkite, Gore Vidal, Saul Bellow, and the retired Columbia University history professor Jacques Barzun. "A nation has to have certain things in common," Barzun said, and "these include certain customs, a language and a historical tradition—and the historical tradition can't be picked up without the language." Their plan was to make English the official language in several states, and then push to amend the US Constitution. They said that an overwhelming majority of Americans supported their goals. Many Americans assumed English already was the official language, yet it never had been. California was just the beginning for them, and an important beginning it was, as one of the biggest states with the largest undocumented immigrant population.

US English raised about $2.5 million in the year leading up to the election. This was the money it used to finance the campaign in California. Samuel Ichiye Hayakawa—a former president of San Francisco State University, a former Republican senator from California, and a cofounder of US English—was its lead organizer. The Michigan ophthalmologist John Tanton was also on the ground for much of the California campaign. Together, Hayakawa and Tanton insisted that Proposition 63 had nothing to do with nativism or cultural chauvinism but was concerned only with giving California's non-English speakers their best opportunity for success in the United States. They argued that bilingual education and bilingual ballots mandated by the 1975 amendment to the Voting Rights Act kept immigrants living in "language ghettos."

Generally, US English targeted only government initiatives, but individual supporters of Proposition 63 also waged campaigns to require restaurants to remove Spanish from their menus and to require the Pac Bell telephone company to stop publishing a Spanish-language version of the Yellow Pages, or Páginas Amarillas.

The opponents of Proposition 63—a coalition made up of labor unions, civil rights organizations, and Hispanic advocacy groups—nevertheless considered the English-only law and the anti-immigrant pronouncements of candidates such as Antonovich to be motivated by the same animus. They argued that both Proposition 63 and Antonovich's campaign tapped into the "deep resentments" felt toward immigrants by many California voters who saw them as invaders who took jobs, drained public resources, and trafficked drugs. Hayakawa, though, said, "The racist argument is the argument of desperation by the opposition."

The passage of Proposition 63 by more than 70 percent of the voters in California was a big win for US English. In general, non-Hispanic white Republicans were the most likely group of voters to support Proposition 63, but significant percentages of non-Hispanic white Democrats and African Americans supported it as well. Hispanics voted against Proposition 63 by a margin of 60 percent to 40 percent, which meant that it still gained significant support. Its Hispanic opponents included Hispanic Republicans and Democrats. The "people behind US English and FAIR are a bunch of crazies," Fernando Oaxaca said, "who came out of the environmental movement and think the environment is damaged by people, especially people different from themselves."

As soon as the English-only bill passed in California, several other states, including Texas, Arizona, Colorado, and Florida, announced their intention to follow suit. English-only laws would prove unpopular among Hispanics in those states as well. Just days after the election in California, the Republican Party of Texas announced its intention to propose similar legislation. In response, the state chairman of the RNHA complained that if the Republican Party could offer only "second-class participation," then Hispanics were "in for a terrible time." He said it was the "same old story—they want Hispanics to tend the cattle and pick lettuce" but not to get "involved in politics." Finally, he warned that the Republican Party's support for English-only laws could hurt Republicans at the polls and even wondered whether Democrats were right, that Republicans in Texas cozied up to Hispanics "only when they need them," and, even then, only paid them "lip service." He concluded, "I pray that is not the case."

Linda Chavez was among the minority of Hispanics who supported English-only laws. She had opposed bilingual education for most of her career, with the American Federation of Teachers; the US Commission on Civil Rights; and, whenever she had the opportunity, as the director of the Office of Public Liaison. She argued that bilingual education programs designed to help Spanish-speaking children in fact harmed them. Instead of brief transition periods, during which Spanish speakers were supposed to be learning and improving their English, they ended up spending years learning in both English and Spanish, segregated from their non-Spanish-speaking peers. But in the wake of her failed bid for a US Senate seat, a period that she recalled as the low point of her political career, she opposed bilingualism in a more focused and official capacity as the president of US English.

The organization founded by Tanton and Hayakawa hired Chavez in early 1987. She had a friend on the board who was familiar with her long history of opposition to bilingual education and thought she'd be perfect for the job. Her appointment as president helped the members of US English claim that their organization wasn't racist. Chavez said she didn't know a lot about the individuals who had spawned the movement at the time she accepted the job. But she was "very concerned about the semi-official status" that "Spanish was acquiring in many parts of the country."

After her failed Senate race, Chavez became something of a public intellectual, offering commentary on National Public Radio and in a weekly column for the *Chicago Sun-Times*. As president of the organization, she hoped that she could "elevate the level of debate" about language policy and "give reassurance that there was nothing racist or anti-Hispanic about promoting a common language in this nation of immigrants."

But accepting the position, she wrote, made her the "Most Hated Hispanic in America." She came in for rough treatment everywhere she went as a spokesperson for US English. Hispanic protesters in California, Texas, Colorado, and elsewhere threw things at her and waved signs in her face that called her a "Tia Taco" and "La Malinche," the Aztec wife of the Spanish conquistador Hernán Cortés, who, according to one version of the stories told about her, was a traitor to her race for serving as a translator and giving Cortés intelligence about the Aztecs.

Chavez's time as president of US English was brief but intense. Not long after she accepted the position, a reporter confronted her with some of Tanton's more "unsavory" writings, letters written in October 1986 that demonstrated his eugenicist thinking. Hispanics, he claimed, were hyperbreeders who would have a degrading cultural influence, and Americans were asleep while Hispanics infested the country with their offspring. It was too much for Chavez. She resigned immediately. So did Tanton.

WHILE CHAVEZ TRAVELED THE COUNTRY SPEAKING ON BEHALF OF US ENGLISH, FERNANDO Oaxaca cruised through East Los Angeles in his Jaguar, stopping at the homes of Mexican immigrants to talk with them about the new immigration law. It was his job to let undocumented immigrants know that beginning on May 5, 1987, they could apply for legalization if they had lived in the United State since before January 1, 1982.

At the time, Oaxaca was an executive at the Los Angeles firm Coronado Communications, which specialized in Spanish-language media. He had founded the company that became the lead partner of a joint venture called the Justice Group, which won a multimillion-dollar contract from the INS to advertise and explain the new immigration law. Oaxaca was president of the Justice Group. He had been a stalwart Republican for two decades. Reagan needed him to sell an immigration law that almost everyone had reason to dislike.

In the days, weeks, and months after the law was signed, many Americans expressed their doubts that it would actually work. They thought it might prevent some immigrants from crossing illegally into the United States, but not all of them, and not for long. A month after the signing ceremony, early reports claimed that the number of undocumented immigrants congregating in border cities, waiting for employment, had dropped by half. Because of the new sanctions in *la ley Simpson* (the Simpson law), immigrants were uncertain that employers would hire them. The fears of those who opposed employer sanctions were confirmed. Acting out of "confusion and an abundance of caution," the *Chicago Tribune* reported, employers had already begun to lay off workers they suspected of being in the country illegally.

Criticisms of the law grew only louder, not just because employers were, in fact, found to discriminate against Hispanic employees, but also because the law—not the amnesty provision, not the sanctions, not the increased enforcement, and not the increased number of visas—didn't prevent the entry of undocumented immigrants in the long run.

The skepticism about the law made Oaxaca's job all the more difficult. He disliked parts of the law himself. Many were surprised that he signed on to promote a law he had spent two years lobbying against. Oaxaca said that he, like most Hispanics, had "always believed in legalization," or amnesty, particularly for immigrants who had "lived here for years." But he opposed employer sanctions, and this issue divided Hispanics. A little more than half of Hispanics opposed sanctions, as Oaxaca did, but others saw them as a way of curbing undocumented immigration. Because of how controversial the law was and because of how little immigrants actually knew about it, Oaxaca said it would be a "monstrous challenge" to communicate its complexities to them and to employers.

The main difficulty he faced, reported the *Wall Street Journal*, was convincing undocumented immigrants to trust an agency that many believed only wanted to chase them "through factories or from homes," or separate them from their families and handcuff them "for a trip to a detention center." Herman Baca, an immigrant rights advocate from San Diego, said that asking immigrants to change their views of the INS would be like "asking the Indians to trust Custer." He wondered, "Would the blacks have allowed the KKK to administer the Civil Rights Act?"

Compounding his problem, Oaxaca also had to teach employers how to avoid penalties and discrimination against job applicants. They had to fill out a new form for all employees—the I-9, now part of the E-Verify process—to demonstrate they had made a "good-faith effort" to confirm their legal status. Oaxaca said he hadn't been "real interested" in promoting the law, but now he was "absolutely on the INS team." He had a job to do and he intended to do it.

In large part because of Oaxaca's promotional efforts, the INS claimed victory when two million undocumented immigrants initiated the legalization process in the yearlong window between May 5, 1987, and May 4, 1988, when application processing centers closed.

OAXACA'S CAMPAIGN WAS IN MANY WAYS THE FINAL HALF MILE OF A FIFTEEN-YEAR marathon—the implementation of an immigration bill that ultimately contributed to a division between Hispanics and the Republican

Party. Instead of ending a debate over undocumented immigration that began in the early 1970s, though, Oaxaca's campaign can be seen as another beginning. Complaints about the law were expressed— and remedies proposed—almost before the ink from Reagan's pen had dried.

As Oaxaca sought to convince undocumented immigrants to apply for amnesty, critics pointed out that the law applied to different groups of Latin American immigrants unequally. Because IRCA said that undocumented immigrants had to live in the United States continuously since the last day of 1981, the hundreds of thousands of Central Americans, in particular, who arrived in 1982 or later were ineligible.

The US-backed president of El Salvador, José Napoleón Duarte— who was elected on the promise that he would occupy a middle ground between left and right, represented by the FMLN and death squads embedded in the military—appealed to Reagan to permit the four hundred thousand to six hundred thousand Salvadorans seeking refuge in the United States to stay there. Duarte acknowledged that they were ineligible under the law, but after seven years of violent conflict and a devastating earthquake in October 1986 that left three hundred thousand Salvadorans homeless, he asked for special consideration, and added that it might ruin El Salvador's economy if they were to return. When she was still a Florida state senator, Ileana Ros-Lehtinen similarly complained that the law didn't cover Nicaraguans forced to leave their country after the 1981 cut-off date. Like Salvadorans, Nicaraguans who couldn't take advantage of the law's amnesty provision were defenseless.

Ros-Lehtinen, who assumed a position as the Hispanic Republican most engaged with the Central American refugee issue, may have seen the fundamental similarity between Salvadorans and Nicaraguans, but the immigration officials evaluating their applications often saw them as distinct. Thousands from both countries applied for political asylum. The vast majority of their applications were denied,

yet they were rejected at different rates. About a quarter of the Nicaraguans who applied were granted asylum—not an overwhelming percentage compared with the 75 percent of Soviets whose applications
were successful—while only 2 percent of the applications by Salvadorans and Guatemalans were approved. Nicaraguans were freedom
seekers, whereas Salvadorans and Guatemalans fled countries whose
governments were propped up by the Reagan administration.

It would have put Reagan in an awkward position, indeed, to
grant asylum to the citizens of countries whose leaders he backed.
That would have been tantamount to acknowledging that his administration had helped create the dangerous conditions that refugees
were fleeing. The White House argued instead that most Nicaraguans
and the vast majority of Salvadorans and Guatemalans were economic
migrants looking for jobs, rather than refugees. They didn't meet the
criteria for asylum laid out in the Refugee Act of 1980.

Some prominent Hispanic Republicans supported this position.
Congressman Luján agreed with Reagan that most of the Central
American immigrants arriving in the United States were economic
immigrants. He said that it just didn't "make sense" to him that all
of them—hundreds of thousands—would be political refugees. Luján said he sympathized with Central American immigrants, but he
could not tell Americans to "violate the law" by offering them refuge
and safe haven. Like his opposition to amnesty, his position that Central Americans were economic immigrants ran counter to the belief
of many Hispanic Republicans, such as Ros-Lehtinen, who saw them
as political refugees.

Luján's statement came at a moment when the national conversation about Central America and Central Americans was especially
tense. Since 1982, Sanctuary Movement volunteers had provided
safe haven in churches and communities across the United States
to thousands of Central American refugees, and between 1984 and
1986 several of them were prosecuted by Reagan's Justice Department for smuggling and harboring undocumented immigrants. Their

conviction led to a high-profile, nationally televised debate between church and state over the relative strength of legal and moral authority. Reagan's lawyers said sanctuary workers broke the nation's immigration laws, while sanctuary workers guided by religious conviction said they obeyed a higher authority. They were convicted in a court of law, but they triumphed in the court of public opinion. The prosecutions brought them favorable press coverage, including comparisons with the Underground Railroad that led slaves to freedom in the nineteenth century. By 1987, the Sanctuary Movement had spread and included hundreds of declared sanctuary cities, two sanctuary states, and more than four hundred churches.

As sanctuary volunteers went on trial, reports circulated that some of the Contra leaders had used millions in US aid for their self-enrichment. Rather than the freedom fighters that Reagan made them out to be—he called them the "moral equal of our Founding Fathers"—the Contras were a twenty-thousand-man army in the business of gunrunning, drug trafficking, and assassination. Meanwhile, Contra supporters in Miami sold guns to Central America and smuggled cocaine from Colombia in seafood containers. The charges came to light just as Congress debated another $100 million in aid to the Contras. The congressmen who opposed Reagan's Central America policies, and even some who supported them but were disturbed by the allegations against the Contras, were in no spending mood. Reagan blamed Congress for any misspent funds and said that the charges were part of a vicious propaganda campaign against the rebels.

Then there was the question the reporter sneaked into the immigration law's signing ceremony, about whether there was an Iran deal. Over the next several months, Americans learned that there was, indeed, a deal with Iran, and it shook Reagan's presidency to the core, leading to the indictment and conviction of almost a dozen administration officials. Americans were again glued to the television to watch the unfolding of another presidential scandal. They watched Colonel Oliver North, who at the time was a National Security Council staff member, explain his secret arms sales to Iran, and use of the proceeds to fund the Contras in Nicaragua.

Incidentally, the televised hearings on the Iran-Contra scandal began on May 5, 1987, the same day that undocumented immigrants could begin applying for legalization. Oaxaca therefore promoted the new immigration law at the same time that millions of Americans learned of the Reagan administration's illegal efforts to stamp out communism in the hemisphere. Most Americans probably didn't see the connection between these simultaneous occurrences, but Hispanic Republicans across the country must have felt their worlds converging. In the end, Reagan avoided the same fate that befell Nixon only because investigators never learned whether the president knew about North's scheme.

THE IRAN-CONTRA SCANDAL PUT REAGAN'S HISPANIC SUPPORTERS IN THE DIFFICULT PO-sition of defending him, just as Nixon's Hispanic supporters defended him after Watergate. And just like Nixon's Hispanic supporters, Reagan's were loyal.

Iran-Contra, wrote a reporter for the *El Paso Times*, had "soured the atmosphere in Congress and eroded support for aid to the contras." But even as the US House of Representatives voted to cut off aid to the Contras after the current funding ran out at the end of September 1987, Ros-Lehtinen urged the federal government to continue offering assistance to the seventy thousand Nicaraguan refugees living in Florida, 90 percent of them in Miami-Dade County. They had established the Nicaraguan-American Foundation to lobby for the interests of Nicaraguan refugees, including their applications for asylum. It was based on the model of the CANF. Ros-Lehtinen proposed her own bill in the Florida Senate, S788, which would "establish local councils to distribute special state aid to local groups to feed, clothe, shelter, and educate Nicaraguans."

Dozens of Cuban Americans from southern Florida—some of

whom may have been Ros-Lehtinen's constituents—were deeply involved in the Contra war in Nicaragua. Cuban Americans in Miami proudly noted how their adopted city became a "rebel base" for the Contras, just as it had been for Cuban exiles training for the Bay of Pigs invasion. Members of Somoza's National Guard moved to Miami after the Sandinistas took power. One umbrella Contra organization—the United Nicaraguan Opposition—set up headquarters in an unmarked office building in Miami, most US government aid to the Contras was funneled through Miami banks, and Miami hospitals treated Contras injured in the fighting.

Some Cuban Americans were pessimistic about the revolution's potential for success, seeing it as a "slow-motion Bay of Pigs." Nevertheless, dozens of Bay of Pigs veterans bunkered in Central America, where they served as "military advisers" to the Contras. Their Miami-based supporters funded their ventures and raised more than $1 million for humanitarian aid for shoes, uniforms, compasses, and canteens. The powerhouse lobbyist Mas Canosa continued his verbal assault against the Sandinistas. He enlisted the support of Jeb Bush, who at the time was the chairman of the Miami area Republican Party. Bush said he supported the Contras but that he hadn't "been involved in aiding them directly." It was good enough for Mas Canosa, who said of Bush, "He is one of us." Along with Mas Canosa, Bush, and others, Ros-Lehtinen led the calls to support the Contras and the thousands of Nicaraguan refugees who landed in Florida.

Congressman Manuel Luján Jr. was another strong backer of Reagan's Central America policy, even after the Iran-Contra revelations. Luján said he didn't condone the means North had used to accomplish his goals, but he nevertheless considered him a "national hero" for trying to secure the release of hostages in Iran while also supporting the Contras in Nicaragua. He said that both of these were goals he supported as well. When the US House passed a measure to freeze aid to the Contras, Luján voted against it, while fellow New Mexican Bill Richardson voted for it. A month into the televised Iran-Contra hearings, when one of his constituents sent him a letter to say how important they were as a public service, Luján acknowledged that most Americans agreed, but he himself thought they were an "expensive dog and pony show."

In his correspondence with his constituents in and around Albuquerque, Luján always supported the president, even if it meant disagreeing with the voters who could remove him from office. He received more than four hundred calls about the Contra war, the majority to oppose the continuation of military assistance. When constituents wrote him to voice their opposition to continued funding for the Contras, and instead suggested that humanitarian aid be given to Nicaraguans in need, Luján wrote diplomatically but insistently that, like them, he wanted peace in Central America, and the only disagreement was over how to achieve that goal.

By supporting the continued funding of the Contras, Hispanic Republicans such as Ros-Lehtinen and Luján implicitly encouraged violence against Latin American leftists. They didn't seem overly concerned that some Hispanics expressed uneasiness about sending Hispanic soldiers to fight in Central America, where they might kill, or be killed by, others of Latin American descent. Opposition to the Contra war, though, ran deep enough to make Luján's support for military aid to Nicaragua a talking point against him during his 1988 reelection campaign. He wouldn't be an elected official much longer anyway. The next Republican president had other plans for him.

Toward the end of 1987, Luján and other Republicans were encouraged by the recent round of peace talks in Guatemala City. The foreign ministers of five Central American nations—Guatemala, El Salvador, Nicaragua, Honduras, and Costa Rica—discussed the possibility of a cease-fire. More broadly, they discussed national reconciliation, economic cooperation, refugee assistance, arms control, and the ending of support for paramilitary groups such as the Contras. The president of Costa Rica, Óscar Arias, had drawn up the basic framework, and for his efforts he received the 1987 Nobel Peace Prize.

The Iran-Contra scandal largely sidelined the United States in the negotiation process, but Reagan, Luján, and other Republicans were relieved anyway. Congressional funding for the Contras was about to run out, and Congress was unlikely to vote for another aid package that included military support. Reagan preempted the negative vote by saying that the peace plan meant the United States no longer had to send aid to the Contras. According to the plan, the Organization of American States would oversee and confirm the cease-fire. Once that happened, Cuba and the Soviet Union would stop sending aid to the Sandinistas, and the United States would lift the embargo on Nicaragua. In a letter to a constituent, Luján wrote of the peace agreement, "It is my hope that this finally brings about the much needed improvements in that trouble spot, without hanging the Contras out to dry."

REAGAN'S SECOND TERM WAS A DIFFICULT PERIOD FOR HISPANIC REPUBLICANS, WHO were divided over IRCA and had to defend President Reagan's Central America policies in the face of the sanctuary movement and the Iran-Contra scandal. They also didn't know how to respond when Cuban refugees who had arrived during the Mariel boatlift rioted in prisons in Georgia and Louisiana, setting fires and taking hostages. They had been detained for committing felonies in the United States. Should they be sent back to Cuba, or should they be allowed to remain in this country? Mas Canosa and the CANF didn't answer that question clearly or consistently. Right after the riots, Mas Canosa said he favored deportations only of the "most hardened criminals," but then he opposed all deportations and said he was working behind the scenes on behalf of the detainees.

Yet Mas Canosa and other Cuban Americans weren't sure of their position. What if one of the detainees, after being released, killed someone? What would the "public's perception" be of the Cuban Americans who fought so hard for them to stay? After the 1988 election, one Cuban American considering this dilemma said, "Look at the price" the Democratic contender Michael Dukakis "had to pay over Willie Horton." The Bush campaign had smeared Dukakis by arguing that his leniency toward criminals such as Horton—a convict serving a lifetime sentence in Massachusetts, where he was furloughed for a weekend; failed to return to prison; and assaulted, robbed, and raped a woman—enabled heinous acts against innocents.

The RNHA didn't know how to answer the question about the detainees either. They sympathized with the plight of the Cuban prisoners who didn't want to return to Communist Cuba, but they said that they weren't lobbyists, so they would stay out of it.

After some trying years, Hispanic Republicans looked ahead to the 1988 presidential election. When Vice President Bush visited Miami in October 1987, where he was accompanied by his son Jeb, the increasingly influential Cuban American politician Ros-Lehtinen wrote, "We're at the gates of an impressive victory." Floridians, she said, would again show their strong support for Republicans. More generally, as part of her effort to position herself as an expert on Hispanic politics nationally, not just Cuban exile or Central American refugee politics in Florida, she wrote that Hispanics had a bright future politically because of the increasing number of Hispanic elected officials and the growing Hispanic population, which was predicted to hit twenty-five million by the year 2000. Like others before her, she wrote expectantly about the moment in the near future when Hispanics—and Hispanic Republicans, in particular—would claim genuine political power.

She wrote another article about the imminence of Hispanic political power right before the 1988 convention, which cited a recent study by the National Association of Latino Elected and Appointed Officials (NALEO) reporting that the number of Hispanic voters had increased by 50 percent between 1976 and 1986, and three million Hispanics were expected to vote in the 1988 election, mostly, Ros-Lehtinen wrote, Cuban Americans from South Florida, Puerto

Ricans from New York, and Mexican Americans from California, Texas, and New Mexico.

Hispanics, Ros-Lehtinen wrote optimistically, were trending toward the Republican Party. Reagan had won 35 to 40 percent of their votes by highlighting ideological issues such as family values, work ethic, and patriotic pride. He had captured their imagination and enthusiasm. Hispanic Republicans expected Bush to build on Reagan's success. Ros-Lehtinen predicted that two issues would be of particular importance: the ongoing struggle against communism in the Caribbean and Central America, and the efforts by US English to make English the official language of the United States. The first issue spoke to her position as a prominent Cuban American politician and daughter of a leader of the Cuban exile community. The second was a sentiment she shared with most other Hispanics in the United States, that they were under attack by the nativist wing of the Republican Party.

Some Cuban Americans had begun to rumble about the fact that Reagan had promised to rid Cuba of Castro. They recalled his famous speech on Cuban Independence Day in 1983, when he exclaimed, "*¡Cuba sí, Castro NO!*" Yet Castro remained in power at the end of his second term. They also said that the embargo against Cuba had been a "complete joke," since there were still economic ties between Miami and Havana. But these sins, Ros-Lehtinen wrote, paled in comparison with how much worse it would be with one of the Democratic candidates in office. All of them—Dukakis, Al Gore, Dick Gephardt—had opposed aid to the Contras. They would turn a blind eye to the Caribbean and Central America, cut defense budgets, and let Marxism spread throughout the region. Cuban American voters, she argued, had to prevent this from happening.

Hispanic Republicans rallied around Bush, but cracks had appeared in the facade of unified Hispanic Republican support. In addition to IRCA and support for the Contras, they were even divided over how well Reagan had followed through on his promises to them. Reagan and Bush still saw the Republican Party as inclusive of Hispanics. But powerful currents of American conservatism pulled in the opposite direction.

The Federation for American Immigration Reform (FAIR) opposed the rising numbers of undocumented Mexican immigrants, and they seized upon the failures of IRCA to keep building their movement for immigration restriction. The law hadn't stopped the flow of undocumented immigrants, and it gave amnesty to two million Mexicans. US English sought to capitalize on the momentum they had gained in California by taking their movement to other states. Conservative but mainstream Republicans were increasingly persuaded. They considered the Immigration and Nationality Act of 1965 to be one of the worst pieces of legislation in US history for how it changed America's complexion, and they didn't think that the new immigration law was much better.

Even though Reagan ultimately forged the immigration compromise that had eluded his predecessors, IRCA unleashed a firestorm that caused many conservatives to claim that the president had caved to Hispanics by giving them amnesty. Other debates that were central to Hispanic Republican politics also turned on the issue of migration from Latin America, by Cubans during the Mariel boatlift, Central Americans who fled civil wars throughout the region, and undocumented Mexicans. These issues caused tension between Hispanic Republicans and the Republican Party. Anti-immigrant and seemingly anti-Hispanic conservatives were building power. Reagan couldn't control them, and sometimes they felt emboldened by him. Hispanic Republicans hoped the party would swing back toward moderation, and that's what it seemed to do with the selection of the vice president as its standard bearer. Hispanics were simultaneously courted and pushed away by the Republican Party. But over the previous two decades they had become loyal Republicans, so they stuck with the party that, they still believed, would allow them to shape its future.

PART IV

LOYALTY

CHAPTER 9

LA FAMILIA BUSH

Vice President George H. W. Bush's trip to New Orleans began with a gaffe. He was there for the convention, to become the Republican Party's nominee for president. Ronald Reagan greeted Bush on the morning of August 16, 1988, after his plane touched down at Louisiana's Belle Chasse Naval Air Station, the same base where Cuban exiles trained for the Bay of Pigs invasion.

Reagan had delivered his farewell address the evening before, just fifteen miles away at the Superdome. Ignoring the controversies of his eight years in office, he celebrated his triumphs and the fact that "not one inch of ground" had "fallen to the Communists." He praised his vice president and threw barbs at the Democrats, who had met in Atlanta the month before.

The meeting between Bush and Reagan at Belle Chasse was part of a delicately choreographed passing of the baton. The trick for Bush, the *Chicago Tribune* reported, was to "emphasize his relationship with the immensely popular President without appearing to be a second banana." Belle Chasse seemed the perfect place for this high-wire act. Access to the military base was limited, so crowds couldn't form to cheer Reagan while ignoring Bush. The base also offered the perfect backdrop for photographers to document the meeting.

The most reported-upon moment of their encounter came when

Bush introduced Reagan to his half Hispanic grandchildren. They belonged to Jeb and Columba Bush (née Columba Garnica de Gallo), Jeb's naturalized Mexican American wife who was born in León, Guanajuato, Mexico. George P. Bush, who recited the Pledge of Allegiance at the convention that evening, was thirteen at the time, Noelle Bush was eleven, and Jeb Bush Jr. was five. "These are Jebby's kids from Florida, the little brown ones," Bush said.

Democratic leaders pounced on the statement, claiming it was a sign of Bush's insensitivity toward Hispanics. Ruben Bonilla, the general counsel for the League of United Latin American Citizens said, "It's vintage George Bush." It was a "cavalier, aloof attitude that thrives on reinforcing the differences among Americans." Bonilla wanted to know, did Bush call his children "my little white ones"? Bush's statement also seemed to reflect the rising anti-Hispanic and anti-immigrant sentiments expressed most forcefully by the Republican Party's right wing, which had begun to influence its moderate leaders as well.

Bush countered their outcry with his own. Seated beside his running mate, Dan Quayle, Bush sounded defensive at the next day's news conference. To wit: "This heart knows nothing but pride and love"; "For anyone to suggest that that comment of pride is anything other than what it was, I find it personally offensive"; and "Those grandchildren are my pride and joy, and when I say pride, I mean it." Nobody doubted Bush's sincerity, but surely beginning on defense wasn't how the vice president's campaign team drew up their plan for the convention.

Hispanic Republicans were forced to defend another misstep by a Republican leader. They gave a series of superficial statements. His remark was a sign of "sincere affection" for his grandchildren, they said, and there was "nothing at all vulgar or insidious or disrespectful about being brown or being black." It was "just a description." Bush's remark was a bit like Gerald Ford's bite into an unshucked tamal: an unfortunate incident for a man whose insensitivity was only skin deep. Bush's Hispanic supporters didn't let it ruin their week. They were loyal to Bush and would make sure their support got across.

A Hispanic Republican wasn't given the keynote slot in New Orleans, as Ortega had been in Detroit, but several Hispanics took turns at the speaker's podium. A Hispanic teen spoke as a representative of the Georgia Teenage Republicans. "The Hispanics and the youth of this great nation are going to work to elect George Bush president," he said. The Hispanic bishop from the Catholic Diocese of Corpus Christi, Texas, offered an invocation: "Bless the members of the Republican Party, the delegates to this convention, and the candidates they select."

Leaders of the Hispanic Republican movement spoke as well. Catalina Vasquez Villalpando spoke as chairperson of the Republican National Hispanic Assembly (RNHA). The Mexican American banker had reclaimed leadership of the organization from the Cuban American Del Junco. The GOP stood "for progress, for opportunity, and for strength, both here and in Latin America," she said, and Bush was a "strong supporter of Hispanic literacy, jobs for progress, and economic development." Florida state senator Ileana Ros-Lehtinen introduced General Alexander Haig, who had run against Bush in the Republican primaries. She praised both men for their strong stand against communism in Nicaragua and Cuba.

Through the column she wrote for Miami's *Diario las Américas*, Ros-Lehtinen shared her impressions of the convention and its host city. It wasn't every day that a "Cuban refugee" had the chance to speak at her party's most important gathering. Ros-Lehtinen praised the conservative party platform, which staked out strong positions on right to life, anticommunism, family values, and the inherent worth of hard work. She also praised the cultural and linguistic diversity of New Orleans. "Cajuns," she wrote, spoke their French-derived dialect freely, "without suffering discrimination or creating resentment."

Her words, written in Spanish, were ostensibly about the convention's host city, but Hispanic readers would have grasped that she was also writing about the virtues and fact of bilingualism, as well as the tolerance and even admiration with which non-English tongues should be heard. They were her stand against the ongoing campaign to make English the official language of the United States.

The longtime Republican congressman from New Mexico Manuel Luján Jr., who had recently announced that he would retire after the current session, introduced Pete Wilson, the US senator from California, who was a rising star in the Republican Party. Meanwhile, the longtime leader of Puerto Rico's prostatehood movement, former governor Luis Ferré, in his mideighties at the time, wasn't given a formal speaking slot, but he got to deliver Puerto Rico's fourteen delegate votes for Bush. Ferré proclaimed to everyone listening that Bush had said on various occasions that the time for statehood was "now." He had led the march for statehood for more than twenty years, promoting it both in Puerto Rico and the fifty states.

If there was any doubt that Bush had great affection for Hispanics, or that his remark about the "little brown ones" had been anything other than an expression of love, or an innocent mistake, Bush's daughter-in-law, Columba, sought to lay it to rest. Significantly, most of her speech was in Spanish. She began, *"Buenas noches, queridos hermanos"* (Good evening, my dear brothers and sisters). She was only there, she said, to say that *"la familia* Bush" (the Bush family) "had opened its arms and its heart to her," much like "this generous country" that was "full of opportunities" for her and her children. The vice president had been like a father to her. He was an honest, intelligent, and capable man.

Here was a Mexican immigrant who had become a naturalized Mexican American, whose native language was Spanish, and who had been brought into the inner circle of the Bush family. The love she had for Bush, and the love Bush had for her, her husband, and her children, was a message to Hispanic voters that they could have faith in this man and his party.

THE REMARKS BY HISPANICS, AND OVERTURES TO THEM AT THE CONVENTION, WERE BOTH symbolic and strategic. In the months leading up to the beginning of the general campaign season, the national media was once again predicting that Hispanic voters could determine the outcome of the election. Ninety percent of all Hispanic voters lived in a handful of states that included California, Texas, Florida, New York, and Illinois. California and Texas alone had a quarter of the electoral votes a candidate needed to win.

The 1988 election would be the first since 1976 without a candidate from California in the race, and Ford had won there by only 1.7 percent. Twelve years later, there were more Hispanic voters in the state, and if they voted for Dukakis, a Democrat might win there for the first time since Lyndon Johnson in 1964. It was the kind of political calculus that had become commonplace over the previous two decades as the Hispanic population grew and as states with large Hispanic populations became increasingly important electorally.

Bush and his connections with Hispanics would be important to the Republican Party's ability to hold on to the White House. The vice president was to temper the rising anti-immigrant and anti-Hispanic sentiment within the Republican Party. For the Bush family, Columba's speech was an important reaffirmation of Republican principles at a time when rivals were trying to take the party in a different direction.

The Hispanic Republican network lifted Bush in Texas and Florida, where he was well known, but also in California, New York, and Illinois, all states he won in 1988. It was the last time a Republican would accomplish that feat. Hispanic Republicans placed their faith in him not only because of what he had done for them over the course of his career, but also because they saw him as the last best hope to save the party from its more conservative insurgents.

WHEN JEB BUSH WAS STILL A TEENAGER, IN 1970, HE HAD TRAVELED TO GUANAJUATO, Mexico, on a trip with classmates from Andover, as part of the class they were taking on "Man and Society." While there, he met his future wife, Columba. Unlike his brother George W. Bush, who followed his father's footsteps to Yale, Jeb went to the University of Texas at Austin, where he majored in Latin American Studies. He married Columba at the Newman Catholic Center on campus. He was an important part of his father's appeal to Hispanics.

Jeb leveraged his Latin American Studies major and marriage to a Mexican immigrant to claim familiarity and kinship with Hispanics and Latin America more broadly. His employer after college, the Texas Commerce Bank, sent him to Caracas, Venezuela, to open a branch there. Jeb then campaigned tirelessly for his father in Puerto Rico, where the Bush family would form close friendships that lasted for decades with the island's leading Republicans. After the Puerto Rico primary, Jeb stumped for his dad in every state with a large Hispanic population. At the end of the campaign, he settled in Miami, where he worked with the Cuban American real estate developer Armando Codina and mapped out his own political future.

By 1988, Jeb had developed an extensive web of connections with Hispanics that spread across the hemisphere, from Mexico to Texas and from Venezuela to Puerto Rico to Florida. His father had taken to calling Jeb and Columba his "secret weapon" with Hispanic voters. It was at the end of the primary campaign trail in 1988 that Jeb and his grandkids joined the patriarch of the family in New Orleans. But his father's victory would be his victory, too. The relationships with Hispanics he developed with his father's help became important to his—and his brother's—political career.

Despite the Bush family's long-standing connections with Hispanics, Bush had to work hard to shed his patrician background when talking to Hispanic audiences. He couldn't appear to be part of the East Coast elite. That would make him seem like the least ethnic American imaginable, as one critic described him in comparison with the Greek American Dukakis. Bush also couldn't be the Texas oilman whose early career benefited from family loans. Instead, Bush recounted his brave service in World War II and how he got married to his high school sweetheart right when he got home.

As a student at Yale, he worked hard and played baseball, he said, just like José Canseco and Fernando Valenzuela. Canseco was Cuban and Valenzuela was Mexican, and both athletes were heroes to many Hispanics. After college, instead of staying on the East Coast and becoming a banker or lawyer—the expected path for someone with his pedigree—he and Barbara wanted to "get out on our own," to blaze a trail toward the frontier, a metaphor that had been an animating force in the lives of so many in his and earlier generations of Americans. They didn't want to be "so-and-so's" daughter or son. They wanted to go to a place such as Texas, where every American could make his or her own name.

It was equalizing rhetoric designed to put him on the same plane as the Hispanics whose votes he sought. He brought them into his vision of America, in which there were no "hyphenated Americans"—Mexican-Americans, Irish-Americans, African-Americans, Asian-Americans, and so forth. There were "only Americans." This inclusive spin on America's ethnic past, along with Bush's probusiness, proreligion, profamily message, constituted the core of his appeal to Hispanics.

LIKE EVERY SEPTEMBER, IN 1988 HISPANIC HERITAGE MONTH—WHICH HAD EXTENDED TO a month from a week in 1987—saw Republican and Democratic candidates actively recruiting Hispanic voters. The "Hispanics for Bush" campaign launched on September 14, at Moreno's Mexican Restaurant in Orange, California. The candidate stood for a photo in front of a "Viva Bush" banner, then told the crowd of five hundred that he would "fight for the vote of Hispanic Americans, all the way from the East Coast to the West." He then sat in the ample shade of three large pepper trees and devoured a breakfast of beans, potatoes, corn tortillas, and *machaca*, a dried and rehydrated beef dish. Mariachis serenaded him inside the walls of the courtyard, while protesters heckled from across the street.

Bush's Democratic opponent, Dukakis, also planned an all-out push to win the votes of Hispanics. He touted the fact that both he and his running mate, Senator Lloyd Bentsen of Texas, spoke Spanish. They used their linguistic capabilities and Bentsen's hometown advantage to win the primary in Texas, where they spoke in Spanish to the "poor farm workers" of the Rio Grande Valley, where Bentsen had grown up. Dukakis also highlighted his Greek heritage and, according to an article in the *Washington Post*, "virtually wallows in his status as a first-generation American." A Dukakis campaign official called the candidate's ethnic appeal an opportunity to lure back the Hispanic voters who supported Reagan in 1980 and 1984.

Bush's Hispanic surrogates were having none of it. It didn't matter that Dukakis spoke Spanish because, one said, "Hispanics will prefer to be told the truth in English rather than lied to in Spanish." He and Bush sparred on several issues of importance to Hispanics, including deficit reduction, the defense budget, US-Soviet relations, and US–Latin American relations, including what to do about the political left, immigration, drug trafficking, and debt crises in several Latin American countries aggravated by unfavorable terms from US lenders. Hispanic Republicans thought their candidate—a former ambassador to the United Nations, special envoy to China, and director of the CIA—was uniquely qualified to handle such consequential matters. Dukakis paled in comparison, and might even lead the country down a dangerous path.

Hispanics gave Bush about 30 percent of their votes. In an interview with Univision—by then the preeminent Spanish-language media outlet in the United States—Ros-Lehtinen claimed that Bush's victory was a win for anticommunism and conservatism generally. Addressing Cuban Americans in South Florida, she wrote, "here we have voted against communism because the inhabitants of this state know that the correct thing to do is continue with the conservative, anti-liberal movement." The "Latin American masses of Florida"—by which she meant Cuban Americans, Puerto Ricans, and the growing population of Nicaraguan Americans—came to the United States to escape communism. They knew its dangers, and by voting for Bush they supported a leader who would demonstrate strength in Latin America. Florida was a "Latino state," she said, and it was Hispanics who helped Bush win there.

Bush, too, believed that he was indebted to Hispanics. In December, he invited a hundred of his strongest Hispanic supporters—those who had campaigned for him for more than a year—to a summit in Washington, DC, at the offices of the vice president. On the agenda were bilingual education, unemployment, high school dropout rates, crime, US–Latin American relations, and other issues. Ros-Lehtinen was there. She had become one of Bush's closest Hispanic advisers. The Mexican American secretary of education, Lauro Cavazos, who had been confirmed during Reagan's last month in office, was there, too. So were Ferré and RNHA chairperson Villalpando.

They met with Bush for three hours. Cavazos focused on bilingual education. Ferré addressed the interests of Puerto Ricans on the mainland and on the island, especially statehood. Ros-Lehtinen lamented the "tragic situation" of Nicaraguan refugees, whom she urged Bush to help settle in the United States, and she warned Bush against reestablishing relations with Castro, which he said he wouldn't do as long as the Cuban regime was antidemocratic.

BUSH WAS INAUGURATED ON JANUARY 20, 1989, AND ACCORDING TO A REPORT IN THE *LOS Angeles Times*, several of the inauguration celebrations highlighted his support not only among Hispanics, but also Asian Americans, Native Americans, and African Americans. Two Hispanic Democrats attended the Hispanic American Inaugural Ball at the Washington Hilton. They said they went to see what "the competition" was up to. They were impressed, and acknowledged that the Republicans were "doing damn well" with Hispanic outreach. Dispirited by the November loss, they now understood what happened. Flipping through the program for the festivities, they exclaimed, "Look at these names!" The Republicans were organized and had successfully gone after the "upwardly mobile" Hispanics. The inaugural events were just the beginning of their celebrations.

In Bush's first year in office, Hispanic Republicans notched major victories. Back in July, Bush had promised to appoint the first Hispanic to his cabinet if he were elected. In early August, Reagan preempted him by nominating Cavazos, who at the time was serving as the president of Texas Tech University, and before that had been a professor and dean at Tufts University's Medical School. As could be expected, Reagan denied that Bush's announcement had influenced him in any way, or that he nominated Cavazos because he was Hispanic, or because Bush needed to win Texas, where Cavazos was from.

Cavazos's nomination hearing was chaired by Massachusetts senator Edward Kennedy, who praised the nominee but criticized Reagan for slashing education spending and proposing to eliminate the Department of Education. Bush's promise to become the "education president" was a welcome shift in rhetoric, Kennedy said, but he called it an "election year conversion." Then addressing Cavazos, he joked, "As they say in the circus, it is a big job cleaning up after a big elephant," but he could count on Congress's help.

Senators variously hailed Cavazos for his commitment to expanding Hispanic access to higher education and offered lukewarm criticisms, noting that he wasn't "universally loved" by the faculty who worked for him or by Hispanics in Texas. But all acknowledged that he was qualified. The Senate confirmed him unanimously, by a vote of 94–0. Cavazos swore on his family Bible when he took the oath of office. His wife, Peggy, was by his side, along with their ten children. Hispanics finally had their first cabinet member, after decades of lobbying by the RNHA and others.

Los Angeles Times reporter Frank del Olmo wrote that the appointment was "symbolic at best, and downright cynical at worst." But many other Hispanics said they were happy to have Cavazos represent them. Right after the election, Bush asked him to stay on as secretary of education and promised that "minority representation would not stop" there.

Before his inauguration, Bush named a second Hispanic to his cabinet, selecting the outgoing New Mexico congressman Manuel Luján Jr. as his interior secretary. Luján had been short-listed for the position during the Reagan years but wasn't selected. When Bush's transition team first gauged Luján's interest, he said he'd rather return home to the Southwest. But when Bush pressed him about it personally, he relented. At the Department of the Interior he would be a "steward" of the natural resources of the United States; "safeguard" public lands and wildlife; "preserve" the natural environment; and manage the economic development of resources such as coal, oil, gas, and renewable energy. These were issues that had been on his mind for a long time, as a westerner and a New Mexican.

For most of his time in Congress, Luján had served on the Committee on Interior and Insular Affairs, the Joint Committee on Atomic Affairs, and the Committee on Science, Space, and Technology. Therefore, as the senator chairing the hearing put it, he was "well acquainted" with the work of the Interior Department. Luján, as a member of Bush's cabinet, would become "a role model for Hispanics in America." After the Senate confirmed Luján by a vote of 100–0, one Democrat called him Bush's "least controversial nomination." His biggest job was overseeing the cleanup of the *Exxon Valdez* oil spill, which occurred on March 24, 1989, only a couple of months after he was confirmed.

Bush appointed more Hispanics still. Catalina Vasquez Villalpando stepped down as head of the RNHA to become treasurer of the United States. Her nomination didn't receive the same amount of press as the nominations of Cavazos and Luján, but she succeeded

Romana Acosta Bañuelos and Katherine Ortega as the third Hispanic woman to hold the post. Bush also nominated Antonia Novello, a Puerto Rico–born pediatrician and expert on children with AIDS, to succeed C. Everett Koop as surgeon general of the United States. Her nomination sparked controversy among pro-choice advocates because she opposed abortion except for cases of rape, incest, or when the mother's life was in danger.

AFTER BUSH'S MEETING WITH HISPANICS IN DECEMBER, FERRÉ ANNOUNCED PUBLICLY that Bush had told him he would urge Congress to organize a referendum on the island's status. It would state that Congress would be bound to implement the results of the vote, whatever they were. Ferré said he asked Bush to state his support for statehood publicly, not just to him.

About halfway through his first State of the Union address, on February 9, 1989, Bush, with his squinting, soft eyes, and a slight, barely noticeable smile, shoehorned a statement about statehood between paragraphs on caring for the homeless and the benefits of personal savings. "There's another issue that I've decided to mention here tonight," he said. "I've long believed that the people of Puerto Rico should have the right to determine their own political future," he said. "Personally, I strongly favor statehood," he added, but ultimately it would be up to Puerto Ricans and then to Congress. He urged Congress to allow Puerto Ricans to organize a referendum on their status.

With Bush's endorsement, Puerto Rican Republicans believed statehood was imminent. José Celso Barbosa and other founders of the Republican Party of Puerto Rico had fought for statehood since the early twentieth century. Now the president of the United States offered his full-throated support with millions of Americans watching. The inroads that Puerto Rican Republicans had made with Bush were about to pay off. They heard Bush's words as a kind of policy recommendation that would lead to a referendum, and then the incorporation of Puerto Rico as the fifty-first state.

The State of the Union address unleashed a flurry of activity in Puerto Rico and the District of Columbia. Ferré's New Progressive Party (Partido Nuevo Progresista, or PNP) was the main prostatehood party on the island, but there were dozens of smaller prostatehood groups. After Bush's address, they all came together to form the so-called Statehood Caucus. Ferré and other prostatehood leaders, including former governor Carlos Romero Barceló, began traveling back and forth between Puerto Rico and DC to lobby members of Congress. To shore up the sitting president's statement, Ferré asked former presidents Richard Nixon, Ford, and Reagan for letters expressing their support for statehood. They happily obliged. The leaders of the RNHA voiced their support for statehood, bolstering the arguments of statehooders on the island.

The members of the PNP also worked to convince Puerto Ricans that statehood was the best outcome. To those who argued that statehood would lead to the erasure of Puerto Rican culture, Ferré said he envisioned an *estadidad jíbara*, which meant something like a "creole statehood," in which Puerto Ricans would remain Puerto Rican; they would continue speaking Spanish and maintain traditional customs. He also claimed that Bush's statement put the lie to the "completely invented" antistatehood "propaganda" that the United States didn't want Puerto Rico as a state.

In relatively short order, Congress proposed and scheduled hearings on two separate bills, one in the Senate and one in the House of Representatives. The Senate proposed S.712, the Puerto Rico Status Referendum Act, and the House proposed H.R.4765, the Puerto Rico Self-Determination Act. They held public hearings, and Ferré was a witness at each. He stressed that Puerto Rican Republicans had fought for statehood for half a century; that statehood would lead Puerto Rico away from economic dependence; and that equal rights, responsibilities, and opportunities had worked well for the other fifty states, so wouldn't they for Puerto Rico?

It didn't take long, though, for Congress to begin casting doubt on statehood by arguing that incorporating Puerto Rico would be a financially ruinous decision. Bush had made a campaign issue of lowering the federal debt, but many in Congress, both Democrats and Republicans, argued that statehood would require a significant federal layout, especially through various social service programs. Once American companies no longer benefited from the tax exemptions provided under Section 936 of the Internal Revenue Code, which provided a "tax credit equal to the tax liability of certain kinds of businesses" in Puerto Rico, the companies would leave the island. Certainly that wouldn't be good for Puerto Rico.

For about fifteen years, the prostatehood party had embraced Romero Barceló's argument that "statehood is for the poor," primarily to convince other Puerto Ricans that it wasn't a design by and for the rich. But when Congress began to question how much statehood might cost, they downplayed how it might help the poor and focused instead on the value that Puerto Rico would add to the Union.

Congress began to back away from the guarantee that they would comply with the results of the vote and automatically transition to whatever status Puerto Ricans favored. Instead, they said that Puerto Ricans should organize the referendum themselves, and then make a specific request that Congress would consider once they knew the results, mimicking the Republican Party's position since the 1940s, rather than the new direction that Puerto Rican Republicans heard in Bush's State of the Union remarks. They would have to wait and see how debates over the referendum proposals would play out.

AS CONGRESS DEBATED PUERTO RICAN STATEHOOD, HISPANIC REPUBLICANS RECEIVED unexpected good news when Ileana Ros-Lehtinen was elected to the US House of Representatives in the special election held to replace longtime Democratic congressman Claude Pepper. Pepper had died in office on May 30 at age eighty-eight, and a week later Ros-Lehtinen declared herself a candidate for the seat that Pepper had held since 1962, when Florida's Eighteenth District was created. She chose Jeb Bush as her campaign manager, and enlisted the president himself to host a fund-raising lunch that netted $500 per plate and a total of $400,000. The Republican National Committee (RNC) and the Republican Party of Florida were willing to spend big on the election because it would be an important pickup for them. For decades, Florida had been a Democratic stronghold, but Republican candidates—for president, governor, and Senate—were on a winning streak. As the *Los Angeles Times* put it, Florida was "prime territory for Republican expansion."

The election of Ros-Lehtinen would signal victory for Cuban Americans, who still didn't have a national representative. Mas Canosa and the Cuban American National Foundation worked hard to get Ros-Lehtinen elected, and claimed their fair share of the credit.

Racial tensions defined the campaign. The candidates talked about US-Cuban relations, U.S.-Israel relations, crime prevention, elder care, and abortion, which Ros-Lehtinen opposed and her Democratic opponent Gerald Richman did not. But above all, the race pitted Hispanics who supported Ros-Lehtinen against non-Hispanic whites, Jewish Americans, and African Americans who supported Richman. During the "short, tense race," as the *Los Angeles Times* described it, Richman's non-Hispanic white supporters said there were too many Hispanic immigrants. They sounded more like the Democrats of an earlier era than Democrats in the late twentieth century, who increasingly saw the embrace of multiculturalism as key to their future success. In Miami, national immigration politics got turned upside down as Democrats argued for restriction and Republicans argued for inclusion.

In the end, Ros-Lehtinen defeated Richman in the August election by 53 to 47 percent. She won almost 90 percent of the Hispanic vote, 13 percent of the non-Hispanic white vote, and 3 percent of the African American vote. What's more, almost 70 percent of eligible Hispanic voters turned out. The day after the election, campaign manager Jeb Bush reflected, "Ethnic pride is what we saw in the Cuban American community here." During one of his Hispanic Heritage Month speeches his dad spoke of Ros-Lehtinen's victory, saying, "She is going to be a tiger, a force to be reckoned with on Capitol Hill." Beyond what the victory meant for Ros-Lehtinen herself, it was another sign that Cuban American Republicans had arrived. Ros-Lehtinen was the first Cuban American elected to Congress, and the only Hispanic Republican since Luján. But more were on the way.

ROS-LEHTINEN'S VICTORY, SEVERAL HIGH-LEVEL APPOINTMENTS FOR HISPANICS, AND Bush's statement in support of statehood during his first State of the Union address all made 1989 a pivotal, overwhelmingly positive year for Hispanic Republicans. So much of what they had worked toward seemed to be coming together: they had elected their first Cuban American to Congress, seen multiple Hispanics appointed to Bush's cabinet, and appeared to be on the brink of securing statehood for Puerto Rico. During the next couple of years, though, many of these successes would come undone.

The momentum for Hispanic Republicans ground to a halt just as Bush's presidency gained steam. A Gallup poll taken in January 1990, one year after his inauguration, gave Bush an 80 percent approval rating, the highest for any president after a year in office. He began as a "second fiddle, a loyal cheerleader" for Reagan. But in a quick twelve months, he went "from wimp to winner, from lapdog to leader," as one article put it. While he received mixed reviews on domestic issues— from cleaning up the environment to labor relations to the war on drugs—in the arena of foreign affairs he presided over the collapse of communism in Eastern Europe and improved relations with the Soviet Union. In December 1989, right before the poll results were released, US troops also invaded Panama, ousted Manuel Noriega, and helped install his replacement, Guillermo Andara. It was a show of strength, his supporters said, that protected Americans in the Canal Zone, defended democracy in Panama, combated Noriega's drug trafficking, and maintained the neutrality of the Panama Canal.

Hispanic Republicans were pleased with the result in Panama, but then 1990 began with mounting criticism at home of the first Hispanic cabinet member: Secretary of Education Cavazos. He was criticized as a weak leader who hadn't made progress on Bush's campaign promises to reduce illiteracy, expand education programs, and increase federal spending on education. When there were whispers about his poor performance, Bush at first defended him, saying he was doing an "outstanding" job and that the post was his "as long as he wants it." But when Bush replaced several cabinet members later in the year, Cavazos was among the first to go.

Bush's resolve on Puerto Rican statehood also withered in the face of mounting criticism. The president's State of the Union address marked the beginning of the war between prostatehood and antistatehood forces in Washington, DC, and Puerto Rico. In the final days of 1989, the Subcommittee on Insular Affairs called public hearings on the island's status; the hearings began in March 1990. From the time the Senate and House bills were proposed, support for statehood faced resistance from House Republicans. They were emboldened by a Congressional Budget Office report concluding that statehood would lead to a significant reduction in economic growth in Puerto Rico. Social welfare programs may provide an immediate stimulus, but later would contribute to slow growth because of decreased investment, lower production, and rising unemployment in the manufacturing sector. Many of these negative effects would come about because US companies would no longer receive the tax benefits of Section 936.

Ferré complained bitterly about the budget report because he believed it made a lot of unfair assumptions. First, he said the report acknowledged that the economic effects of a status change would be hard to predict, but then went on to do just that. Second, he said the report overestimated the unemployment that would result because it didn't account for the new industries that would come to the island. Puerto Rico's new congressional representatives would lobby companies to move there. Third, he added that Puerto Ricans were eager to contribute their fair share in taxes and would not be a drain on the federal government. The taxes they would pay would offset the cost of social welfare programs. Fourth, American companies wouldn't leave the island, even after their Section 936 benefits ended, because they would still make more money in Puerto Rico than if they returned to the mainland.

He argued that statehood would be revenue neutral, if not financially advantageous for the United States. But the opponents of statehood were unmoved, and they had in their corner one of the leading ideologues of the far right, the person who, as much as anyone else, worked to pull the Republican Party in a rightward direction.

Pat Buchanan, a former assistant to Republican presidents from Nixon forward and emerging as a representative of the far right of the conservative movement, weighed in on the Puerto Rico statehood issue in early 1990, as the public hearings were taking place. The main argument of his columns in the *Washington Times* was that Puerto Rico had a strong independence movement bolstered by radical Puerto Ricans on the mainland. The United States did not need a new state full of separatists who would tear apart the United States from within, as had happened in Québec and Northern Ireland. As he colorfully put it, "The last thing America needs now is to clasp to her bosom, forever, three hundred thousand embittered Hispanics who yet dream of an independent country." He reminded his readers of the attempts by Puerto Rican nationalists on Harry Truman's life in 1950, and the torrent of bullets they fired in Congress in 1954. Ten percent of Puerto Ricans still favored independence. Eight thousand supporters of independence had recently demonstrated in San Juan. They wouldn't be incorporated easily.

The rest of what Buchanan had to say was even more controversial. He wrote that half of all Puerto Ricans would be eligible for welfare, including Aid to Families with Dependent Children, Medicaid, and food stamps. "Can we afford this?" he asked. He added the line of attack that Puerto Rico, the fifty-first state, would send six new Democrats to the House of Representatives, and two liberal senators who would "overturn the Reagan Revolution." He also wrote that Puerto Ricans would bring a "fundamental change in the character of our Union." Sixty percent of them, he claimed, didn't understand English. With Puerto Rico as a state, the United States would become bilingual, he said, as if it were ever anything else.

Ferré responded immediately, also in the pages of the *Washington Times*, as an insulted, wounded fellow Republican. "Seldom has the image of the United States of America been soiled with greater injustice and prejudice," he wrote. Buchanan had launched an attack against 6 million American citizens, 3.5 million in Puerto Rico, and 2.5 million in the fifty states. He refuted the idea that half of Puerto Ricans would be on welfare. Puerto Ricans weren't looking for welfare. They would work and save and pay their taxes. Ferré added that 200,000 veterans of the US military were Puerto Ricans. They had fought in every war since World War II and included Congressional Medal of Honor winner Fernando Ledesma. Hundreds of thousands of other Puerto Ricans were factory workers, professors, teachers, and businessmen.

He reminded Buchanan that Puerto Ricans weren't asked in 1898 whether they wanted to join the "American family" but rather were brought in "by unilateral action of the United States" because Puerto Rico was seen as essential to the defense of the country. And insofar as Spanish was concerned, Puerto Ricans wouldn't constitute any more of a threat than the fifteen million Hispanics already living in the United States. This probably wasn't Ferré's most convincing argument, since Buchanan and the hundreds of thousands of other Americans who had made donations to US English did indeed believe that all Spanish-speaking Hispanics were a threat. Nevertheless, his basic point, to which he returned at the end of his essay, was that Puerto Ricans were "loyal Americans" who were eager to assume their "full responsibilities in the Union."

Puerto Rican Republicans living in the United States carried on the attack against Buchanan, whose essay they called a "perversely xenophobic attack." Ferré was extremely popular on the island, they said, so it wasn't necessarily the case that Puerto Rico would become a Democratic state. In fact, Republicans, they claimed, would "have a darn good chance of being a majority in Puerto Rico's congressional delegation." Finally, they argued, Puerto Ricans were bilingual and understood English perfectly, and statehood would be revenue neutral, because federal expenses and the closing of tax loopholes would offset one another.

Unfortunately for the prostatehooders, Buchanan's words had hit their mark, and he got in the last word in another essay, written in May. Like a skilled debater, he reiterated his main argument that Puerto Rico was a "nation in utero." It therefore would be an "invitation to chaos" to invite it into the United States, with its "built-in secessionist movement" that was "determined to break the Union apart." He also reiterated what he'd written earlier about Spanish. Puerto Ricans would soon demand that all official documents be published in Spanish—as many government documents already were, because of the 1975 Voting Rights Act—and from there the demand would spread to Texas, California, New Mexico, and other states with large Hispanic populations. Finally, welfare would ruin Puerto Ricans, just as it had already ruined African Americans and Native Americans.

Why, Ferré wondered, had the Republican Party supported statehood in almost every party platform since 1940, only to turn against it now? Was it that they never really intended to incorporate Puerto Rico as a state? Was it that growing concerns about the national debt made it the wrong moment? Was it the organization of proindependence forces in response to Bush's State of the Union address? Bush administration officials, including Secretary of the Interior Luján, had attended several prostatehood rallies in Puerto Rico, and each time they did, the president was criticized for intervening in the affairs of the island. After one visit by Luján, a Democrat from Puerto Rico wrote, "President Bush has thrown the full weight of the White House behind the subtle and not-so-subtle efforts to influence the people of Puerto Rico to favor statehood."

Bush continued to make statements in support of statehood, but the push was over and the moment had passed. Because Congress remained divided, they said they would respond to any petition by Puerto Ricans but wouldn't push for bills supporting a status change. Puerto Ricans continued to plan for their plebiscite, but, as one historian of Puerto Rico has described it, it would be a "plebiscite of uncertainties," because any decision by Puerto Ricans would still be subject to approval by Congress.

To express their disappointment at the fizzling of their longtime dream, at least for the moment, the Puerto Rico chapter of the RNHA—called the "Luis A. Ferré" chapter—issued a statement in February 1991 that said, "Republicans in the United States should unequivocally know that Republicans in Puerto Rico will not sit idly before this monumental display of ignorance and intellectual obscurantism regarding historical, social, and economic issues related to Puerto Rico." Their faith was shaken. Perhaps the statement that they wouldn't "sit idly" was a reminder that they could rethink their support for the Republican Party, but, as if still responding to Buchanan (surely, "intellectual obscurantism" wasn't a phrase often used by people who didn't understand English), they also implied that a proper understanding of the island's history would lead to their incorporation.

EVEN CUBAN AMERICANS DEVELOPED MIXED FEELINGS TOWARD BUSH. THE FALL OF THE Soviet Union made them excited about Castro's imminent demise.

As Bush and Russian leaders Mikhail Gorbachev and Boris Yeltsin grew closer, Cuba, they hoped, would become increasingly desperate. Russia would need the help of the United States as it transitioned to a "free-market economy," one Cuban American speculated, and this would require them to leave Castro in the lurch. Cuban Americans now had in Ros-Lehtinen a strong leader in Washington who had extracted a promise from Yeltsin that he would not help Castro. She had also fought for aid to the Contras and served as a strong advocate for Miami's business and banking interests. When she launched her reelection campaign in the fall of 1991—with Jeb Bush again serving as her campaign manager, and his father again campaigning on her behalf—there was little doubt that she would win.

But Bush's "declarations on Cuba," wrote one of Bush's conservative Cuban American critics, were "vague and imprecise." While they still wanted to topple Castro, Bush seemed to be moving more toward coexistence. Bush made pronouncements about the need for a democratic Cuba and an end to Castro's human rights abuses, but he wasn't actively working toward his overthrow, as they wished. Ros-Lehtinen and Jeb Bush supported their cause and clearly understood the threat posed by communism. Hopefully, the Cuban American wrote, Ros-Lehtinen and Jeb could persuade the president to stand for them. If not, they had "tremendous electoral power" and could elect other leaders who saw that the interests of the Cuban exile community and of the United States were one and the same.

Cuban Americans and Nicaraguan Americans also felt they had lost the momentum of the Reagan years. The election of Violeta Chamorro as the president of Nicaragua in 1990 unseated Daniel Ortega, marking the end of Sandinista rule and apparently signaling the rise of democratic governance. Bush lifted the sanctions imposed by Reagan and promised hundreds of millions in aid to Nicaragua: $300 million in 1990 and more than $200 million in 1991. But Ros-Lehtinen and other Cuban Americans argued that Chamorro wasn't in control of the country and that the Nicaraguan government was still infested with Communists. She joined the fight to reclaim properties confiscated by the Sandinistas, criticized the continued violations of human rights in Nicaragua, and advocated for Nicaraguan exiles to be granted asylum in the United States, even though they now fled the democratically elected Chamorro regime.

The United States seemed to lose interest in Nicaragua after Chamorro's election and the end of the Contra War. Chamorro herself traveled to Washington to appeal to Congress for more aid. Whereas the floor of Congress was packed during debates over Nicaragua in the 1980s, very few congressmen showed up for her speech this time. Their attention had shifted from Nicaragua back to El Salvador. Ros-Lehtinen sought additional funding for the Salvadoran military in their ongoing campaign against the Farabundo Martí National Liberation Front (FMLN). Congress also shifted its attention to the Middle East, and from communism to terrorism.

THE GROWING SKEPTICISM AMONG HISPANICS LED TO THE CONCLUSION BY *HISPANIC BUSIness* magazine that "Hispanic enthusiasm" for the president and the Republican Party "may be stagnant." Bush still earned high marks for his handling of the war in Iraq and for his foreign policies in general. He also had the approval of Hispanic entrepreneurs and industrialists, who supported his trade policies with Mexico and won defense contracts during the Gulf War. But many "political professionals" claimed that the 1990 midterm elections pointed to a dark future for the GOP given the rising nativist backlash against immigration and the nation's growing non-white population generally. They anticipated that the loyalty of Hispanic Republicans would be tested.

Ferré had just written Bush a couple of months earlier, to congratulate him on the "astounding success of Operation Desert Storm," but his letter, too, came with a note of dissatisfaction. It was clear that statehood wasn't going to happen, at least not then, so Ferré linked the Gulf War and Puerto Rico's struggle in one more effort to achieve his goal. Ferré set up his ask in a paragraph that was ostensibly about the Gulf War. He wrote, "You have shown the world that the cause of democratic self-determination and the principles of human dignity and liberty can triumph over the most severe threats to their existence." Bush must have known where the letter was headed. Ferré had used words such as "self-determination," "dignity," and "liberty" to describe Puerto Rico's desires as well. He transitioned to Puerto Rico by talking about the Puerto Rican soldiers who had "served so bravely" in the Middle East. Then he wrote that Republicans needed to understand that their opposition to statehood was "inconsistent with the principles the United States just defended in the Persian Gulf." Puerto Rico also had to be liberated from the "shackles of second-class citizenship" they had worn for nearly a century.

The president was being pressured on other fronts as well. Claims of job discrimination had spiked as a result of the employer sanctions in the Immigration Reform and Control Act. Rather than face a stiff fine for hiring an undocumented immigrant worker, many employers chose not to hire Hispanics at all. The law had also done little to discourage undocumented immigrants from attempting to enter the United States. Moreover, Hispanic high school dropout rates were rising despite the fact that Bush promised to be "the education president," and drugs continued to negatively affect Hispanic communities despite Bush's promise that he would win the war on drugs, begun by Nixon and waged even more aggressively by Reagan. With considerable Hispanic support, Democrats had won the governorships in key states such as Florida and Texas.

Bush's personal reputation seemed not to suffer. In September 1992, just two months before he was voted out of office, Hispanic survey respondents considered Bush to be "the man they most admired." He edged out Julio Iglesias and Pope John Paul II.

Despite his popularity, the general dynamics that doomed Bush's reelection campaign also hurt him among Hispanics. There was the infamous 1988 pledge "Read my lips: no new taxes," a promise on which he reneged. Bush also never saw himself as part of the conservative movement whose ideologues were enamored with Reagan, and they never saw Bush as one of them. Many conservatives promised to withhold their votes when he increased taxes, even though Reagan had done the same not once but five times during his presidency. As one of their own, Reagan could do no wrong. They didn't feel the same way about Bush. Emerging leaders such as Newt Gingrich turned on him.

Bush also contended with challengers Buchanan and Ross Perot. Buchanan's repugnant views appealed to a narrow but committed slice of the primary electorate. He called Congress "Israeli-occupied territory"; wished that "Englishmen" and not "Zulus" would be let into the country; and said that Latin American immigrants represented an "assault" on Western culture. He nevertheless won 37 percent of the vote in the New Hampshire primary, and from there gained a loyal following of like-minded culture warriors. There was no love lost between Ferré and Buchanan. After his second-place finish in New Hampshire, Ferré, trying to galvanize opposition against Buchanan in the upcoming Puerto Rico primary, said he was a "moral and political dwarf on stilts." Perot, meanwhile, would peel 14 percent of the Hispanic vote from Bush and Clinton, primarily because of Perot's opposition to the North American Free Trade Agreement (NAFTA).

Republicans thought free trade would be a home run for them because it would lead to a "trade-fueled economic boom." Instead, many Mexican Americans in the border region were more afraid than excited because they believed it would cost them their jobs. Cuban Americans also opposed entering into a free trade agreement with Mexico as long as Mexico continued to do business with Cuba, which needed trading partners after the fall of the Soviet Union. In one of her columns, Ros-Lehtinen noted that commerce between Cuba and Mexico reached $184 million in 1991, and Mexico predicted it would reach $284 million in 1992. Puerto Rican businessmen were skeptical as well, because they thought the trade deal would make the tax exemptions Puerto Rico offered American companies less attractive.

Then there was Bill Clinton, the man from "a place called Hope." Hispanics met the Democratic nominee with mixed reactions, in part due to the efforts of the Bush campaign to hammer his record as governor of Arkansas. "Governor Clinton has not been a friend to the Hispanic community," said José Sosa, a Puerto Rican Republican assemblyman from New Jersey. He and other Hispanic surrogates noted that Clinton had signed a bill in 1987 declaring English the official language in Arkansas, and had endorsed a policy allowing police to search cars matching a certain "drug-deal profile" that many Hispanics claimed discriminated against them.

Bush's treasurer of the United States, Catalina Vasquez Villalpando, delivered some off-the-cuff remarks about Clinton that got her in trouble. On the eve of the Republican national convention in Houston, Villalpando met with members of the New Jersey delegation. The former Mexican American mayor of San Antonio, Henry Cisneros, had just joined the Clinton campaign to help him recruit Hispanics. Rumors about Clinton's sexual indiscretions had dogged him for years, and Cisneros had retreated from public life three years earlier after acknowledging he had an extramarital affair. Villalpando asked, "Can you imagine two skirt chasers campaigning together?" George Stephanopoulos, Clinton's communications director, redirected the accusation of improper behavior toward Bush, whose team, he argued, had entered the "business of sleaze" by stooping to personal attacks.

Villalpando said she was "deeply" sorry, but her apology was only the beginning of a rough period for her. Over the next few months, the Department of Justice would open an investigation into accusations that she had improperly used her influence as treasurer of the United States to secure federal contracts for her former employer, Communications International. Based in Norcross, Georgia, the company had been granted almost $70 million in noncompetitive contracts since the early 1980s, including a recent deal, inked while Villalpando was treasurer, to provide communications services to the US military in the Middle East and to help rebuild Kuwait after the Gulf War. Her home was raided days before the election, and she was placed on leave by the Treasury Department. She was later convicted of tax evasion and the destruction of evidence in a criminal investigation and sentenced to four months in prison.

Despite the efforts of Bush's Hispanic supporters to undermine Clinton, the Democratic nominee created problems for the president. According to Clarence Page of the *Chicago Tribune*, Clinton showed "greater appeal across racial and class lines than any other presidential candidate since Robert F. Kennedy." Clinton also didn't cede issues to Bush that had been Republican Party strengths for more than a decade. Like Republicans, Clinton took hard-line positions on Cuba and immigration. Clinton supported the Cuban Democracy Act proposed by a Democratic representative from New Jersey, Robert Torricelli, which tightened the US embargo on Cuba. He would also propose his own restrictive immigration reform law, to go along with other conservative legislation including welfare reform and mandatory minimum sentencing.

Beyond Clinton, the fact that the Republican Party was taking a hard right turn, especially on immigration and border control, also troubled the waters for Bush and for the Republican Party. The president's differences with movement conservatives included their divergent views of these increasingly divisive issues. Gingrich's split from Bush gestured toward rifts within the party generally, but the biggest threat to Bush and moderates would come from Buchanan and figures such as the governor of California, Pete Wilson, who had spoken in support of Bush at the 1988 convention and was regarded as a presidential contender himself. In New Orleans, he had celebrated California's diversity, but he and Buchanan now pushed the Republican Party in a nativist, anti-immigrant direction. Hispanic Republicans would remain loyal, but for how long?

CHAPTER 10

BUILDING THE WALL

"Structures." The word spooked Hispanic delegates at the 1992 Republican national convention in Houston. It appeared in the official party platform establishing the GOP's current and future direction. The offending line said, "We," the Republican Party, "will equip the Border Patrol with the tools, technologies and structures necessary to secure the border." Many Hispanics read it as a sign that the GOP supported the construction of more border fencing between the United States and Mexico. By the early 1990s, there were chain-link fences and newer corrugated steel walls—made from repurposed Vietnam War–era chopper landing mats—along some stretches of the border. But these were hardly impregnable. The prospect of more fencing precipitated a brawl within the Republican Party that became increasingly heated.

President Bush's right-wing challenger Pat Buchanan and his supporters sparked the controversy. Buchanan had captured almost a quarter of the votes cast in the Republican presidential primaries, seriously weakening the president. Securing the support of Buchanan's followers was essential for Bush to win the general election. That compelled him to placate the right-wing populist by giving him a prominent speaking slot at the convention, despite his promise that he wouldn't "swear fealty to King George." Bush hoped Buchanan would encourage his supporters to fall in line, but instead Buchanan inflamed tensions by telling them that structures did indeed refer to the "physical barrier" often referred to as a "tortilla curtain."

That Republicans were even discussing a border wall in 1992 marked a sharp break with the past. The 1980 platform said that Republicans "have opened our arms and hearts to strangers from abroad" and that "we favor an immigration and refugee policy which is consistent with this tradition." On the campaign trail that year, Ronald Reagan had declared that undocumented workers should be documented, that they should be allowed to stay for however long they wanted to stay, and that there shouldn't be a fence along the border. But times were changing, and the anti-immigrant forces within the Republican Party were growing stronger. This shift was already evident in the 1984 and 1988 platforms, which insisted upon "our country's absolute right to control its borders." Bush's voice was supposed to be the voice of moderation, but anti-immigration hard-liners had begun to drown him out, forcing his supporters to assuage fears sparked by the harsh language in the party platform.

Bush's people insisted that "structures" referred to the maintenance and improvement of existing fences, which had been constructed along the border sporadically throughout the twentieth century. But Buchanan's people, one newspaper reported, countered that it meant "building walls . . . to repel illegal border crossings." A campaign spokesman bluntly bragged that "the GOP is going to build the Buchanan fence."

Hispanic Republicans agreed that such a fence would be "offensive and discriminatory." But despite Buchanan's affirmation, they debated whether a fence was what he actually wanted. Gaddi Vasquez, a Mexican American delegate who chaired the subcommittee that wrote the immigration plank, said he told the members of his committee that "structures" referred to "organization structures"—whatever that meant—or beefed-up highway checkpoints for use by the Border Patrol. Secretary of the Interior Manuel Luján Jr. claimed that structures simply meant legal structures, or "new laws to prosecute organized illegal immigration." Rita DiMartino, the Puerto Rican former head of the Republican National Hispanic Assembly (RNHA) in New York, didn't have a specific explanation but said that it certainly didn't refer to a wall. The language in the platform "just flowed," she said. José Manolo Casanova, the Cuban American chairman of the RNHA—leadership of the organization would bounce back and forth among Mexican Americans, Cuban Americans, and Puerto Ricans during the 1990s—agreed that Buchanan was only "making mischief."

Lou Gallegos, a Mexican American delegate from New Mexico, wasn't satisfied. Gallegos dismissed as excuses the explanations given for the meaning of "structures." He thought that by allowing the language, the GOP was "pandering" to rising anti-immigrant sentiments among the party's conservative right wing. Gallegos said he was "as good a Republican as anybody," a loyal Republican, but he took issue with the feeling that "you've got to fence people out."

Many non-Hispanic Republicans were upset as well. The divides between them and the anti-immigrant wing of the GOP suggested the growing tensions within the party. An oilman from Roswell, New Mexico, said it didn't make sense for Bush to create the economic free-trade zone established by the North American Free Trade Agreement (NAFTA) while simultaneously considering the construction of "some kind of wall." Jack Kemp, Bush's secretary of housing and urban development, didn't "buy this idea that we have to set border guards to shoot people coming across." To Kemp, "people don't come to America for welfare. They come for their family." William Bennett, Bush's drug czar, acknowledged that "the American people expect to be protected" from drugs, but "they don't want the country to turn into a totalitarian state with borders that have hundred-foot walls."

Republican anxiety and opposition prompted delegates to pass a separate resolution clarifying that whatever "structures" meant, it didn't mean walls. Nevertheless, the language forced into the 1992 platform by the Buchanan brigades was becoming mainstream. Former Klansman David Duke, who at the time held a seat in the Louisiana Senate, also ran in several primary races. Buchanan was a longtime admirer of Duke, and had famously stated that the Republican Party should borrow from his "portfolio of winning issues" those that didn't conflict with its own, including stands against affirmative action, "reverse discrimination," illegal immigration, and welfare, whose recipients, he and Duke claimed, were people of color. Their populist nationalism appealed to a significant minority of Republicans, pointing to the growing divide within the Republican Party.

Even though the 1992 convention began with controversy for Bush and Hispanic Republicans, Hispanic leaders did their best to give Bush momentum. Ileana Ros-Lehtinen, Al Cardenas, Jeb's son George P. Bush, and RNHA chairman Casanova were some of the featured speakers. Ros-Lehtinen argued that the Cold War was still very much alive in Cuba and Nicaragua and that Bush would be the only one to stand up to the Communists in these countries. Cardenas said the quest for freedom was the same in all Hispanic communities, "from the barrios of LA to Chicago's South Side, from New York's Spanish Harlem to Miami's Little Havana." Casanova said the Bushes had a "special relationship" with Hispanics that began in their own family but also stemmed from the values they shared: democracy, free enterprise, free trade, religion, family, a strong work ethic, loyalty, and patriotism.

The 1992 Republican primary campaign showed that there was a war between establishment leaders such as Bush and insurgents such as Buchanan. Because of the internal divisions within the Republican Party, the third-party candidate Ross Perot, and Bill Clinton's own talents, Democrats won the White House for the first time in twelve years. Over the next four years, during Clinton's first term, the conservative insurgents seemed to be winning, pushing the Republican Party—and the Democratic president—to support hard-line stances on immigration, border control, welfare, and crime. But after a resounding defeat in 1996, which swept Clinton into a second term in office with the largest share of the Hispanic vote since Carter beat Ford, Republicans moved back toward the center, at least on the border and immigration. The rise of another Bush—George W. Bush, the governor of Texas—gave them hope that the Republican Party still represented them and that leaders like him would define the party's future.

IF THERE WAS UNCERTAINTY IN HOUSTON ABOUT THE IMMINENCE OF FENCE CONSTRUC-tion, it was cleared up when stretches of border wall were built during Clinton's first term. By the second decade of the twenty-first century, calls to build a border wall had become a rallying cry for millions of Republicans. Segments of the border wall got built, but Republicans increasingly called for its completion, from the Gulf of Mexico to the Pacific Ocean.

Clinton didn't begin with fence construction. He started with more personnel. In the months after his inauguration, Clinton implemented Operation Hold the Line, a new border security measure in El Paso, one of the main points of illegal entry. Hundreds of Border Patrol agents were sent to apprehend illegal immigrants along a twenty-mile stretch of the Texas border. According to one journalist, they "stopped illegal immigration in its tracks." Before Hold the Line, he wrote, thousands of immigrants were crossing the border illegally every day. But as a result of the initiative, "record highs" had become "record lows." Within just a couple of months the number of illegal entries was "negligible." Hoping to repeat its success, Clinton ordered the Immigration and Naturalization Service (INS) to implement Operation Gatekeeper in California in 1994, and Operation Safeguard in Arizona in 1995.

It didn't take long, though, for fence construction to become seen as a necessary complement to increased personnel. Much of the fencing along the border needed fixing, and much of the two-thousand-mile border didn't have fencing at all. When the Clinton administration announced hundreds of millions of dollars of funding for border security in early 1994, the Border Patrol announced that they were considering using some of the funding for the construction of two ten-foot-tall border walls in Arizona, where many illegal immigrants began crossing after the fortifications at El Paso and San Diego. The proposal immediately stirred controversy, pitting Republicans against Democrats, Republicans against Republicans, Democrats against Democrats, and Hispanics against Hispanics.

Debates over fence construction had become highly symbolic for Hispanics and nativists alike. To Hispanics, they were representative of the growing anti-immigrant sentiment nationally, which often was conflated with anti-Hispanic attitudes in general. Most of the insensitivity toward immigrants came from Republicans, but they needed Clinton's support to enact their restrictive immigration and border proposals. For Hispanic Republicans in particular, the wall was an affront, a diminishment of their role within the party. Even though conservatives claimed that the wall was meant only to deter illegal immigrants, Hispanic Republicans knew they could be caught up in the dragnet of discrimination as well.

For nativists, meanwhile, the border wall symbolized their desire to hold the line against cultural and demographic change. They said it was necessary to preserve the rule of law, defend national sovereignty, and prevent crime. But these were canards. The borderlands were safer than other regions, undocumented immigrants committed crimes at lower rates than citizens, and some of the louder calls for a border wall came not from border communities—which would experience the effects of illegal immigration most immediately—but from areas hundreds or thousands of miles away. What nativists really feared was the browning of the United States and the long-feared reconquest of the Southwest by Mexico.

Regardless, the path from Reagan's rejection of a border wall, to debates about the meaning of "structures," to the growing feeling among Republicans that a border wall would solve the nation's immigration problems and slow the progress of broader cultural and demographic shifts, marked a turning point in the Republican Party's relationship with Hispanics. The GOP became seen as the anti-immigrant and anti-Hispanic party. It was the party that was most responsible for the rising tide of discrimination against them nationally, making it increasingly difficult—but not impossible—for Hispanic Republicans to justify their continued support.

Puerto Rican Republicans were further aggravated by the members of Congress in their own party who opposed their bid for statehood. They ended up organizing their referendum anyway. The new prostatehood governor, Pedro Roselló, oversaw the process. Luis Ferré, now in his nineties and having gone through four surgeries in the previous year and a half, remained statehood's greatest cheerleader. In the November 1993 plebiscite, Puerto Rican voters favored commonwealth status over statehood by the slim margin of 48 to 46 percent. Even though it was the greatest show of support for statehood ever, it nevertheless demonstrated a division over the issue that caused Congress to continue sitting on its hands.

Cuban Americans had their own reasons for frustration, because they saw their favored immigrant status slipping away. For decades Cubans arriving in the United States were seen as refugees from communism. The Cuban Adjustment Act, still in effect, had eased their settlement, assimilation, and upward mobility. But this, too, was changing as a result of the end of the Cold War and growing anti-immigrant sentiment. In 1994, thousands of rafters fled Cuba for Florida, and, like those who arrived during the Mariel boatlift in the early 1980s, they were met with discrimination instead of open arms. One Cuban American employee at a nursing home in Miami said she overheard a doctor and a nurse having a laugh at their expense. "What's the Cuban national anthem?" one asked. The answer? "Row, Row, Row Your Boat." One Cuban American journalist wrote, "Cubans are facing the overt hostility that others have fought for years."

What she meant was that Cubans felt the same hostility that

Mexicans had combated for years, and continued to fight against in the early 1990s. The arrival of Cuban refugees had become a national issue, but not of the same magnitude as the challenges posed by the US-Mexico border and unauthorized Mexican immigration, which increasingly were discussed as national crises. California became the model for the Republican Party's approach to dealing with the border and illegal immigration, and many political analysts would argue that the path California Republicans chose ultimately doomed the GOP there.

FOR MOST OF 1994, A CALIFORNIA FLAG HUNG UPSIDE DOWN IN FRONT OF GLENN SPENcer's Sherman Oaks home northwest of Los Angeles, as a symbol of the extreme distress he believed his state suffered due to illegal immigration from Mexico. Along with an accountant from Tustin and a crime analyst with the Anaheim Police Department—both in the conservative stronghold of Orange County—Spencer led the grassroots campaign in support of the Save Our State (SOS) initiative. They agitated against undocumented immigration through organizations called Voice of Citizens Together and the California Coalition for Immigration Reform, whose name nodded to Tanton's Michiganbased Federation for American Immigration Reform (FAIR). When Save Our State gained hundreds of thousands of signatures and qualified for the November ballot, it became known as Proposition 187.

Explaining the need for Proposition 187, which would have denied nonemergency health care, public education, and other services to undocumented immigrants, Spencer said, "We have people who are flooding across our borders with a very high fertility rate and a very low educational level." He added, "Unless something is done, the state has nothing to face but fiscal havoc." While Spencer focused on the bill's financial impetus, the crime analyst from Anaheim highlighted Proposition 187's cultural logic. In 1991, she had walked into a social services center "with babies and little children all over the place," speaking Spanish. They qualified for the same public benefits as US citizens, she learned, which began her conversion from "political neophyte to fiery crusader" against demographic change in California. No matter how much supporters emphasized the state's inability to cover the expense, most Hispanics believed that fears of demographic change and racism were behind the bill.

Once it became evident that Save Our State had widespread grassroots support among California voters, Governor Pete Wilson, who was running for reelection in 1994, made Proposition 187 and anti-immigration sentiment more broadly the main plank of his platform. A former mayor of San Diego and US senator from California, Wilson aired inflammatory campaign commercials that placed him at the center of the immigration debate, as well as efforts to combat illegal entry. "They keep coming!" said a voice in one of his ads about undocumented immigrants. There were two million of them in California. The federal government did nothing to stop them from entering, yet required the taxpayers of California to "pay billions to take care of them." Wilson said he was fighting bravely to address the issue. He had deployed the National Guard to help the Border Patrol, sued to force the US government to control the border, and supported the denial of state services to undocumented immigrants. "Enough is enough," Wilson said as he looked into the camera.

To conservative Republicans across the country, Wilson was a hero finally taking a stand, an elected official with national aspirations at the forefront of the immigration debate. Talk radio hosts including Rush Limbaugh and G. Gordon Liddy, one of the planners of the Watergate break-in, heaped praise on Wilson. The once and future candidate for president Buchanan also lauded his efforts. In many ways, Wilson's support for the measure revived his campaign. Before he embraced immigration as his main issue, he was down by 20 points to his Democratic opponent. By late October, just a few weeks before the election, he was ahead by 10 points. As for Proposition 187 itself, it passed by a vote of 59 percent for and 41 percent against.

Hispanics in California were the group most opposed to the measure. They voted against it by a margin of 77 to 23 percent. Political scientists concluded that Hispanic support or opposition didn't necessarily break down along partisan lines. Instead, the greatest predictors of Hispanic support for Proposition 187 in the run-up to the election had to do with whether they were citizens who spoke English or noncitizens who spoke Spanish. The Hispanics who supported Proposition 187 were those who didn't think it would affect them, since they were citizens who spoke English, while the Hispanics most likely to oppose it were those who feared it would have negative consequences for them personally.

The longest-lasting consequence of Proposition 187 would be the damage it did to the Republican Party in California, in large part because of the political mobilization of Hispanics who opposed it. A majority of Californians voted for the Republican presidential candidate in every presidential election from 1968 to 1988, but never since. In the run-up to Wilson's reelection in 1994, hundreds of thousands of Hispanics marched against Proposition 187 in the largest protests in California history. Even more troubling for the Republican Party was how debates over the bill divided conservatives. Republican stalwarts Bennett and Kemp, who two years earlier had opposed the word "structures" in the Republican Party platform, wrote an op-ed for the *Wall Street Journal* that encouraged Republicans "not to support an anti-immigration movement that we consider, in the long run, to be politically unwise and fundamentally at odds with the best tradition and spirit of our party." They argued that anti-immigrant sentiment would drive Hispanics from the Republican Party.

When the head of the California chapter of the RNHA publicly criticized Proposition 187, the bill's coauthor called him a "fat burrito eater." Because of such sentiments, one Hispanic Republican called the bill a "poison pill" swallowed by the GOP. Another said Proposition 187 demonstrated Wilson's belief that Hispanics "don't work hard," and another still said Hispanics felt like they were "being maligned as a group." An editorial in the *Chicago Tribune* called Proposition 187 an "alarm bell" that alerted Americans to what immigration debates could begin to look like nationally, since Californians "often set trends for the rest of the country."

Things got worse before they got better. The Republican Party seemed to be coming apart at the seams because of debates over immigration. Moderate Republican leaders spent the rest of the nineties apologizing to Hispanics for Proposition 187.

Although a US district court ultimately declared Proposition 187 unconstitutional because immigration was the purview of the federal government and it violated federal law, it nevertheless influenced national debates in Congress about immigration and border enforcement. Operations Hold the Line, Gatekeeper, and Safeguard, and proposals to build sections of border fencing, were immediate and partial solutions intended to slow or halt undocumented immigration. But President Clinton and Republican lawmakers, in particular, believed that a more comprehensive immigration reform was also necessary. They felt compelled to respond to the "national backlash against immigrants," so they began considering various measures in 1994 that ran parallel to California's debates over Proposition 187.

REPUBLICANS WON A MAJORITY IN CONGRESS IN THE MIDTERM ELECTIONS OF 1994, AND one of the first things the Republican-led Congress did was get to work on restrictive immigration legislation. Their proposals went even farther than Proposition 187. They would deny public services not only to undocumented immigrants, but also would bar legal immigrants from collecting Social Security and welfare. Ros-Lehtinen and Lincoln Díaz-Balart, who was elected in 1992 and served alongside Ros-Lehtinen as the two Cuban Americans in Congress, rejected the proposals because of the negative impact they would have on immigrant communities in South Florida. They tried to gain permanent resident status for the tens of thousands of undocumented immigrants in their districts who faced deportation, agreeing with the head of the Miami-based Fraternidad Nicaragüense—Nicaraguan Brotherhood—that immigrants worked hard and deserved status and benefits.

But Ros-Lehtinen wasn't confident that she could do anything for them. She and Díaz-Balart had met with the commissioner of the INS, Doris Meissner, who told them that her agency planned to grant work permits only to refugees whose asylum claims had already been approved, and they were going to take into consideration that Nicaragua had been a stable democracy since the election of 1990. Ros-Lehtinen left the meeting frustrated with Meissner's response. It was a clear sign to her that Nicaraguans, too, would have to contend with the nativism spreading across the United States.

The rise of anti-immigrant sentiment coincided with the shifting politics of the post–Cold War period, as demonstrated by Meissner's remark about Nicaragua's stable democracy. For Cuban Americans such as Ros-Lehtinen and Díaz-Balart, the Cold War hadn't ended. In their view, Sandinistas controlled Nicaragua even after the democratic election of Violeta Chamorro, who replaced the Sandinista leader Daniel Ortega, and Castro still posed a national security threat, even if high-ranking military officers concluded otherwise. They supported any measure that further tightened the economic embargo against Cuba, even after the collapse of the island's greatest benefactor, the Soviet Union. They became outraged when entrepreneurs and elected officials traveled to Cuba to explore possible avenues for economic and diplomatic relations.

Therefore, Díaz-Balart was ecstatic after an early-vote approval of the Helms-Burton Act, which strengthened the embargo in part by penalizing third countries that did business with Cuba. He said, "they're going to have to choose between collaborating with Castro or participating in the US market." Opponents of Helms-Burton argued that it would increase hardships for Cubans, and that policies of isolation hadn't worked so far, so there was little reason to believe they would work then. But supporters, including Jorge Mas Canosa and the Cuban American National Foundation (CANF), felt Castro couldn't hold on much longer because of how desperate conditions on the island had grown.

When Helms-Burton passed through both chambers of Congress and was signed by President Clinton in March 1996, the European Union, Brazil, Mexico, and other US allies that did business with Cuba rejected it. Their opposition was only one of the reasons why Helms-Burton and the politics of immigration put President Clinton in a difficult position. Mas Canosa, Ros-Lehtinen, and Díaz-Balart pushed the president to maintain a hard-line stance against Cuba, which he did. But they opposed the deportation of Cubans and Nicaraguans, who were victims and veterans of earlier Cold War struggles. Therefore, they protested when Clinton lifted the ban on the deportation of Nicaraguans and negotiated a secret agreement with Castro to repatriate Cuban rafters. The INS stated that they wouldn't begin a campaign of mass deportation, but Nicaraguans and Cubans felt increasingly uncertain about their prospects of being welcomed by the United States.

DURING THE SAME MONTH WHEN CLINTON SIGNED THE HELMS-BURTON ACT, THE HOUSE OF Representatives voted on early versions of an extremely harsh immigration bill that they knew would have to be "softened" before Clinton signed it. The proposal nevertheless represented the desires of the conservative wing of the Republican Party. Among other things, the House bill, like California's Proposition 187, proposed to permit states to deny public education to the children of undocumented immigrants. It also proposed to hire five thousand additional Border Patrol officers over five years; increase penalties for "smuggling" undocumented immigrants; make it easier to deport "criminal aliens"; allow the Justice Department to return undocumented immigrants to the interior of Mexico instead of just across the border; fund the construction of a triple fence along the San Diego–Tijuana border; and block undocumented immigrants from receiving benefits including welfare, Medicaid, and Medicare.

As a disincentive to hiring foreign workers, the House bill would have required employers to pay noncitizens 110 percent what they paid US workers, and give employers in states with large immigrant populations access to a toll-free call service to verify the immigration status of potential employees. Finally, it proposed a lifetime immigration ban on anyone caught entering the United States illegally, and to grant local and state law enforcement officers the authority to detain illegal aliens and hand them over to the INS.

The 1996 law left important legacies that lasted into the twenty-first century, including making it a crime to immigrate without papers, often referred to as the criminalization of undocumented immigration; its E-Verify program, which required employers to verify the ability of their employees to work in the United States; and 287(g), which deputized local authorities to carry out immigration enforcement.

The most controversial provision of the House bill would have denied the children of undocumented immigrants access to public education. In its 1982 *Plyler v. Doe* decision, the US Supreme Court ruled that Texas could not deny education to the children of undocumented immigrants, so many opponents deemed the denial of education to be unconstitutional. But they also argued that it would make teachers "adjunct officers" of the INS and that it would make immigrants a permanent underclass, since they would be blocked from their best shot at social and economic upward mobility. As an editorial in the *Miami Herald* put it, immigrant children had "only one childhood, one chance to get an education to prepare, perhaps, for citizenship one day." Nevertheless, Republican leaders including Newt Gingrich supported the measure, noting that California spent an estimated $1.7 billion per year, and Florida $424 million, educating the children of undocumented immigrants. They should not be allowed to "live off law-abiding taxpayers," he said. His critics said he was politicizing "public anxieties over illegal immigrants."

It also became clear that the immigration bill would do little to help the growing number of undocumented Nicaraguans in Florida, despite consistent pressure from Ros-Lehtinen and Díaz-Balart to act on their behalf. Ros-Lehtinen's fears after the midterm elections in 1994 were confirmed. She wouldn't be able to help them. She had asked Clinton to grant them permanent residency, but he said no. The most the Clinton administration would concede was allowing some sixteen thousand Nicaraguans in Miami to reopen their asylum cases, to give them another shot at success if their applications had already been denied. If they were denied again, in the best-case scenario they wouldn't have access to public services, and in the worst-case scenario they would be deported. Ros-Lehtinen therefore became intensely interested in the elections in Nicaragua that year, sending members of her staff to Central America as election observers to look for evidence that the Sandinistas were meddling to prevent a free and democratic election because they were scared of losing influence. She believed that would give her further ammunition to argue on behalf of asylum applicants.

Ros-Lehtinen and Díaz-Balart at first opposed the early version of the House bill because children, Ros-Lehtinen said, didn't choose to immigrate illegally, so they didn't deserve punishment. Her argument preceded the one offered in support of DREAMers—the beneficiaries of the Development, Relief, and Education for Alien Minors Act, which granted resident status to immigrants who entered the United States as minors—in the twenty-first century. In the end, this early version of the House bill that included the denial of access to education passed despite their opposition, by a vote of 257 in favor and 163 against. The *Miami Herald* said it was a reflection of "anger," "frustration," and the "groundswell of opposition" to benefits for undocumented immigrants. It was a bill straight out of the playbook for California's Proposition 187, and then some.

The final House bill, passed on June 13, 1996, did not contain the provision that would deny access to public education to the children of undocumented immigrants. Neither did the version that passed in the Senate on July 18, nor the version signed by President Clinton on September 30, as the Illegal Immigration Reform and Immigrant Responsibility Act of 1996 (IIRIRA).

ASTONISHINGLY, ALL OF THE DEBATES ABOUT IIRIRA WERE TAKING PLACE IN THE MIDDLE of the 1996 presidential election, during a summer that featured the Olympics in Atlanta, the Republican national convention in San Diego, and the Democratic national convention in Chicago. On the one hand, the willingness of Congress to debate such massive legislation in the middle of the campaign demonstrated bipartisan support for doing something about undocumented immigration. But that didn't prevent heated partisan debate over which party did and did not support immigrants, and which did more or less to control their entry. IIRIRA's harshest provisions were proposed by Republicans, but about half of the Democrats in the House and the Senate supported it as well, and the bill was signed into law by a Democratic president. More words in the Republican Party platform were spent on the ills of unauthorized immigration and the president's failure to do anything about it (the conventions were held before he signed the law), but both party platforms recognized the need for action.

The platform approved by the Democrats said, "We cannot tolerate illegal immigration and we must stop it." It borrowed an idea oft spoken by Republicans, that the border had been so wide open that it "might as well not have existed." But then came along Clinton, who was "making our border a place where the law is respected and drugs and illegal immigrants are turned away." He had increased the Border Patrol by 40 percent. Border agents in El Paso were "so close together they can see each other." Clinton had deported thousands of undocumented immigrants, and filed felony charges against those who returned unlawfully. The far right not only moved into the mainstream of their own party, but influenced Democrats as well.

The truth was that both parties were divided. Leading Republicans, such as Ros-Lehtinen, Díaz-Balart, New York mayor Rudy Giuliani, and New Mexico senator Pete Domenici, opposed the harshest measures supported by members of their own party. The Democratic Party platform even singled out Giuliani and Domenici for their stand against the "mean-spirited and short-sighted" proposal to "bar the children of illegal immigrants from schools." Meanwhile, Kemp had opposed Proposition 187, but when he became the vice presidential pick of Kansas senator Bob Dole, who won the Republican primary in 1996 after two unsuccessful attempts, he toed the party line and said he supported "denying the children of illegal immigrants a public education." He called the shift part of his "metamorphosis."

Hispanics were divided as well. One poll showed that Hispanics in California, Texas, New York, and Florida supported reduced levels of legal immigration but opposed the building of a border wall and the denial of government services to legal or illegal immigrants. A Cuban American talk show host added that he didn't agree with "cutting off legal aliens" from receiving welfare benefits, but generally agreed with what the Republicans were "doing on immigration" because "this country can only take so many people, and you cannot just keep taking these immigrants and letting them get on welfare."

The former chairman of the RNHA of California, Xavier Hermosillo, said the push for immigration reform wasn't about preventing the entry of undocumented immigrants, but rather about discriminating against Mexicans and Mexican Americans. He said he was disgusted with the Republican Party for supporting Proposition 187, which set their party back twenty-five years. He explained how he had become a Republican when Nixon was president and the party was still a "redneck party." Over the years, he had witnessed it become more inclusive. It was "a big tent," but "now the Neanderthals have come in and cut all the ropes and the tent is caving in."

Hispanic Republicans didn't like the direction in which Republicans were headed on immigration, but they wouldn't abandon the GOP. They felt like they didn't have anywhere else to turn. State party chairmen were no longer returning their calls. Their concerns went "in one ear and out the other." But they didn't believe that Democrats represented their interests either. They still believed in small government, economic development, and "self-empowerment," so even though they were dispirited, they believed most Hispanic Republicans would "rally to the party in November." As the conservative firebrand Linda Chavez continued to insist, there was a "natural affinity" between Hispanics and the Republican Party.

THE ANTI-IMMIGRANT TIDE ROSE AT JUST THE WRONG TIME FOR THE REPUBLICAN PARTY, when the Hispanic population of the United States had grown by 20 percent in the previous five years and when many predicted that by 2020 Hispanics would represent 16 percent of the US population. Republicans, one article stated, seemed to be on the brink of "alienating" one of the "fastest-growing blocs in America just as they are emerging as a formidable political force."

And indeed, Hispanics were an increasingly formidable political force, especially because of their rush to become naturalized in the nineties. In response to increasingly discriminatory rhetoric, Proposition 187, and IIRIRA, the numbers of Latin American immigrants applying for naturalization skyrocketed. Applications for citizenship increased 500 percent in Los Angeles between 1994 and 1995. Ros-Lehtinen led a naturalization drive in Miami. On one weekend, thousands of immigrants became citizens at the same convention center where Reagan had delivered his speech on Cuban Independence Day in 1983. All across the United States, applications for citizenship were expected to increase from 555,000 to 760,000. Instead of surging patriotism, these immigrants applied for citizenship because they were afraid of being deported, losing their welfare privileges, or both.

Immigrants without papers who did not apply for citizenship didn't return home, in part because they feared that they wouldn't be able to come back to the United States if they did. They were pushed further underground. The restrictive reforms of the 1990s therefore ended the pattern of circular migration in which immigrants moved frequently back and forth between the United States and Mexico for work. It kept them trapped inside the United States. Those who did apply to become citizens hoped they would get their papers in time for them to vote in 1996, or in the midterm elections in 1998.

Unfortunately for Republicans still interested in courting Hispanics, the Republican nominee, Dole, wasn't an impressive campaigner. Many suspected that he won the nomination because he was next in line, not because he was an inspiring leader. Buchanan again entered the race and again was a formidable opponent, picking up 20 percent of the primary vote and winning four states—Alaska, Louisiana, New Hampshire, and Missouri. But he didn't win as many votes in 1996 as he did in 1992, and he did worse in almost every state with a large Hispanic population, especially California, where his 26 percent in 1992 dipped to 18 percent in 1996.

Observers of the race argued that the anti-immigrant laws passed between 1992 and 1996, combined with their increasing rates of application for citizenship, signaled that Hispanics were reluctant to support Republicans.

A campaign visit to Miami in early March 1996, just a few days before the Florida primary, was telling. Dole still had his supporters, including Ros-Lehtinen, Díaz-Balart, and Al Cardenas. Ros-Lehtinen said that Dole had "always been a friend of anti-communism" and that he "is a man of principle who has always supported Freedom." Dole paid the requisite visit to the Bay of Pigs Monument to lay a bouquet of flowers at its base as 150 onlookers sang the Cuban national anthem. He later had dinner at the famous Versailles Restaurant. Despite his performance, the *Miami Herald* pointed to troubles on the horizon. Dole spent most of his time at his vacation home in

Bal Harbour, lounging by the pool. He canceled a fund-raising event, even though his campaign was in terrible financial shape. He hadn't defined himself as a candidate and he seemingly lacked motivation, perhaps, the paper insinuated, because of his age. He was seventy-three at the time, which would have made him the oldest president ever elected were he to win.

The Republican Party's backing of statehood also seemed to have stalled after Republicans in Congress pulled their support for the bills debated between 1990 and 1992, as well as the mixed results of the 1993 referendum on the island. Republican voters in the Puerto Rico primary nevertheless supported Dole overwhelmingly, in part because they believed that Buchanan was a "threat to everything we stand for."

One of the new leaders of the Republican Party of Puerto Rico was Zoraida Fonalledas, and Puerto Rican Republicans counted on her to renew the push for statehood in advance of the next plebi-scite vote, planned for 1998. She was the granddaughter of Rafael Martínez-Nadal, president of the Senate of Puerto Rico in the 1930s and a founder of Puerto Rico's Republican Party. She attended a Ro-man Catholic school in New Jersey—the College of St. Elizabeth—then studied law at the Interamerican University Law School in Puerto Rico and received a master's degree in labor law at NYU. Af-ter returning to Puerto Rico, she worked as the assistant director of San Juan Legal Services and married fellow attorney Jaime Fonalle-das. Together they ran Empresas Fonalledas, a prominent commer-cial development company and owner of the largest shopping mall in the Caribbean, Plaza las Américas. She knew that Dole supported statehood and was gratified by the usual platform statement, that the Republican Party supported it if Puerto Ricans so desired. But she also knew that statehood would be an uphill battle.

Other signs of trouble surfaced at the convention. Pointing to the subsurface feud between them since the early 1980s, Cuban Ameri-cans expressed disappointment that Mexican Americans were given prominent speaking slots while they were not. Cuban Americans voted for Republicans more reliably than Mexican Americans and Puerto Ricans. They had given more money to the Republican Party. Yet Mexican Americans were rewarded with more appointments, despite the fact that several Mexican American appointees were disgraced. They pointed to Romana Bañuelos, whose firm was raided by the INS during her confirmation process, and Catalina Villalpando, who wound up a convicted felon. If the Republican Party continued to "slap" Cuban Americans, one author concluded, "many angry Cubans" might "punish with their silence the Republicans in the November Elections."

Punish Republicans they did. Even before the election, reporters concluded that it didn't look good for Dole. Record numbers of Hispanics were registering to vote. Members of the media misunderstood who these Hispanics were, sticking with the oversimplified idea that Cuban Americans were Republicans while Puerto Ricans were Democrats, along with "all" Mexican Americans. But they were right that Hispanics were motivated to vote because of the anti-immigrant bills passed in California and by Congress, the ongoing movement to make English the official language of the United States, and welfare reforms that affected even legal aliens. Antonio Gonzalez, president of the Southwest Voter Registration and Education Project, argued that the "Reagan-Bush era," when Republicans successfully appealed to Hispanics, was over, and now Republicans had decided to go after the "white 'bubba vote.'"

Even staunch allies such as Ros-Lehtinen, who had opposed her party's positions on immigration, welfare, and bilingual education, were disheartened. The Republican Party had "drifted" from its "natural base" of Hispanic voters. "I wish our party would be more aggressive in courting the Hispanic vote," she said, "but because of welfare and immigration reform and English-only issues we are afraid to try and solicit their support." She still believed that Hispanics could be a "gold mine" for the Republican Party, but first Republicans had to "shed our fears." In the end, Dole received the lowest percentage of the Hispanic vote since Ford in 1976. Dole was the first Republican to lose Florida since 1968, and won a paltry 22 percent of the Hispanic vote in California. The relationship between Hispanics and the Republican Party was damaged. Hispanic Republicans for decades had fought for greater influence within the party, and they believed they had made significant progress. But after the 1996 election they weren't so sure.

ONE OF THE MAIN TAKEAWAYS OF DOLE'S DEFEAT—PERHAPS THE MAIN TAKEAWAY—WAS that the Republican Party had taken the wrong stance on border control and immigration reform. President Clinton had signed IIRIRA into law, but the Republican-led Congress took the blame for it. Republicans had catered too much to the conservative wing of the party, and in doing so had alienated Hispanics, a constituency they had to appeal to more forcefully if they didn't want their party to become extinct. Ros-Lehtinen had warned them at the convention, when she delivered her veiled rebuke of the anti-immigrant sentiment pulsing within her party. "The America I know is caring," she said, "The America I know is inclusive, is nurturing."

The new governor of Texas, George W. Bush, who spoke at the convention before her, anticipated her remarks. In his speech, he said, "What we need in America is a renewal of spirit, a return to selfless concern for others." He added, "I know we can create a more decent and compassionate society," and "We Republicans and we Americans must reach out for the hands of those who are not as strong and not as fortunate as we are." Listeners could hear almost anything they wanted in his words, which didn't explicitly mention the border or immigration. But he spoke about being a good neighbor, and he called for a more compassionate society. He reiterated these sentiments in the coming years, on bigger and bigger stages.

One year after Clinton's reelection, Republicans seemed to have

taken Ros-Lehtinen's and Bush's words to heart, concluding that they
had overreached on border control and immigration. In early 1997,
many Hispanics continued to protest laws that would deny legal immigrants access to public services. Some of them were elderly folks
who had lived in the United States for decades, but for a number of
reasons had not become citizens. Public benefits such as food stamps
and Social Security were the only income they had. Once the new
law was announced, they applied to become citizens, but they didn't
know if their applications would go through before they were either
cut off or deported from the country. Hearing their complaints, the
Clinton administration responded by extending aid until late August,
but Ros-Lehtinen said these half measures would soften the blow for
only a brief period. She and Díaz-Balart both introduced bills that
would repeal some of the harshest aspects of the laws by continuing
aid to certain groups of immigrants, including those who were in the
process of applying for naturalization, or who became permanently
disabled after their arrival in the United States.

By the fall of 1997, some Republicans were still trying to push
through conservative measures such as getting rid of bilingual education. One leader of the RNHA even denied that the Republican Party
had embraced anti-immigrant positions. It was only the "negative
media" attention that had "created a perception" that Republicans
were against immigrants. Nobody denied that the Republican Party
was still engaged in a debate over whether "record rates of immigration should be slowed or continued," as one reporter described their
dilemma.

There were also fractures among Hispanic Republicans. The longtime head of the CANF, Mas Canosa, died. His passing sent shock
waves through Miami's Cuban American community and created a
power vacuum at the top of that powerful organization. The RNHA
itself seemed to be breaking apart as well, either, as different Hispanic
Republicans have suggested, because Mexican Americans and Cuban
Americans disagreed on Cuba policy; because the organization had

successfully incorporated Hispanics into the Republican Party, so it was no longer necessary; or because electoral laws had changed and grassroots groups affiliated with the RNC and DNC could no longer receive funds from the parties and as a result, had a harder time sustaining themselves. Regardless, by the mid-1990s, the RNHA no longer represented the Hispanic Republican movement as a whole, and its members articulated many different political viewpoints. Despite these fractures, it seemed that the Republican Party changed after the 1996 election in ways that were favorable to Hispanics.

Many Republicans worried that more bills like Proposition 187 would doom the party forever. These voices of moderation once again seemed to represent the majority opinion. The head of the liberal National Immigration Forum expressed his surprise by saying in late November, "Wow," the "proimmigrant wing of the party may be in control." Nobody would have predicted it even six months earlier and, he added, "no one would have believed you if you did." A few of the Republicans held up as the visionaries of this move back to the middle were leaders such as Jack Kemp, Dole's running mate, who tacked back to the center after the election; New York mayor Rudy Giuliani; and Los Angeles mayor Richard Riordan. Both Giuliani and Riordan, one reporter wrote, "publicly embraced the huge influx of immigrants" into their cities.

The Republican Party seemed to have taken a turn. They "dismantled" much of the 1996 immigration law, and the Hispanic Republicans in Congress were thrilled. IIRIRA contained some extremely harsh immigration controls, but now the Republicans in Congress wanted to restore welfare benefits, ease the threat of deportation, and provide undocumented immigrants with a pathway to citizenship. Díaz-Balart said he believed that the Republican Party had "turned a corner." He and others had worked to educate the Republican leadership, and their new posture would "break the misperception" that Republicans were "anti-immigrant, and to some extent, anti-Latino." Even Gingrich, who had supported tough border and immigration policies before the election, got in on the action and said he had turned over a new leaf. He visited Miami in late 1997, and after stating his support for Cuban and Nicaraguan immigrants, he saw banners that said "Hispanics Love Newt" and heard chants of "*Gracias*, Newt." When he returned to Washington, he attended a breakfast held for Hispanic Republicans, where he said, "If we extend *un gran abrazo*" (a big hug) "to everyone, they will extend it back to us and we will be a big American family."

But to some, this change in approach was about politics rather than a meaningful transformation. To Mark Krikorian, the head of the conservative Center for Immigration Studies, a spin-off of FAIR, the preelection positions of Republicans was the real ruse. They only "made a show of being tough on illegal immigration," he said, because they knew it would play well with conservative Republicans. But once they suffered a defeat at the ballot box, they reverted to their soft-on-immigration selves. Liberal Hispanics believed the opposite to be true, that their tough-on-immigration positions from 1992 to 1996 were reflections of the beliefs held by most Republicans, and that they had reversed course only after their election loss because they hoped to do better in the 1998 midterms and the 2000 presidential election. The head of the National Council of La Raza said, "The best we can say is that we're now getting mixed messages from the Republicans instead of a consistently ugly message." The political scientist Rodolfo de la Garza, a professor at the University of Texas at Austin, said, "It's too easy to say there has been a turnabout." He added, "I don't think the battle is over."

GEORGE W. BUSH'S CANDIDACY FOR PRESIDENT WAS TO BE A CONTINUATION OF THIS NEW turn for the Republican Party. He delivered a message of inclusiveness that appealed to moderates, political consultants, and pollsters. Two years after Dole's defeat, some conservatives still believed that Republicans shouldn't waste their time trying to recruit Hispanic voters and shouldn't bother "pandering" to the "Latino ethnic lob-bies." Even some Hispanic Republicans weren't fully on board with the GOP's shift in course. The former Reagan administration official, now columnist, Linda Chavez believed that immigration should be curtailed until the immigrants already in the United States learned English and became "Americanized." Growing numbers of immi-grants, she believed, would only make their assimilation more dif-ficult. But the majority of Republicans agreed with the head of the California Republican Party that Hispanics, including immigrants, were "the future of the conservative movement" because they were "self-sufficient, family-oriented, and entrepreneurial."

The Republican Party went back to the well and again sought the counsel of Stuart Spencer, who had helped candidates from Nixon to Reagan court Latino voters, first in California and then nationally. He said that Republicans would be committing "political suicide" if they did not "mend fences" with Hispanics. They also heeded the advice of a new counselor, Ralph Reed, the head of the Christian Coalition, who argued that Republicans should back Puerto Rican statehood to court Puerto Ricans. He noticed the growing Puerto Ri-can population not only in Florida but also throughout the US South, and believed that the best way for the Republican Party to reach them was through their churches. So he advocated bringing pastors from Puerto Rico, who could migrate easily because they were already citi-zens and who could begin organizing the Puerto Rican community religiously and politically. The Bush team, including Karl Rove and Ken Mehlman, liked Reed's idea and put it into action.

Bush had already demonstrated that he could successfully court Hispanics. He had done well among Mexican Americans in Texas during his 1994 and 1998 campaigns for governor, winning re-election with a reported 50 percent of their votes. In 1994, he de-nounced Pete Wilson's support for Proposition 187, saying he would never support such a measure in Texas. He supported bilingual education—or "English plus," as he called it—and, after winning the

election, he appointed Mexican Americans as the Texas secretary of
state, the Texas commissioner of insurance, and to the University of
Texas Board of Regents. He appointed Harvard Law grad Alberto
Gonzales to the Texas Supreme Court. On the campaign trail, Bush
talked about personal responsibility, inclusion, education, and friendship with Mexico. He said he would be a compassionate conservative, a "different kind of Republican." That is, he echoed his father
and brother. Mexican American Republicans (and many Democrats)
bought in.

Beyond his appointments and talking points, much of his appeal
among Hispanics in Texas was attributed to his personal charm and
charisma. He spoke Spanish, ate Mexican sweetbread in border cities,
and for Christmas he made enchiladas and tamales that he, unlike
President Ford, shucked before eating. Precisely because of his charm,
Bush's Hispanic critics said he was all symbol and no substance. They
pointed to persistent poverty and poor health care along the Texas-
Mexico border, as well as educational inequalities despite his attention
to education. His Hispanic supporters brushed aside these criticisms
and expected his success to carry over to his national campaign for
president. His support for Hispanics was heartfelt, they said.

When Bush launched his campaign, he gestured toward the importance of Hispanics by spending more than $10 million on out-
reach and appointing some of his closest Hispanic associates in Texas
as leaders of the effort, including Tony Garza, his secretary of state;
Raul Romero, one of his appointments to the University of Texas
Board of Regents; Lionel Sosa, the adman from San Antonio who
worked for John Tower, Reagan, and Bush's father; and Joe Fuentes,
the former attorney general of Puerto Rico whose wife's family knew
Bush from Andover. He also asked for support from longtime family
friends including Ferré, who responded, "Your message of embracing
cultural diversity" was the right one for "America and the world as we
head into the new millennium."

It could be said that Bush went on to win the 2000 presidential election because of support from Hispanics including his Mexican American troops in Texas and Puerto Rican Republicans such as Fuentes and Ferré. But more precisely, it could be said that he won because a Cuban boy's mother died in the Florida straits on Thanksgiving Day in 1999. Like Desi Arnaz's mother, Elián González's mother was trying to take her son to Miami. After she drowned, her son was picked up by two fishermen who turned him over to the US Coast Guard. They brought him to Miami, where he lived with relatives until the US government determined whether he could remain in the United States or would be returned to Cuba.

According to Bill Clinton's "wet feet, dry feet" policy, Cubans apprehended in waters off the Florida coast would be returned to Cuba, but if they made it to the mainland they could apply to become legal permanent residents. In April 2000, Attorney General Janet Reno ordered that González would be returned to his father in Cuba, since he had been pulled from the water. When his relatives in Miami defied Reno's order, a nine-day standoff began that ended when INS agents, armed with automatic rifles, forcibly entered the house where González was staying to remove him. It was after midnight on April 22, Easter Sunday. *Time* magazine printed a picture of González a few hours after he was reunited with his father at Andrews Air Force Base. The boy smiled widely. The caption said, "Papa!"

The whole episode sparked a controversy for Democratic and Republican presidential candidates alike, especially in Florida. It became a kind of litmus test for the candidates. Democrats thought González should be returned to his father, per the policy agreed upon by Cuba and the Clinton administration. In opinion polls, a majority of Americans agreed, believing it was in González's best interests to be returned to his father. Republicans took the opposite position, following the lead of Miami's Cuban exile community. To them, it would be best for González to remain in the United States, in part because returning him would mean caving to Castro. The González case became a call to action for Cuban American voters. They embraced the slogan *Libertad*

Elián (Freedom for Elián). Bush sided with them. It would be a "wonderful gesture" to give González citizenship, he said, while also extending an invitation to González's father to visit the United States so he could "get a taste of freedom," too. Bush's brother Jeb was the governor of Florida, and he fought alongside them for González to stay.

Recognizing how politically fraught the issue had become for Democrats, Al Gore, even though he initially supported the Clinton administration's decision to return González to Cuba, decided by the end of March to break ranks with his boss and other Democrats. He was the only Democratic candidate to back US citizenship for González, but his departure from the party line wasn't enough to earn the support of Cuban Americans in Florida. More than 80 percent of them voted for Bush in an election that he—now famously, with help from the US Supreme Court—won by only 537 votes from a total of more than 5.8 million cast. Some considered the González case the greatest mishandling by a Democratic president of US-Cuban affairs since Kennedy's Bay of Pigs disaster. In November, they punished Gore for Clinton's decision to return the boy. They called it *el voto castigo* (the punishment vote). The *Atlantic* ran an article about it the following spring, titled "Elián González Defeated Al Gore." From the time that one Bush left office until another entered, it had seemed that an insurmountable wall was being built between Hispanics and the Republican Party. But with the election of George W. Bush, Hispanic Republicans had their faith in the Republican Party restored.

THE FUTURE FOR HISPANIC REPUBLICANS

The first months of George W. Bush's presidency helped Hispanic Republicans feel that the Republican Party would return to an era of inclusion, as it had been under Nixon, Ford, Reagan, and Bush Sr. There were setbacks during their administrations—seemingly callous remarks by these leaders, tensions between different Hispanic groups, scandals involving leading Hispanic Republicans—but most Hispanic Republicans didn't waver in their belief that their influence and power within the Republican Party were rising. As a result, it became customary for a significant minority of Hispanics to support Republicans, and it was more remarkable when they didn't than when they did. With George W. Bush's election, it looked like this pattern would continue, and maybe even grow stronger.

On Inauguration Day, Texas Supreme Court justice Alberto Gonzales became the eightieth attorney general of the United States. It was the highest executive position ever held by a Hispanic. Bush also nominated Linda Chavez to become the secretary of labor. She withdrew after fallout from an ABC News report that an undocumented Guatemalan woman had lived with her, cleaned her home, and cared for her children. Bush also appointed representatives of the next generation of Hispanic Republicans, including Héctor Barreto, son of the founder of the Hispanic Chamber of Commerce, as administrator of the Small Business Administration; Alfonso Aguilar, who became the first chief of the US Office of Citizenship; and Daniel Garza, who became the deputy director of external and intergovernmental affairs in the Office of the Secretary at the Department of the Interior, then associate director of the Office of Public Liaison.

They were the "Bushies," and they became leaders of the Hispanic Republican movement in the early twenty-first century. Barreto became chairman of the Latino Coalition, an advocacy group for Hispanic small-business owners. Aguilar became president of the Latino Partnership for Conservative Principles. Garza became president of the Koch-funded LIBRE Initiative, which promoted free enterprise in Hispanic communities. Their service in the Bush administration was a pivotal moment in their careers. They all remembered a photograph Bush arranged with his Hispanic appointees standing in front of the White House. They said the line stretched from one end of the building to the other.

The picture was taken in early September 2001, right before Hispanic Heritage Week, when Bush hosted his Mexican counterpart, Vicente Fox, the first leader of the National Action Party (Partido Acción Nacional, or PAN) ever elected as president. One of Bush's priorities was improving the relationship with Mexico and working—again—on immigration and drug trafficking.

Bush and Fox have both recalled that they were making good progress on an immigration plan that included more guest worker visas and a reduction in tensions over border issues. A statement by Bush's secretary of state, Colin Powell, summed up Bush's approach. "Our common border is no longer a line that divides us," he said, "but a region that unites our nations, reflecting our common aspirations, values and culture." Hispanics appreciated the sentiment. It sounded like something Reagan or Bush's dad would have said.

The momentum Bush and Fox had built ground to a halt with the terrorist attacks on September 11, 2001. Not only were the Bush administration's attentions diverted, but also many Americans reverted to a state of anti-immigrant hysteria and calls for heightened border restrictions. Islamic terrorists were the new enemy, but immigrants of all backgrounds suffered. Bush's replacement as the governor of Texas, Rick Perry, used the new War on Terror to once again make the case for restricting immigration and reinforcing the southern border. He said that the United States was vulnerable to attacks by terrorists who could easily enter that country carrying missiles. Perry, Trump's future energy secretary, repeated this idea for more than a decade, saying in 2014 that it was a very real possibility that ISIS crossed into the United States through Mexico.

The vast majority of Hispanics supported Bush's War on Terror—90 percent of them, or one percentage point more than the 89 percent of non-Hispanic whites who supported it. Their support was part of a general pattern of patriotism during wartime, and many had family members in the military. But they had been drawn to Bush in the first place because of his moderate stances on immigration, border policies that favored cooperation over conflict, and what they saw as his genuine empathy with Hispanics.

In his 2004 reelection, Bush set a new standard for success among Hispanic voters, earning more than 40 percent of the Hispanic vote according to most reports. Bush benefited from the patriotic zeal of most Americans in the years after the 9/11 terrorist attacks, but also from his high-level Hispanic appointments and focus on issues that Hispanics said mattered to them, including education and jobs.

BUT WHEN IMMIGRATION MADE ITS WAY BACK ONTO THE AGENDA IN BUSH'S SECOND TERM, the most serious proposals, as during the early 1990s, focused on border enforcement instead of leniency toward undocumented workers. The Secure Fence Act of 2006, for example, provided for the construction of hundreds of miles of fencing at a cost of more than $2 billion.

The wave of anti-immigrant sentiment after 9/11 was compounded by the financial crisis of 2008, with the result that John McCain during his run for president that year sounded more like Pete Wilson and Buchanan than Bush. In one campaign ad, McCain yelled, "Build the danged fence!" McCain supporter Fernando de Baca, the onetime special assistant to President Ford, was the chairman of the Bernalillo County Republican Party in New Mexico. He was forced to resign when he exposed the ugly side of historical tensions between Hispanics and African Americans. He said, "The truth is that Hispanics came here as conquerors," while "African-Americans came here as slaves." He should have stopped before he even began, but he continued, "Hispanics consider themselves above blacks," so they "won't vote for a black president," referring to Illinois senator Barack Obama, who defeated Hillary Clinton in the Democratic primary campaign.

One of the less high-minded questions asked about Obama's candidacy was whether the United States was ready for its first black president. De Baca answered that question from the perspective of one Hispanic Republican. But as a bald expression of some of the racial animus that greeted Obama, it was a sign of the times.

Four years later, in a primary campaign debate in Florida, Republican candidate Mitt Romney, months before he secured the nomination, called for the "self-deportation" of immigrants, a policy of making life in the United States so difficult for them that they would decide to leave on their own. Just a couple of months earlier, one of his primary opponents, the African American businessman Herman Cain, had suggested that an electrified border fence could be a good way to curb undocumented immigration. The argument wasn't about wall or no wall, structure or no structure, but about whether the fence would have lethal powers.

The sense of hope for an inclusive Republican Party that Hispanic Republicans had felt as a result of Bush's election was fading quickly. In retrospect, it seemed representative only of Bush's personal appeal and his family's relationships with Hispanics. At the national level, the Republican Party in the twenty years between 1992 and 2012 was undeniably more inhospitable to Hispanics than the Republican Party in the twenty years between 1972 and 1992 had been. Hispanic Republicans supported Bush, but the strong currents of nativism and xenophobia flowing beneath his presidency were upsetting. It had always been true that Hispanic Republicans cared about issues other than immigration and the border—foreign policy, jobs, education, health care, homeownership—and it had always been true that Hispanic Republican opinions on immigration and the border were not uniform. But never before did they have to state so forcefully their support for the Republican Party despite its hard-right turn on these issues, as a justification of how they could continue to vote for Republicans in a manifestly unwelcoming environment.

THE 2016 ELECTION MIGHT HAVE ENDED UP LIKE THE 2000 ELECTION. AFTER ANOTHER four-year period of anti-immigrant sentiment and legislation—SB1070 in Arizona, HB56 in Alabama, Cain's electric fence, Romney's calls for self-deportation—followed by a devastating loss in the 2012 election, the Republican Party might have returned to the inclusiveness of the Bush years. That's what many Republicans thought would happen after Romney's defeat. That's what RNC Chairman Reince Priebus expected to happen.

Priebus believed that the Republican Party would learn a tough lesson after the elections of 2008 and 2012. Perhaps McCain's bowing to anti-immigration conservatives in 2008 didn't much matter, since his prospects dimmed as news about the unfolding financial crisis became increasingly grim. But Romney's odds seemed better; many dynamics pointed in his favor, including not only Republican disapproval of Obama but also the success of Hispanic Republicans at the state level in 2010, among them Nevada governor Brian Sandoval, New Mexico governor Susana Martinez, and Florida senator Marco Rubio. Romney ended his campaign in Boston on the morning of the election, confident that he was headed toward victory. When he lost, a campaign adviser said, Romney was "shellshocked." Republicans spent the next several months soul-searching as they watched Obama's second inauguration from the sidelines.

Particularly stinging was that Hispanics in key battleground states such as Colorado, Florida, New Mexico, and Virginia seemed to have tipped the scale in Obama's favor. To help Republicans process and learn from their defeat, the RNC, in March 2013, released its now-famous hundred-page *Growth & Opportunity Project* report, also known as the "autopsy report." It offered conclusions about the reasons for Romney's loss and the shortcomings of the Republican Party more generally. Zoraida Fonalledas, national committeewoman for the Republican Party of Puerto Rico, was one of the five members chosen to participate in this project. She worked, in particular, on the sections that related to Hispanics throughout the United States and its territories, including Puerto Rico.

The sections that got the most attention had to do with immigration and how the Republican Party should reach beyond its base to be more inclusive of Hispanics and others. On immigration, the report concluded, "we must embrace and champion comprehensive immigration reform," including a path to citizenship. "If we do not," the report stated, "our Party's appeal will continue to shrink to its core constituencies only," by which it meant non-Hispanic whites. In its analysis of the report, the *Washington Post* anticipated a "backlash from those appealing to anti-immigration exclusionists."

The report noted that "Hispanics made up 7 percent of the electorate in 2000, 8 percent in 2004, 9 percent in 2008, and 10 percent in 2012." As a percentage of American voters, they were steadily growing. But as always, their influence went beyond the number of potential voters. The views of undocumented immigrants were reflected in the votes of their family members who were citizens, and Puerto Ricans and Latin Americans helped shape the views of their relatives who could vote. The Republican Party had to find a way to reach Hispanic voters or Republicans would be left behind. The report quoted George W. Bush, the gold standard for the successful recruitment of Hispanics. "Family values don't stop at the Rio Grande and a hungry mother is going to try to feed her child," he had said. Such expressions, the report concluded, made Hispanics willing to listen to the other proposals he asked them to support.

The report also cited Dick Armey, a conservative Republican from Texas, House majority leader, and leading voice of the emergent Tea Party movement. He, too, understood the Republican Party's dilemma. "You can't call someone ugly and expect them to go to the prom with you," Armey had said, adding, "We've chased the Hispanic voter out of his natural home." To begin repairing the relationship, the report proposed the creation of an RNC Growth and Opportunity Inclusion Council, which would be tasked with outreach to Hispanics in addition to African Americans, Asian Pacific Americans, and Native Americans.

In the wake of the RNC's autopsy report, it made perfect sense that candidates such as Jeb Bush and Rubio were early front-runners to win the nomination in 2016. Jeb's appeal was clear. He was the governor of an important swing state. He selected the powerful Cuban American lawyer and lobbyist Al Cardenas as a senior campaign adviser and fund-raiser. Jeb spoke Spanish, was married to a Mexican American woman, had valuable Latin America experience, and had spent decades cultivating Hispanics in Puerto Rico, Florida, along the US-Mexico border, and in the increasingly purple mountain West. He also had a campaign slogan—Jeb!—that simultaneously marked him as independent of his brother and father and lent itself to lighthearted mockery in the twitterverse. After the *New York Times* reported that Jeb had listed himself as "Hispanic" on a 2009 voter registration application, many joked that Jeb was actually pronounced "Heb."

Meanwhile, Rubio was the first Hispanic to run for the Republican nomination since Benjamin "Boxcar Ben" Fernandez. Bill

Richardson, whose mother was born in Mexico, ran in 2008, but as a Democrat. Like Jeb, Rubio was from the critical swing state of Florida. He was a Cuban American who channeled the Cuban exile, Cold War, anti-Castro origin story. In interviews, he told Florida reporters that his family had come to Miami after the Cuban Revolution in 1959. When the *St. Petersburg Times* presented naturalization records demonstrating that his family in fact arrived in 1956, before the Cuban Revolution, it became an embarrassing fact to be explained, since it undermined an important aspect of the story about his political evolution, that his family had been forced out by Castro.

An older generation of Cuban Americans in Florida, whose political identities got forged in the crucible of opposition to Castro, nevertheless continued to embrace him as one of their own. They sometimes called him *joven viejo* ("young fogey," as the *New Yorker* article about Rubio translated it), evoking his kinship with his politically powerful elders. They were his mentors, and he connected with them more than he connected with Cuban American youth who didn't grow up during the Cold War and who had supported Obama.

Ted Cruz ran, too, but his Hispanic Republican credentials were a little bit different. He was from Texas and half Canadian, half Cuban. Cruz's father had compelling stories about his escape from Castro's Cuba, and the important lessons about freedom that he carried with him for the rest of his life and recounted to his son, a Princeton University and Harvard Law School grad, associate attorney general at the Department of Justice, member of the George W. Bush administration (another "Bushie"), US senator from Texas, and presidential candidate. Nevertheless, Cruz didn't grow up in the same mix of South Florida's Cuban exile politics that Rubio did. More important, Cruz held anti-immigration positions that were out of step with what many Hispanic Republicans believed. It probably didn't help that Rubio mocked his Spanish in a primary debate in South Carolina.

Indeed, it seemed as though the next Republican nominee would be the candidate who most appealed to Hispanic voters. In addition to Bush and Rubio's efforts, the Democratic nominee, Hillary Clinton, highlighted her experience in the early seventies canvassing for George McGovern in South Texas. She also noted that some two thirds of Hispanics supported her instead of Obama in the 2008 primaries. It seemed even more likely that Hispanics would be key to deciding the next president after Donald Trump, in a speech announcing his candidacy, called Mexicans rapists, drug dealers, and thieves. He went on to say that Mexico would pay for the border wall he wanted to build. Then he said that Judge Gonzalo Curiel's Mexican ethnicity made Curiel biased against him. Then he gave a divisive speech in Phoenix in which he detailed gruesome crimes committed by undocumented immigrants against US citizens, whose mostly white relatives joined him onstage. It went on and on.

Trump's nativism was much like Buchanan's had been twenty years earlier. In the 1990s, nativism appealed to more Americans than many at the time suspected—about 20 to 25 percent of them, if we can judge by the support Buchanan received in 1992 and 1996. Many Americans were again drawn to a nativist Republican in 2016, when Trump reprised Buchanan's "America first" slogan. As Trump won primary after primary, it became clear that his divisive message resonated, and the Nixon-Ford-Reagan-Bush eras of Republican Party seemed to be over. The profile of the segment of Trump's base that got the most attention in many ways resembled Buchanan's: non-Hispanic white voters whose anti-immigrant views stemmed from their fears that they were being replaced by the brown hordes that got all the benefits of living in the United States while they got left behind. But as we've learned, his base also included more suburbanites, women, and, yes, Hispanics than many imagined possible.

THE HISPANIC REPUBLICAN MOVEMENT OF AN EARLIER ERA WAS, IN FACT, COMING TO AN end, or entering a new phase, as many of its founders passed away. A new generation of Hispanic Republicans succeeded them. The RNHA also had new life, as some of its chapters were reconstituted, and others were established for the first time. But it wasn't clear what a national Hispanic Republican movement would look like going forward.

Benjamin Fernandez died in the first year of the new millennium, on April 25, 2000. Perhaps more than anyone else, Fernandez defined the Hispanic Republican movement that ultimately led a reliable base of Hispanics to vote for Republicans. Yet hardly anyone noticed his passing. There wasn't a single obituary in a major paper for the man who had cofounded the RNHA and was the first Hispanic to run for president. It took almost five months for a newspaper to commemorate him at all. Finally, on September 15, the first day of Hispanic Heritage Month, Fernandez's longtime acquaintance Carlos Conde wrote an article for the *El Paso Times* that eulogized him as a "memorable presidential candidate."

Conde, a journalist from Texas and fellow Brown Mafia member, said that Fernandez was an "original Hispanic Republican," one of the first to clearly state that Hispanics were Republicans because of their "self-reliance, free enterprise, and family values." According to Conde, Fernandez helped usher in a time when "a Hispanic no longer need feel like a stranger in this land." He was a millionaire with a "wunderkind reputation" who helped other Hispanics become rich, too. A Republican to the end, Fernandez attended a Las Vegas meeting of the RNHA the month before he died. Nobody even knew he had colon cancer. "Reportedly, he died alone in a hospice," Conde wrote.

Others of Fernandez's generation passed as well, but received greater tributes. Fernando Oaxaca died four years after Fernandez, in June 2004. He cofounded the RNHA and then served in the Nixon, Ford, and Reagan administrations. He led the National Association of Latino Elected and Appointed Officials (NALEO), the Hispanic Council on Foreign Affairs, and the Mexican American Opportunity Foundation. An obituary in the *Los Angeles Times* quoted one of Oaxaca's colleagues—the editor of HispanicVista.com, for which

Oaxaca wrote a weekly column—who said that Oaxaca was "first and above all else a staunch Republican, who could not be swayed from his mission of bringing the Republican philosophy to the Latino community, and while at it, became one of the most ardent defenders of Latino rights."

Two more lived into the Trump era, and witnessed what became of the movement they helped to found.

The death of Romana Acosta Bañuelos in January 2018 received national attention. Obituaries appeared in the *New York Times* and the *Los Angeles Times*. The Los Angeles paper said of Nixon's former treasurer of the United States, "In a career that stretched from a small Arizona town to the heights of the business world," Bañuelos "helped open doors" to Latinos in America. At the time of her passing, the company she founded, Ramona's Mexican Food Products, was still a family business and earned more than $10 million every year. A picture of her with President Nixon and the "first printouts" of dollar bills with her signature on them still hung in Ramona's headquarters in Gardena, California.

Manuel Luján Jr. died on April 25, 2019, nineteen years to the day after Fernandez's passing, but he lived long enough to offer his assessment of Hispanic and American politics in the twenty-first century. The Democratic governor of New Mexico, Michelle Luján Grisham, a former head of the Congressional Hispanic Caucus, said he was the "picture of a statesman." Luján had already begun to lament that the country had become so divided. He said that if politics in the sixties, seventies, eighties, or early nineties had been like politics in the early twenty-first century, he never would have served. "Back then," he said, perhaps nostalgically, "Republicans and Democrats all got along." But Washington, Luján concluded, was "no longer a friendly place."

Still others from the first generation of Hispanic Republicans lived on, promoting Hispanic political involvement as they had done for all of their adult lives. Albert Zapanta headed the US-Mexico Chamber of Commerce. Linda Chavez worked at a Washington, DC, think tank. The former chairman of Nixon's Cabinet Committee on Opportunities for Spanish-Speaking People, Henry Ramirez, organized an event at the Richard Nixon Library in Yorba Linda, California, to discuss Nixon's efforts to include Hispanics in the Republican Party. Ramirez starred in another event there after the publication of his memoir, *A Chicano in the White House: The Nixon No One Knew*, which offered another fawning account.

The memories they've shared were a kind of eulogy for the movement they built. They had cultivated a loyal base of Hispanic Republicans who would give any Republican candidate a greater share of the Hispanic vote than any time before Republicans began courting Hispanics more than half a century earlier. Hispanics had demonstrated that they could play decisive roles in state and national elections, which gave them political clout. They looked back on their accomplishments with some satisfaction, but also expressed some unhappiness with the current situation.

BEFORE THE 2016 REPUBLICAN NATIONAL CONVENTION, LIONEL SOSA WROTE IN HIS HOMEtown paper, the *San Antonio Express-News*, that he would leave the Republican Party, to which he had belonged since the 1950s, if Trump were the nominee. He didn't acknowledge Trump as a "real conservative." He said Trump was "just a shark, a self-promoter out to see how far his out-of-control ego can take him." He was sad about what the resonance of Trump's message said about the state of the GOP. "Instead of 'Tear down this wall'"—a reference to Reagan's famous speech in Berlin—"The party promotes a new and bigger wall." A "thousand points of light"—a phrase Bush used in his 1988 nomination acceptance speech, as praise for volunteerism—"has been replaced by a thousand points of anger." In the fall, closer to the election, Sosa told reporters that he would vote for Clinton.

During the general election campaign, after more divisive remarks, some of Trump's Hispanic supporters—including several members of his Hispanic Advisory Council—abandoned him, even if they didn't make the same decision Sosa made to leave the Republican Party. Jacob Monty, a lawyer from Texas, said Trump "used us as props." Ramiro Peña, a pastor from Texas, said the Hispanic Advisory Council had always been a "scam." Alfonso Aguilar withdrew his support after the candidate's speech in Phoenix in August 2016. He had hoped that Trump would pivot away from his candidacy announcement ripping Mexicans and his comments about Judge Curiel, but his anti-immigration speech in Phoenix made clear that it wasn't going to happen. Aguilar said he gathered from that speech that Trump's position was "either self-deport or be deported."

Other leading Hispanic Republicans stuck with Trump, hoping they might be more influential with a President Trump if they remained by his side instead of jumping ship. For some, Trump was their favored candidate from the beginning. In the Puerto Rico primary, for example, the restaurateur John Regis supported Trump over Rubio, who won the votes of Puerto Rico's delegates. Regis said he wanted to support the winner rather than offer symbolic support to Rubio just because he was Hispanic.

Very few of the Hispanic Republicans I talked to said that they would vote, or had voted, for Hillary Clinton. Sosa was the only one. Even if Hispanic Republicans didn't love Trump, they weren't prepared to turn their backs on him or the Republican Party. They had been Republicans for most or all of their lives, so to vote for a Democrat was against their nature. Trump has been a polarizing figure for Hispanic Republicans, as for everyone else, but he hasn't turned many of them away.

As Daniel Garza put it, he wasn't going to let one man—Trump—ruin the movement that he and others before him had built. Garza's idea that Hispanics would vote for the Republican Party because of their conservative principles, even if they didn't like Donald Trump marks a 180-degree reversal compared with the position they had taken some forty years earlier, when they began to articulate their support for Ronald Reagan. After Reagan announced his candidacy, Hispanic Republicans made the case to other Hispanics, Republicans and Democrats alike, that until recently the Republican Party hadn't cared about them, but in Reagan they had a man who understood them. They should vote for the man, not the Republican Party.

This shift—from man-not-party to party-not-man—helps explain why even a candidate such as Trump won almost 30 percent of the Hispanic vote. The pollsters at Latino Decisions, who worked for the Clinton campaign, expected Hispanic voters to reject Trump by historic margins, leaving them and everyone else shocked when he won a little less than the usual third of the Hispanic vote, and performed better than the establishment candidates McCain and Romney before him. One main aim of this book has been to make that result seem less surprising when seen in in historical perspective, and to explain how it could happen again.

AS IN THE RUN UP TO THE 2016 ELECTION, FEW HISPANIC REPUBLICANS HAVE STATED that they won't support Trump in 2020. Linda Chavez told me that she has worked with Bill Kristol to find a Republican to challenge Trump in the 2020 primaries. Another longtime conservative Mexican American I talked to said, "I'll be blunt, I think Trump is an idiot." He said he was sad for the country and for the Republican Party. A longtime Puerto Rican Republican told me he thought Trump was bad for the Republican Party. The Hispanic Republicans who were most critical of Trump wanted their remarks to remain off the record.

But others have been more supportive. At the Latino Coalition's annual leadership summit in Washington, DC, in March 2019, several speakers repeated the idea Trump was the best thing that had happened to Hispanics in a long time. Ted Cruz told a story about the fight for American freedom that drew a straight line between Castro, the embroiled Venezuelan socialist leader Nicolás Maduro, and the rise of socialism in the United States, embodied by the first-term representative from New York Alexandria Ocasio-Cortez. The danger was very real, and only Republicans would protect us, he argued. A South Florida Puerto Rican I spoke with echoed Cruz. Could I believe, she wondered, that it's 2019 and we're facing the real threat of socialism in this country?

One Puerto Rican Republican told me that if it weren't for Trump's big mouth, he'd have the highest approval ratings of any president in history because of his accomplishments. He added that even if the president continues to make anti-immigrant and anti-Hispanic statements, it wouldn't matter because the RNC had been spending millions of dollars for more than twenty years on grassroots Hispanic outreach, so whatever Trump said about immigrants and border walls, these grassroots campaigns had convinced Hispanics that the Republican Party would put them to work and help them buy homes.

Moreover, most of Trump's Hispanic supporters don't believe that Trump is racist. He's an unrefined speaker, and sometimes careless with his words, as they could be as well—here was an expression of empathy with him—but they didn't believe that made him a bad person. There was the president's public persona, and then there was what he did behind the scenes, entrusting Hispanics in his administration, such as Jennifer Sevilla Korn, special assistant to the president in the Office of Public Liaison, to answer their calls. Such grassroots organizing over a long period of time and maneuvering within the administration are two more reasons why Republican candidates still win a significant share of Hispanic votes, even if Hispanic voters disagree with a particular candidate's positions or style.

HISPANIC REPUBLICAN SUPPORT FOR AND OPPOSITION TO TRUMP POINT TO AN UNCERTAIN future for the Hispanic Republican movement. Hispanic Republicans would, of course, prefer that their party be more inclusive, and openly so, instead of quietly, through behind the scenes pressure. But it's far from certain that they will leave the Republican Party if that doesn't happen. They've built a reliable base of Hispanic Republican support that has been able to withstand attacks from within the party, and even from party leaders. But they cannot be sure that this loyal base won't begin to crumble in the future.

There has been dissent among Hispanic Republicans in the past, and there will be more going forward. Critiques by Hispanic Republicans are yet to be reflected by dramatic shifts in how they vote, but doubts about their place in the Republican Party, fears that Republicans have ignored them, and the assault by Republican leaders today suggest that the foundation may falter at some point. But then again, it might not. The main lesson of history—its very definition—is that things change over time. We just don't know the direction of the change.

Even as the Trump administration has tried to slow demographic change by curtailing illegal and legal immigration, the United States, for now, is still hurtling toward a majority-minority future. Hispanics are still poised to play an important role in the United States. Electorally, that could happen in 2020 or beyond. It's only a matter of time. But one important lesson of the Hispanic Republican movement is that Hispanic political influence won't necessarily lead to the Democratic, liberal majority many have imagined. This book could become a requiem for a moment of inclusion that seemed possible from the 1960s through the 1980s, and for a brief period in this century, but then ended. Or it could be a story about how Republicans maintained power despite the changing demographics of our country, in part thanks to support from Hispanics.

It used to seem that an aging non-Hispanic white base and the growth of the Hispanic population would force the Republican Party to choose one group or the other—and that if it chose the former, it might die. But given the recent shake-up of American politics and how dug-in partisanship has become, we can't be certain. Regardless, we should refocus political arguments on policy rather than what too frequently gets described as Hispanic nature. Hispanics themselves and the political parties who seek their support have to account for the totality of their individually and communally held beliefs, especially on issues that have most marked their history in the United States, or in Latin America and the United States. I write this for loyal Hispanic Republicans, too, who are curious about the history of the movement they're participants in, and who've long been committed to the Republican Party, but who may, at this very moment, be imagining the different paths they might follow in the future.

ACKNOWLEDGMENTS

This is the second book I've written inspired by my grandfather Geraldo Cadava Jr. I thank him for his example and for everything he is. I thank the rest of my family, too, including my parents, grandparents, siblings, aunts, uncles, cousins, and friends for all their support over the years.

Archivists and librarians in Puerto Rico, Mexico City, Arizona, California, New York, Texas, Florida, Michigan, Kansas, New Mexico, and elsewhere have been infinitely patient and helpful. Heidi Iverson at New Mexico State University, Fabiola Barreto at the University of Texas at Austin, and Alex Moisa and Isabella Soto at Northwestern University helped me with research. Dozens of individuals who were actors in the history I've written generously shared their memories with me. The efforts of these archivists, librarians, researchers, and interviewees to recover and preserve the past have been indispensable.

Students and colleagues at universities across the United States listened to me talk about this project when I didn't know what it was about. Thanks to audiences at Duke, Florida International University, Grinnell, the Huntington Library, Marquette, Michigan State, New York University, Stanford, Texas A&M, the University of Chicago, the University of Southern California, and the University of Virginia.

An extra special thanks to the Nicholas D. Chabraja Center for Historical Studies for inviting me to workshop a draft book proposal;

to the Stanford Humanities Center's workshop on Critical Orientations to Race and Ethnicity, for inviting me to workshop several chapters; and Nicole Hemmer, Karl Jacoby, and Stephen Pitti, who read the entire manuscript and offered crucial feedback. Thanks also to Lina Britto, Brodwyn Fischer, and Emilio Kourí for reading passages on US–Latin American relations.

A shout-out to students in my courses on Latino Conservatism and Conservatism in the Americas, who discussed my research with me in the project's very early stages. And an extra special thanks to the students I taught in a first-year seminar on Watergate in the fall of 2018, and in a lecture course on Latino History in the spring of 2019, who read parts or all of the book and asked important questions.

I wrote most of this book while at Northwestern, where I've been extremely lucky to work for the past decade. Colleagues, administrators, and students there try every day to make the university a place where we can all thrive and do our best for the world. I had the pleasure of completing the book at the Stanford Humanities Center, and I thank everyone there for the financial support, quiet space, and energetic community they provided. I also thank the Alumnae of Northwestern for their grant to do research in Puerto Rico and Florida, and the Alice Kaplan Institute for the Humanities at Northwestern for helping me pay for image permissions.

For believing in this book and helping me bring it into the world, I thank Edward Orloff at McCormick Literary and the whole team at Ecco.

Most important, I treasure the family I've built over the past ten years with K & O. These have been the best years of my life because of your presence. I love you, and I dedicate this book to you. You make all my days better.

NOTES

INTRODUCTION

x Beginning in the late 1960s: Rigueur, *The Loneliness of the Black Republican*.

xi In their 2014 book: Barreto and Segura, *Latino America*, 233.

xi The National Association: "NALEO: Latino Turnout in Election 2018 Increased 50 Percent from Previous Midterm Election," Latino Rebels, April 25, 2019, https://www.latinorebels.com/2019/04/25/latinoturnout 2018/, accessed on June 18, 2019.

xi Political analysts generally agreed: Jens Manuel Krogstad, Antonio Flores, and Mark Hugo Lopez, "Key Takeaways About Latino Voters in the 2018 Midterm Elections," Pew Hispanic Center, November 9, 2018, https://www.pewresearch.org/fact-tank/2018/11/09/how-latinos -voted-in-2018-midterms/, accessed on June 18, 2019.

xii Ted Cruz did a little better: Ibid.

xii Rick Scott did a lot better: Salena Zito, "Rick Scott Chased the Hispanic Vote and Got It," *Washington Examiner*, January 11, 2019, https:// www.washingtonexaminer.com/opinion/op-eds/rick-scott-chased-the -hispanic-vote-and-got-it, accessed on June 18, 2019.

xii "Thank you," Trump tweeted: Donald Trump Twitter feed, January 22, 2019.

xii Even after the devastating: Asher Stockler, "Parents of Baby in Viral El Paso Photo with Trump Were Supporters of the President," *Newsweek*, August 9, 2019.

xiii In the long run, even if growing numbers: David Byler, "Trump's Support Among Hispanics and Latinos Is Real. Don't Assume It Will

Fade," *Washington Post*, March 5, 2019, https://www.washingtonpost
.com/opinions/2019/03/05/trumps-support-among-hispanics-latinos
-is-real-dont-assume-it-will-fade/?utm_term=.2dc47a409a9d, accessed
on June 19, 2019.

xiv Before Mitt Romney's loss: Micah Cohen, "Why Arizona Isn't a
Battleground State (and Why It May Be Soon)," *FiveThirtyEight*,
at *New York Times*, October 23, 2012, https://fivethirtyeight.blogs
.nytimes.com/2012/10/23/why-arizona-isnt-a-battleground-state-and
-why-it-may-be-soon/, accessed on June 19, 2019.

xiv Right after that election: Ted Cruz in Ryan Lizza, "The Party Next
Time," *New Yorker*, November 11, 2012.

xiv Even though Hillary Clinton lost: Roberto Suro, "Here's What
Happened with the Latino Vote," *New York Times*, November 8, 2016,
https://www.nytimes.com/interactive/projects/cp/opinion/election
-night-2016/heres-what-happened-with-the-latino-vote, accessed December 20, 2019.

xv Four years earlier: "The Hispanic Vote in the 2008 Election," Pew Hispanic Center, November 5, 2008, https://www.pewhispanic.org/2008
/11/05/the-hispanic-vote-in-the-2008-election/, accessed on June 18,
2019.

xv Latinos are "progressive or liberal": Segura, "Latino Public Opinion
and Realigning the American Electorate," 100–105.

xvi But survey findings: Hero, *Latinos and the U.S. Political System*,
64–66.

xvii In 2010, the former Democratic Senate majority leader: Shane D'Aprile,
"Reid: I Don't Know How Any Hispanic Voter Could Be a Republican," Hill, August 11, 2010, https://thehill.com/blogs/ballot-box/senate
-races/113687-reid-i-dont-know-how-anyone-of-hispanic-heritage-can
-be-a-republican, accessed on June 19, 2019.

xvii "The Hispanic-American of the Southwest": Manuel Ruiz, "How to
Get the Mexican American Vote," September 28, 1968, box 1, folder 6,
"Drafts of MR's Writings and Speeches, 1942–1980," Ruiz Papers.

xvii Fifteen years later: Benjamin Fernandez in George Will, "Benjamin Fernandez: A Natural Republican," *Washington Post*, August 23, 1979, A21.

xvii "The dedication to principles": Ronald Reagan, Proclamation 5084
of August 25, 1983, National Hispanic Heritage Week, 1983, *Federal
Register*, August 30, 1983, 39207.

XVII He famously told: Cathy Booth Thomas, "25 Most Influential Hispanics in America: Lionel Sosa," *Time*, August 22, 2005, http://content.time.com/time/specials/packages/article/0,28804,2008201_2008200_2008222,00.html, accessed on June 18, 2019.

XVII In 1989, President Bush said: George H. W. Bush, "Message on the Observance of National Hispanic Heritage Month, 1989," September 11, 1989, *Weekly Compilation of Presidential Documents, September 18, 1989* (Washington, DC: General Services Administration, 1989), 1357.

XIX As Benjamin Fernandez testified: Fernandez testimony, November 8, 1973.

XXII One Hispanic Republican: Garza interview, May 17, 2016.

XXII Moreover, political scientists have argued: Barreto and Segura, *Latino America*, 44–49.

CHAPTER 1: BECOMING REPUBLICAN

3 Desi Arnaz's uninspiringly titled autobiography: Arnaz, *A Book*.

4 One prominent exile: Elaine del Valle, "The Late Desi Arnaz Honored for Bay of Pigs Contributions," *Miami Herald*, May 8, 2004, 3B.

4 He was the cochair: "Desi Arnaz—Ricky Ricardo, TV Mogul—Dies at Home," *UPI NewsTrack*, December 2, 1986.

4 Later, Desi gave: Del Valle, "The Late Desi Arnaz Honored for Bay of Pigs Contributions."

5 They formed groups: Santillan and Subervi-Vélez, "Latino Participation in Republican Party Politics in California," 288.

6 Lionel Sosa, a Mexican American: Lionel Sosa, "Bidding Farewell to My Grand Old Party," *San Antonio Express-News*, June 19, 2016, F3.

6 His father ran a laundry: Sosa, *The Americano Dream*, xi–xii.

6 Eisenhower's acceptance speech: Dwight Eisenhower 1952 nomination acceptance speech, https://www.c-span.org/video/?153098-1/dwight-eisenhower-1952-acceptance-speech, accessed on June 21, 2019.

7 As a Republican himself: Kenneth Burt, "Latinos con Eisenhower: 'Me Gusta Ike' or 'Yo Quiero Ike,'" KennethBurt.com, February 6, 2010, accessed on November 4, 2019.

7 From the Civil War: Gonzales, *Política*, part 5; *Hispanic Americans in Congress*, 69, 193.

7 Many Hispanics therefore credited: Santillan and Subervi-Vélez,

"Latino Participation in Republican Party Politics in California," 287.

8 Estimates of the size: Bean and Tienda, *The Hispanic Population of the United States*, 36–55.

8 Mexican Americans played: Lozano, *An American Language*, "Part 1: A Language of Politics," chapters 1–5.

8 A Republican governor running for reelection: Robert Benitez Robles to the Citizens of Yuma, n.d., box 1, folder 1, Robles Papers.

9 Sometimes the politician: Francis-Fallon, *The Rise of the Latino Vote*, 19–20.

9 As the Hispanic surrogate for a Republican governor: Robert Benitez Robles to the Citizens of Yuma, n.d., box 1, folder 1, Robles Papers.

10 One group of Mexican American businessmen: Republican National Hispanic Assembly website, https://rnhanational.org/index.php/about /history/, accessed on April 26, 2019.

10 Fernando Oaxaca, one of the men: Oaxaca interview, October 23, 1975.

10 Born in El Paso: Ibid.

11 They hired a local member: Stuart Spencer in "Creating Opportunities for Latino Americans," an event at the Richard Nixon Foundation on October 11, 2010, https://www.youtube.com/watch?v=0jSuAzl IWQo&t=3317s, accessed on March 23, 2019.

13 The recently elected governor: Immerwahr, *How to Hide an Empire*, 242–61.

14 In the fall of 1950: Ayala and Bernabe, *Puerto Rico in the American Century*, 162–78.

15 At the same time: "House Records Show Communist Tie with Puerto Rico Rebels," *Daily Boston Globe*, November 3, 1950, 4.

15 Eisenhower wasn't anywhere near: Arch Parsons Jr., "Puerto Rican Rebel Head Here Seized with Two in Collazo's Flat," *New York Herald Tribune*, November 3, 1950, 1.

16 Manuel Machado, for example: Machado, *Listen Chicano!*, xiv–xv.

17 Rita DiMartino, a prominent: DiMartino interview, May 27, 2016.

17 Meanwhile, the family: Guinot interview, February 8, 2019.

17 The Latin Americans who became: Fitz, *Our Sister Republics*, introduction; and De Zavala, *Journey to the United States of America*, conclusion.

18 The Afro-Puerto Rican: Ayala and Bernabe, *Puerto Rico in the American Century*, 54–55.

18 Zoraida Fonalledas, who is today: Fonalledas interview, March 20, 2019.

18 After centuries of domination: Pérez, *On Becoming Cuban*, esp. chap. 1.

19 Henry Ramirez, who served: Ramirez, *A Chicano in the White House*, 14.

20 The Cristero War was: Loaeza, *Acción Nacional*, 60–62.

20 The election of Albizu: 1940 Republican Party platform, https://patriotpost.us/documents/436, accessed on June 25, 2019.

20 Yet Hispanics whose families: Katznelson, *Fear Itself*, chap. 2.

21 Named after its sponsors: Ngai, *Impossible Subjects*, 237.

21 President Truman vetoed the law: "Immigration Bill Vetoed by Truman as 'Infamous,'" *Christian Science Monitor*, June 25, 1952, 7.

21 When Congress overrode his veto: Don Irwin, "McCarran Act Faces a Drive for Change," *New York Herald Tribune*, November 30, 1952, A3.

21 Hispanic views of: Gutierrez, *Walls and Mirrors*, 152–78.

23 Flores also argued: John Flores to Dwight Eisenhower, September 10, 1953, box 406, folder 1-CC F, White House Central Files, President's Personal File, Eisenhower Library.

24 Flores explained: John Flores to Sherman Adams, April 23, 1953, box 1057, folder "Flores, John A.," White House Central Files, Alpha File, Eisenhower Library.

24 Flores wrote Eisenhower himself: John Flores to Dwight Eisenhower, September 10, 1953, box 406, folder 1-CC F, White House Central Files, President's Personal File, Eisenhower Library.

25 Flores congratulated Eisenhower: Ibid.

25 The White House forwarded: A. D. Baumhart to Charles Willis, September 25, 1953; and Charles Willis to Max Rabb, September 25, 1953, box 406, folder 1-CC F, White House Central Files, President's Personal File, Eisenhower Library.

26 Cold War violence: "Guard Increased for Congressmen," *Los Angeles Times*, March 3, 1954, 1.

27 This was the low point: Interview with Ricardo Aponte-Parsi, March 19, 2019.

28 Critics maintained that Eisenhower: Rabe, *Eisenhower and Latin America*, 42–63; Grandin, *The Blood of Guatemala*, 202.

28 After Armas took control: John Harris, "McClellan to Speak in Boston Tomorrow," *Boston Daily Globe*, April 1, 1955, 13.

28 In an age defined: Manela, *The Wilsonian Moment*.

29 The number of unauthorized Mexicans: Hernández, *Migra!*, 184–190.

30 An article in the: Tomme Call, "Operation Wetback: Make It Stick!" *San Antonio Express-News*, August 1, 1954, editorial, A.

30 In the 1950s, Democrats: Hernández, *Migra!*, 184–90.

30 The language of the Cold War: "Jenner Group Sees Red Peril in 'Wetbacks,' " *Washington Post*, May 3, 1954, 2.

30 He endorsed Operation Wetback: McCarran quoted in Foley, *Mexicans in the Making of America*, 143.

31 Another Mexican American from California: Manuel Mesa and Trini Varela to Leonard Hall, November 7, 1956, box 77, folder "Form Letter November 27, 1956," Leonard W. Hall Papers, 1953–1957, Eisenhower Library.

32 Mesa estimated that 80 percent: Leonard Hall to Manuel Mesa, November 27, 1956, box 77, folder "Form Letter November 27, 1956," Leonard W. Hall Papers, 1953–1957, Eisenhower Library.

32 In 1958, he sent Nixon: McPherson, *Yankee No!*, chap. 1.

33 The trip wasn't all bad: "Good-Will Salesman Nixon Takes Case to Latin American People," *Wall Street Journal*, May 7, 1958, 1.

33 But in addition to those who praised him: Ibid.

33 The reception was more or less cordial: Nixon quoted in Rabe, *Eisenhower and Latin America*, 102.

33 He was in Venezuela: Earl Mazo, "Nixon Is on Way Home, Eisenhower Will Lead Huge Welcome Today," *New York Herald Tribune*, May 15, 1958, 1.

34 The journalist Walter Lippman: Walter Lippman, "Pearl Harbor in Diplomacy," *Boston Globe*, May 15, 1958, 1.

34 The United States should be wary: Nixon quoted in Rabe, *Eisenhower and Latin America*, 104.

34 The United States couldn't continue: Milton Eisenhower quoted in Rabe, *Eisenhower and Latin America*, 111.

35 By the time Nixon returned: John A. Flores to Dwight Eisenhower, n.d., box 53, folder "3-A-2 1958 (2)," Central Files, General File, Eisenhower Library.

36 He threw a "Rockefeller fiesta": "Rockfeller Hoisted Up on Shoulders of Crowd," *New York Herald Tribune*, October 25, 1958, 4.

36 Confirming their sense: Horace Sutton, "Rockefeller to Open Hotel in Puerto Rico," *New York Herald Tribune*, November 23, 1958, D8.

37 The Castro brothers: Chasteen, *Born in Blood & Fire*, 284.

38 Mexican president Adolfo López Mateos: Keller, *Mexico's Cold War*, 1–12.

38 Mexico was able: Gillingham and Smith, *Dictablanda*, 23–24.

40 He told a friend: Abe Peña to "Clark," March 1, 1963, box 24, folder 22, Peña Papers.

40 As an outspoken advocate: Minutes of the monthly directors meeting, Grants County Chamber of Commerce, June 13, 1962, box 23, folder 8, Peña Papers.

41 He traveled to the: Ferré, in Baralt, *Desde el mirador de Próspero*, 226.

CHAPTER 2: A NEW GENERATION

43 Goldwater didn't enter: Perlstein, *Before the Storm*, chap. 4.

44 They were more conservative: Díaz interview, April 17, 2019.

44 Many of the most politically active: Oropeza, *¡Raza Sí! ¡Guerra No!*, chap. 1.

45 Goldwater carefully argued: Goldwater, *The Conscience of a Conservative*.

46 His words marked: Rigueur, *The Loneliness of the Black Republican*, chap. 2.

46 On communism, Goldwater used: Goldwater, *The Conscience of a Conservative*.

47 Because of how sweeping: "No. 1 Issue: Civil Rights," *Dallas Morning News*, September 20, 1964, 26.

47 Goldwater had honed: "Barry Calls for Equality Thru Freedom," *Chicago Tribune*, October 17, 1964, 4.

48 To better understand: "No 1. Issue: Civil Rights," *Dallas Morning News*.

48 The League of United: Paul Andow, "Civil Rights 'Quid Pro Quo,'" August 25, 1963, box 1, folder 24, Andow Collection.

48 A teacher from Austin: "No. 1 Issue: Civil Rights," *Dallas Morning News*.

49 Any difference in opinion: Behnken, *Fighting Their Own Battles*, chap. 3.

49 One in California argued: "Padilla Replies," *Daily Review*, October 23, 1964.

49 A Mexican American in Arizona: Radio script by Bob Robles, September 1964, box 1, folder 2, Robles Papers.

50 One historian called: Pycior, *LBJ & Mexican Americans*, 145.

50 Johnson's Mexican American supporters: Ngai, *Impossible Subjects*, 158–166.

50 The greatest pressure against: Harry Bernstein, "Mexican Labor Influx May Continue After End of Bracero Law: 'Cheap Labor' Decried," *Washington Post*, September 10, 1964, F10.

51 On the Bracero Program: Bob Robles to Stanley Ross, September 1964, Robles Papers.

51 One of Goldwater's staunchest supporters: Radio script by Bob Robles, September 1964, box 1, folder 2, Robles Papers.

52 During his campaign: Rabe, *The Most Dangerous Area in the World*, chap. 1.

53 Castro was so aggressive: Ibid.

53 In every debate: Warren Duffee and Alvin Spivak, "Kennedy, Nixon Tangle Head-On over Cuba, State of US Prestige," *Atlanta Constitution*, October 22, 1960, 1.

53 Kennedy stirred controversy: Rabe, *The Most Dangerous Area in the World*, 9–14.

54 Kennedy kept up: John F. Kennedy inaugural address, January 20, 1961, https://www.jfklibrary.org/learn/about-jfk/historic-speeches/inaugural -address, accessed on July 2, 2019.

54 He referred not only: Rabe, *The Most Dangerous Area in the World*, chap. 1.

55 The CIA had charged: Gene Sherman, "Cuba Exile Government Formed," *Los Angeles Times*, March 23, 1961, 1.

55 Cuban exiles back in Miami: García, *Havana USA*, 32.

55 The invading rebels: Peter Beales, "Cuba Awaits Barter for Prisoners," *Washington Post*, December 23, 1962, A1.

56 Leaders such as Manuel Giberga: "Homenaje a Manuel Giberga in Union City," *Nuestra Cuba*, December 6, 1973.

56 Perfectly echoing the Cuban exiles: "Goldwater Hits Kennedy on Cuba 'Bungling,'" *Los Angeles Times*, February 14, 1963, 2.

56 Johnson underestimated the threat: William Kling, "Senator Hits at Johnson on Water Crisis," February 9, 1964, *Chicago Tribune*, 1.

57 He demanded a Senate investigation: "Goldwater Says He'd Train Cubans to Overthrow Castro," *Washington Post*, February 20, 1964, A14.

57 Exaggerating by a hundred miles: Barry Goldwater to Roscoe Gaither, September 2, 1964, box 1, folder 2, Robles Papers.

58 The Hispanic population: Nugent, *Color Coded*.

58 One historian has described: Micaela Anne Larkin, "Southwestern Strategy: Mexican Americans and Republican Politics in the Arizona Borderlands," in Shermer, ed., *Barry Goldwater and the Remaking of the American Political Landscape*, chap. 3.

59 They arrived at the convention: Garcia, *Viva Kennedy*, 33.

59 While there: Francis-Fallon, *The Rise of the Latino Vote*, 56–57.

59 After the convention: Ibid., 53–56.

59 When Nixon campaigned: Jay Rogers, "Alamo Plaza Jam-Packed with Rooters for Nixon," *San Antonio Express*, November 4, 1960, 5A.

59 He also campaigned: Peter Kihss, "City Spanish Vote at a Record High," *New York Times*, November 2, 1960, 30.

60 Still seeking his opportunity: Ruben Salazar, "Doubt Cast on Johnson Parley Here," *Los Angeles Times*, A1.

60 Then, after Kennedy's victory: Ibid.

60 He said he had rallied: "Goldwater Clubs Are Organized for Latin Vote," *Columbus (GA) Ledger*, September 20, 1963, 24.

60 One of Flores's allies: Tom Gavin, "Latin-American Leaders Urge Switch from Kennedy," *Denver Post*, September 19, 1963, 46.

60 His Hispanic critics: Ruben Salazar, "'Latinos' for Goldwater Called 'Publicity Stunt,'" *Los Angeles Times*, 33.

61 The Nationalities Division: Neil A. Martin, "GOP Begins Drive for Ethnic Groups," *Washington Post*, September 15, 1964, A2.

61 Stanley Ross oversaw: Press release, "Stanley Ross Named Coordinator of Spanish-Speaking Sections of the GOP's Nationalities Division," September 18, 1964, box 1, folder 2, Robles Papers.

61 Fernando Penabaz led: "GOP Group Told Missiles Face US," *Evansville Courier*, April 16, 1964, 11.

62 By 1964, Ruiz had earned: Eduardo Quevedo to Manuel Ruiz, June 17, 1963, box 1, folder 1, Ruiz Papers.

62 Goldwater chose Robles: Barry Goldwater to Ignacio Soto Jr., September 8, 1964, box 1, folder 2, Robles Papers.

63 In a speech: Speech by Bob Robles, September 19, 1964, box 1, folder 2, Robles Papers.

63 Goldwater also had support: Francis-Fallon, *The Rise of the Latino Vote*, 94.

64 After Goldwater learned: "Puerto Rico Bid Made by Goldwater," *Baltimore Sun*, June 8, 1964, 10.

64 An expat in Mexico: Roscoe Gaither to Barry Goldwater, August 17, 1964, box 1, folder 2, Robles Papers.

64 Goldwater had developed: Cadava, *Standing on Common Ground*, 124.

64 His Mexican friend also observed: Roscoe Gaither to Barry Goldwater, August 17, 1964, box 1, folder 2, Robles Papers.

64 This last point caught: Barry Goldwater to Roscoe Gaither, September 8, 1964, box 1, folder 2, Robles Papers.

65 The World War II veteran: Francis-Fallon, *The Rise of the Latino Vote*, 98.

65 In his conversations: Bob Robles to Barry Goldwater, August 19, 1964; and Bob Robles to Americans for Goldwater, July 28, 1964. Both letters in box 1, folder 2, Robles Papers.

65 He had invited: Bob Robles to Stanley Ross, September 24, 1964, box 1, folder 2, Robles Papers.

66 Disregarding the precedent: Manuel Ruiz to Carlos Sedillo, October 20, 1964, box 1, folder 10, Ruiz Papers.

66 Their radio ads highlighted: Radio script by Bob Robles, September 1964, box 1, folder 2, Robles Papers.

66 Sedillo said he would: C. B. Sedillo to John Rhodes, August 20, 1964; and C. B. Sedillo to Manuel Ruiz, September 18, 1964. Both letters from box 1, folder 10, Ruiz Papers.

67 A couple of weeks later: Bob Robles to Louis Maniatis, November 30, 1964, box 1, folder 2, Robles Papers.

67 It was created: Manuel Ruiz to Nelson Rockefeller, April 9, 1965, box 1, folder 12, Ruiz Papers.

67 Robles offered a playbook: Bob Robles to Louis Maniatis, November 30, 1964, box 1, folder 2, Robles Papers.

67 In a letter to Goldwater: Bob Robles to Barry Goldwater, October 23, 1969, box 1, folder 2, Robles Papers.

68 In a letter to a group: Bob Robles to Tony Carrillo, Joe Ybarra, Leonard Calderón Jr., David Valenzuela, and John Felix, January 12, 1965, box 1, folder 2, Robles Papers.

68 Nevertheless, Hispanic Republicans: Stephen Shadegg to Bob Robles, December 15, 1964, box 1, folder 2, Robles Papers.

68 They had no choice: Bob Robles to Barry Goldwater, October 23, 1969, box 1, folder 2, Robles Papers.

68 Manuel Ruiz also wrote: Manuel Ruiz to Louis P. Maniatis, October 4, 1964, box 1, folder 10, Ruiz Papers.

69 One group in California: Resolution of the Santa Clara County Republican Central Committee, 1965, box 1, folder 14, Ruiz Papers.

69 Largely seen as a liberal law: Ngai, *Impossible Subjects*, chap. 7.

70 In response to the new arrivals: Public Law 89-732, November 2, 1966.

70 Airlifts from Havana: *Report of the Select Commission on Western Hemisphere Immigration*, 12–14.

71 Even in these years: Bob Robles to Stanley Ross, September 24, 1964, box 1, folder 2, Robles Papers.

71 In New York: Francis-Fallon, *The Rise of the Latino Vote*, 114–15.

72 Also in 1966: Knaggs, *Two-Party Texas*.

72 The key had been: James McCrory, "GOP Woos Mexican-Americans," *San Antonio Express*, April 15, 1966, 10F.

72 With the help of: Knaggs, *Two Party-Texas*.

73 No Democrat had been: Nugent, *Color Coded*, 191.

73 Like other conservatives: Brilliant, *The Color of America Has Changed*, chap. 8.

74 Edward Roybal, a Democratic congressman: "Spanish Voters Ignored, Solon Says," *Albuquerque Journal*, April 3, 1966.

75 Some Hispanics called it: Behnken, *Fighting Their Own Battles*, 112.

75 When Roybal was told: "Spanish Voters Ignored, Solon Says," *Albuquerque Journal*, April 3, 1966.

CHAPTER 3: NIXON'S HISPANICS

79 The hundred or so: Ramirez, *A Chicano in the White House*, 167.

80 The photo of Nixon: Ibid., 166–71.

80 Nixon also knew: Rigueur, *The Loneliness of the Black Republican*, 132.

81 He had the backing: Don Oberdorfer, "Humphrey Has Bilingual Day in South Texas," *Washington Post*, October 24, 1968, A2.

81 Hampering Humphrey's chances: Ruben Salazar, "Latins' Pride in 'Raza' of Benefit to Republicans," *Los Angeles Times*, October 29, 1968, 1.

82 Arnaz sought to help: "Desi Arnaz to Visit El Paso with Nixon," *El Paso Times*, October 28, 1968, 1.

82 The entertainer spent: "Smile Awhile," *San Antonio Express*, October 27, 1968, 16B.

82 Before leaving, Arnaz tuned a guitar: Ramon Villalobos, "Desi Arnaz Says Victory Up to Texas," *El Paso Times*, November 3, 1968, 10A.

83 Barry Goldwater: Francis-Fallon, *The Rise of the Latino Vote*, 171.

83 A Republican congressman: Burt Talcott to Richard Nixon, July 1, 1969, box 8, folder "Inter-Agency Committee on Mexican Affairs/ Cabinet Committee on Opportunities for Spanish-Speaking People— as of 12/30/69," White House Central Files, Subject Files, Federal Government, Organizations (FG-134–149), Nixon Library.

84 But in the months: Ibid.

85 It was a lesson: Phillips, discussed in Lepore, *These Truths*, chap. 15.

85 Nixon's Hispanic supporters: Ramirez, *A Chicano in the White House*, 20–21.

86 "He got to know us": Ibid., 21.

86 One of the first things: Ximenes to Humphrey and McCormack, n.d., box 7, folder "Inter Agency Committee on Mexican American Affairs/Cabinet Committee on Opportunities for Spanish-Speaking Americans—as of 12/30/69," White House Files, Subject Files, Federal Government, Organizations (FG-134–149), Nixon Library.

87 In Johnson's memorandum: Lyndon Johnson, June 9, 1967, box 7, folder "Inter Agency Committee on Mexican American Affairs/Cabinet Committee on Opportunities for Spanish-Speaking Americans—as of 12/30/69," White House Files, Subject Files, Federal Government, Organizations (FG-134–149), Nixon Library.

87 They brought together: Ximenes to Humphrey and McCormack.

88 Because of Nixon's attention: Ramirez, *Nixon and the Mexicans*, 177.

88 On March 5, 1969: Richard Nixon, Executive Order 11458, "Prescribing Arrangements for Developing and Coordinating a National Program for Minority Business Enterprise," March 5, 1969.

88 The SBA ran: Ramirez, *Nixon and the Mexicans*, 176.

89 These sorts of programs: Ibid., 171–78.

89 Johnson's committee: Talcott to Nixon, July 1, 1969.

89 His administration embraced: "Analysis of Legislation to Establish the Committee on Mexican American Affairs," box 7, folder "Inter

Agency Committee on Mexican American Affairs/Cabinet Committee on Opportunities for Spanish-Speaking Americans—as of 12/30/69," White House Files, Subject Files, Federal Government, Organizations (FG-134–149), Nixon Library.

89 Nixon signed the bill: S740 signing statement, December 29, 1969, box 7, folder "Inter Agency Committee on Mexican American Affairs/Cabinet Committee on Opportunities for Spanish-Speaking Americans—as of 12/30/69," White House Files, Subject Files, Federal Government, Organizations (FG-134–149), Nixon Library.

90 The central plank: Ibid.

90 Instead of the criteria: Bean and Tienda, *The Hispanic Population of the United States*, 48.

90 They hired Spanish-speaking: John Ehrlichman to Michael Marmolejo, February 24, 1970, box 7, folder "Inter Agency Committee on Mexican American Affairs/Cabinet Committee on Opportunities for Spanish-Speaking Americans—as of 12/30/69," White House Central Files, Subject Files, Federal Government, Organizations (FG-134–149), Nixon Library.

91 The main engines driving: Ramirez, *Nixon and the Mexicans*, 174.

91 Established in 1970: NEDA monthly reports for March 1972 and April 1972, box 18, folder "National Economic Development Association," White House Central Files, Staff Member Office Files, Robert H. Finch, CCOSS File, Nixon Library.

92 At the end of August 1970: Ruben Salazar, "The Mexican-Americans NEDA Much Better School System," *Los Angeles Times*, August 28, 1970, B7.

92 Hispanic Republicans objected: Ramirez, *Nixon and the Mexicans*, 177.

93 When Congress debated: Clifford Case, in hearing on the extension of the CCOSSP, n.d., box 8, folder "Cabinet Committee on Opportunities for Spanish-Speaking People—as of 1/1/71," White House Central Files, Subject Files, (FG) Federal Government, Organizations (FG-134–149), Nixon Library.

94 In the eight months: Luis Ferré to Richard Nixon, January 29, 1970, box 18, folder "Puerto Rico/Governor [Luis A.] Ferré," White House Central Files, Staff Member Office Files, Robert H. Finch, CCOSS File, Nixon Library.

94 He recommended prominent Puerto Ricans: Guinot interview, February 8, 2019.

94 Before he resigned: Martin Castillo and Hank Quevedo to Robert Haldeman, February 1, 1969, box 7, folder "Inter-Agency Committee on Mexican American Affairs/Cabinet Committee on Opportunities for Spanish-Speaking Americans—as of 12/30/69," White House Central Files, Subject Files, Federal Government, Organizations (FG-134–149), Nixon Library.

94 In May 1971: Ramirez, *A Chicano in the White House*, 297.

94 It would become: Henry Ramirez to Robert Finch, May 21, 1971, box 18, folder "Ramirez, [Henry M.]," White House Central Files, Staff Member Office Files, Robert H. Finch, CCOSS File, Nixon Library.

95 Nixon's chief of staff: Fred Malek to H. R. Haldeman, July 15, 1971, box 5, folder "Mrs. Romana A. Banuelos (Biography) [2 of 3]," White House Central Files, Staff Member Office Files, Barbara Franklin, Series III: Alphabetical Name Files, Nixon Library.

95 Less than ten days: William Marumoto to Fred Malek, July 12, 1971, box 5, folder "Mrs. Romana A. Banuelos (Biography) [2 of 3]," White House Central Files, Staff Member Office Files, Barbara Franklin, Series III: Alphabetical Name Files, Nixon Library.

96 Her message, the administration believed: News clipping, "Ingredients for Success—Dedication and Hard Work," *Gardena Valley News*, n.d., box 5, folder "Mrs. Romana A. Banuelos (Biography) [2 of 3]," White House Central Files, Staff Member Office Files, Barbara Franklin, Series III: Alphabetical Name Files, Nixon Library.

96 When her business: News clipping, "Ingredients for Success."

96 Bañuelos still inhabited: William Marumoto to Fred Malek, July 12, 1971, box 5, folder "Mrs. Romana A. Banuelos (Biography) [2 of 3]," White House Central Files, Staff Member Office Files, Barbara Franklin, Series III: Alphabetical Name Files, Nixon Library.

97 The president was drawn: Jack Robbins, "Romana Banuelos Signing on at the Treasury," *New York Post*, September 25, 1971.

98 Malek reported back: Malek to Haldeman, July 15, 1971.

98 It was a prime: Schedule proposal for Romana Banuelos, July 19–21, 1972, box 5, folder "Mrs. Romana A. Banuelos (Biography) [3 of 3]," White House Central Files, Staff Member Office Files, Barbara Franklin, Series III: Alphabetical Name Files, Nixon Library.

98 Nixon handed her: Richard West, "LA Woman Picked as US Treasurer," *Los Angeles Times*, September 21, 1971, 3.

99 Ramirez described himself: Ramirez, *A Chicano in the White House*, 197, 366.

99 Staff member Robert Finch: Schedule proposal from Robert Finch, July 28, 1971, box 7, Folder "Cabinet Committee on Opportunities for Spanish-Speaking People, 1/1/71–[12/31/72]," White House Central Files, Subject Files, Nixon Library.

99 But Nixon went off script: Richard Moore, August 5, 1971, box 7, folder "Cabinet Committee on Opportunities for Spanish-Speaking People, 1/1/71–[12/31/72]," White House Central Files, Subject Files, Nixon Library.

99 Nixon told Ramirez: Ramirez, *A Chicano in the White House*, 20.

100 As one of Ramirez's first acts: Damaso Emeric to Richard Nixon, August 1971, box 8, folder "Cabinet Committee on Opportunities for Spanish-Speaking People, 1/1/71," White House Central Files, Subject Files, Federal Government, Organizations (FG-134–149), Nixon Library.

100 Meanwhile, a Mexican American: Raul Navarro to Robert Finch, August 5, 1971, box 8, folder "Cabinet Committee on Opportunities for Spanish-Speaking People, 1/1/71," White House Central Files, Subject Files, Federal Government, Organizations (FG-134–149), Nixon Library.

101 With a belly full: West, "LA Woman Picked as US Treasurer."

101 Nixon neglected to mention: Robbins, "Romana Banuelos Signing on at the Treasury."

101 From the moment: "Treasurer Is Named," *New York Times*, September 21, 1971.

102 Everyone involved pointed fingers: News clipping, Ken W. Clawson, "US Nominee's Firm Raided," n.p., n.d., box 5, folder "Mrs. Romana A. Banuelos (Biography) [3 of 3]," White House Central Files, Staff Member Office Files, Barbara Franklin, Series III: Alphabetical Name Files, Nixon Library.

102 Other letters to the editor: News clipping, "The Great Taco Flap," *Chicago Daily News*, n.d., box 5, folder "Mrs. Romana A. Banuelos (Biography) [3 of 3]," White House Central Files, Staff Member Office Files, Barbara Franklin, Series III: Alphabetical Name Files, Nixon Library.

102 One editorial: "Raid Raises Questions About Nixon Choice," n.p., October 11, 1971, 7, box 5, folder "Mrs. Romana A. Banuelos (Biography) [2 of 3]," White House Central Files, Staff Member Office Files, Barbara Franklin, Series III: Alphabetical Name Files, Nixon Library.

103 He reached out to: Bill Marumoto to Fred Malek, October 11, 1971, box 5, folder "Mrs. Romana A. Banuelos (Biography) [1 of 3]," White House Central Files, Staff Member Office Files, Barbara Franklin, Series III: Alphabetical Name Files, Nixon Library.

103 Marumoto urged the presidents: Philip Shandler, "Mrs. Banuelos Backed by 5 Hispanic Groups," *Evening Star*, October 15, 1971, A11.

103 "We urge the": Press release by Alfred Villalobos, box 5, folder "Mrs. Romana A. Banuelos (Biography) [1 of 3]," White House Central Files, Staff Member Office Files, Barbara Franklin, Series III: Alphabetical Name Files, Nixon Library.

103 The raid was an embarrassment: "Treasurer Confirmed Smoothly," *New York Times*, December 7, 1971.

104 In part as a response: "Bill Would Outlaw Employing of Illegal Aliens Knowingly," n.p., n.d., box 5, folder "Mrs. Romana A. Banuelos (Biography) [2 of 3]," White House Central Files, Staff Member Office Files, Barbara Franklin, Series III: Alphabetical Name Files, Nixon Library.

105 These set-asides: Linda Chavez, interview by author, February 18, 2019.

105 According to a Mexican American: Patricia Hitt to Robert Finch, December 3, 1971, box 20, folder "New York/New Jersey [Project Alpha, December 9, 1971]," White House Central Files, Staff Member and Office Files, Robert H. Finch, CCOSS File, Nixon Library.

106 As they campaigned: Robert V. Beier and Pat Kailer, "Administration People Stump in Albuquerque," *Albuquerque Journal*, October 19, 1972.

107 At one campaign stop: Jesse Henry, "400 Attend Republican Fund-Raising Dinner," *San Antonio Express*, October 18, 1972.

107 She logged more: Mark Sanchez, "GOP 'Stand-ins' Laud Gains," *New Mexican*, October 20, 1972.

107 As a local: James McCrory, "Nixon's Minority Help Cited," *San Antonio Express*, October 18, 1972.

107 Not everything was: Pat McGraw, "Hispano Appointees Laud Nixon," *Denver Post*, n.d., box 5, folder "Mrs. Romana A. Banuelos (Biography) [2 of 3]," White House Central Files, Staff Member Office Files, Barbara Franklin, Series III: Alphabetical Name Files, Nixon Library.

107 Bañuelos and Sanchez: Alex Armendariz to Romana Acosta Bañuelos, October 24, 1972, box 25, folder "Banuelos, Romana [2 of 2]," White House Central Files, Staff Member Office Files, Barbara Franklin, Series V: Subject Files, Nixon Library.

107 She autographed bills: Ibid.

107 Going forward, Hispanics: David Manley, "Mexican-Americans Hold Power, Treasurer Says," n.p., n.d., box 25, folder "Banuelos, Romana [2 of 2]," White House Central Files, Staff Member Office Files, Barbara Franklin, Series V: Subject Files, Nixon Library.

108 Bañuelos told a crowd: "Mrs. Banuelos in EP, Campaigns for Nixon," *El Paso Herald-Post*, October 28, 1972.

108 Some Chicanos called: "Nixon Seeking Latins' Support, Says Treasurer," *El Paso Times*, October 29, 1972.

108 The lieutenant governor: "Boosters Labeled 'Showcase' Type," *Las Cruces Sun-News*, October 19, 1972, 17.

108 They made more: Helen Diepenbrock, "Activists Drown Out GOP Rally Here," *Sacramento Union*, November 1, 1972.

108 McGovern campaign pamphlets: "Action Not Promises, McGovern Not Nixon," box 18, folder 1, Reveles Papers.

109 In the days after: Marcelino Miyares to Romana Bañuelos, November 10, 1972, box 25, folder "Banuelos, Romana [2 of 2]," White House Central Files, Staff Member Office Files, Barbara Franklin, Series V: Subject Files, Nixon Library.

109 Ramirez's assistant at: J. Raul Espinosa to Romana Bañuelos, n.d., box 25, folder "Banuelos, Romana [2 of 2]," White House Central Files, Staff Member Office Files, Barbara Franklin, Series V: Subject Files, Nixon Library.

109 The head of: Norman Watts Jr. to Romana Banuelos, November 6, 1972, box 25, folder "Banuelos, Romana [2 of 2]," White House Central Files, Staff Member Office Files, Barbara Franklin, Series V: Subject Files, Nixon Library.

109 It "set the hair": Ramirez, *A Chicano in the White House*, 192.

109 According to Ramirez: Ibid., 188–89.

110 Writing decades later: Ibid.

110 Hearing of their plight: Ibid., 190–202.

110 He had said in 1971: Nixon quoted in "Compassionate Solution to Immigration," *Signal*, June 12, 2012, 5.

111 Ramirez recalled that momentum: Ramirez, *A Chicano in the White House*, 204.

111 The *New York Times* profiled: Michael Kaufman, "Former Representative Peter W. Rodino Jr. Dies at 95; Led House Watergate Inquiry," *New York Times*, May 8, 2005, 30.

CHAPTER 4: THE HISPANIC WATERGATE

112 After pouring himself: Alfred E. Lewis, "Miamians Held in DC Try to Bug Demo Headquarters," *Miami Herald*, June 18, 1972, 1.

112 Once the news: S. J. Micciche, "Court Curbs Power of US on Wiretapping," *Boston Globe*, June 20, 1972, 1.

113 It caused many: Igo, *The Known Citizen*, 264–306.

114 There were three: Hunt, *American Spy*, 118, 211.

115 These and other Cuban exiles: S. J. Micciche, "Barker Sought a Castro Link to Democrats," *Boston Globe*, May 25, 1973, 1.

115 James McCord, the fifth burglar: Hunt, *American Spy*, 211.

115 Martinez wrote an account: Eugenio Martinez, "Mission Impossible: The Watergate Bunglers," *Harper's Magazine*, October 1974.

115 At the time: Hunt, *American Spy*, 222, 194.

116 The break-in at: Ibid., 194–95.

117 Harassment by the Mexican government: Ibid., 116–19.

118 Once reporters began: Ibid., 218.

118 Nixon and his inner circle: Haldeman, *The Haldeman Diaries*, 474; Dean, *The Nixon Defense*, 27.

118 There was plenty: Britto, *Marijuana Boom*, chap. 5.

119 If the "Cuban angle": Dean, *The Nixon Defense*, 30–31.

119 The arrested Cubans became: Francis-Fallon, *The Rise of the Latino Vote*, 263.

120 Five burglars wearing suits: Bernstein and Woodward, *All the President's Men*, 22–23.

120 Woodward called the White House: Bernstein and Woodward, *All the President's Men*, 23–24.

120 When Nixon learned: Dean, *The Nixon Defense*, 49.

120 The $100 bills: "Texas Money Figured in Tapes," *Dallas Morning News*, August 7, 1974, 7.

121 Federal investigators traced: "Magazine Says Reelection Panel Members Aware of 'Bugs,'" *Augusta Chronicle*, August 21, 1972, 17.

121 They sent Ogarrio: "Pennzoil Officials Tell of Transaction," *Dallas Morning News*, September 14, 1972, 38.

121 From there, the Texas oilmen: Philip Warden, "GOP Fund Transfer Told," *Chicago Tribune*, September 13, 1972, 1; "Democratic Break-in Money Traced to Drafts in Mexico," *Dallas Morning News*, July 31, 1972, 7.

122 Ogarrio, of course: "GOP Campaign Check Traced to Break-in Suspect," *Denver Post*, August 1, 1972, 9; "Money Linked to Break-in Is Traced," *Atlanta Constitution*, July 31, 1972, 6A.

123 Reporters, politicians, prosecutors: Arthur Siddon, "Ervin Hints Quiz of Nixon on Legal Role in Watergate," *Chicago Tribune*, June 29, 1973, 1.

123 Their hearings were broadcast live: PBS Video, *Summer of Judgment*.

124 The committee disbanded: S.Res.60, 93rd Cong. (1973–1974).

125 Marumoto nervously read: Marumoto testimony.

126 When he was eight: Ibid.

127 As they went about: Ibid.

127 Under pressure, Marumoto also stated: Ibid.

128 Specifically, the Responsiveness Program: Tony Castro, "Nixon Critics Were Denied Federal Funds," *Charlotte Observer*, July 29, 1974, 17A.

128 In a memo to: "Republican 'Brown Mafia' Efforts Explored by Senate Committee," *El Chicano*, December 27, 1973, 1.

128 The Brown Mafia: Ibid.

128 The reelection campaign strategy: "Capitalizing on the Incumbency," Exhibit No. 262–1, Presidential Campaign Activities of 1972, 5532.

129 If the recipients: "Election Group Meddled in Grant Giving," *Fort Worth Star-Telegram*, January 20, 1974, 10, 12.

129 They gave to: Hartley Hampton, "CREP Helped City Get $3 Million," *Fort Worth Star-Telegram*, January 20, 1974, 10.

130 The Urban Research Group: "Republican 'Brown Mafia' Efforts Explored by Senate Committee."

130 In addition to federal grants: "Pardon Offered for Chicano's Support," *Fort Worth Star-Telegram*, January 20, 1974, 10.

130 The members of: Tony Castro, "Part of President's Reelection Effort May Help Impeach Him," *Del Rio News Herald*, August 4, 1974, 8.

130 During the campaign: "Election Group Meddled in Grant Giving."

130 Sanchez and Marumoto: Ibid.

131 One week after: "SBA Under Investigation," *Odessa American*, April 26, 1974, 2A.

131 Two months later: Marumoto testimony.

131 The case of Leveo Sanchez: Samuel Dash during Marumoto testimony.

131 If the committee focused: "Republican 'Brown Mafia' Efforts Explored by Senate Committee."

131 Similarly, the Southwest Council: "Election Group Meddled in Grant Giving."

132 All of this: Castro, "Part of President's Reelection Effort May Help Impeach Him."

132 Wearing a three-piece suit: Fernandez testimony.

133 The unnamed Cuban American: Ibid.

134 In his testimony: Ibid.

134 Fernandez emphasized his success: Ibid.

135 Linda Chavez, a Judiciary Committee employee: Chavez, *An Unlikely Conservative*, 105–107.

136 Leading up to and during: Hampton, "CREP Helped City Get $3 Million."

136 Meanwhile, Gutiérrez's wife: Ibid.

137 A few years before: Oropeza, *The King of Adobe*, chap. 8.

137 Hoping to take advantage: "Pardon Offered for Chicano's Support."

137 The lone Hispanic: Montoya during Marumoto testimony and Fernandez testimony; and Montoya quoted in José de la Isla, "Effort to Suppress Latino Voters," *Ventura County Star*, October 28, 2010.

139 The "grave consequence": Richard Santillan, "Chicanos and the American Bi-Centennial," *El Chicano*, June 5, 1975, 12.

139 While they were sad: Agenda for Meeting with Hispanic Representatives, April 11, 1975, box 117, folder "Hispanic Americans, April–May 1974," Vice Presidential Papers, Office of Deputy Assistant for Nongovernmental Affairs, Ford Library.

139 Yet despite Nixon's: Frank del Olmo, "1974: The Year Latins Rediscovered Politics," *Los Angeles Times*, December 26, 1974, 30.

CHAPTER 5: POLITICAL POWER

141 The announcement came on: "Statement by the President in Connection with His Proclamation Pardoning Nixon," *New York Times*, September 9, 1974, 24.

141 Only thirty-six years old: Frank del Olmo, "1974: The Year Latins Rediscovered Politics," *Los Angeles Times*, December 26, 1974, 30.

141 De Baca's appointment was historic: Fernando De Baca, interview by Elizabeth Roach, "Ford Aide on Hispanic Status in US," *Washington Star*, October 13, 1975.

142 When he made: Resolution by the Spanish-Speaking Advisory Committee, Spring 1974, box 3, folder "Illegal Aliens—Background Information," Fernando E. C. De Baca, 1974–1976, Ford Library.

142 Three months into: Del Olmo, "1974: The Year Latins Rediscovered Politics," 30.

142 Along the way: Ibid.

142 Finally, De Baca: Ibid.

142 There were screeds: Chris Fox to Gerald Ford, January 14, 1975, box 4, folder "Illegal Aliens—Correspondence," Fernando E. C. De Baca, 1974–1976, Ford Library.

143 On the other hand: "UFW Campaign Against Illegal Aliens—Farmworkers Ask for Citizens' Drive Against Illegal Workers," *¡acción!*, July 1974, in box 3, folder "Illegal Aliens—Background Information," Fernando E. C. De Baca, 1974–1976, Ford Library.

143 De Baca acknowledged: Del Olmo, "1974: The Year Latins Rediscovered Politics," 30.

143 The year 1974: Ibid.

144 Historians have used: Rodgers, *Age of Fracture*, prologue; Grandin, *The End of the Myth*, chap. 12; Kruse and Zelizer, *Fault Lines*, esp. chaps. 1–4; and Perlstein, *The Invisible Bridge*, preface.

144 The number of: Anna Brown, "The US Hispanic Population Has Increased Sixfold Since 1970," Pew Research Center, https://www.pewresearch.org/fact-tank/2014/02/26/the-u-s-hispanic-population-has-increased-sixfold-since-1970/, accessed on July 10, 2019.

145 When corporations recognized: Dávila, *Latinos Inc.*, chap. 2.

145 When many Hispanics became: Mora, *Making Hispanics*, chap. 4.

145 The purpose of: Manuel Luján Jr. to "Dear Colleague," June 23, 1974, box 117, folder "Hispanic Americans, June–August 1974," Vice Presidential Papers, Office of Deputy Assistant for Nongovernmental Affairs, 1974, Ford Library.

145 The first national gathering: Agenda for RNHA organizational conference, July 11–13, 1974, box 117, folder "Hispanic Americans, June–

August 1974, Vice Presidential Papers, Office of Deputy Assistant for Nonovernmental Affairs, 1974, Ford Library.

145 Several of them tried: Francis-Fallon, *The Rise of the Latino Vote*, 275.

146 In a press release: Press release, "RNC's Spanish-Speaking Advisory Committee Auxiliary Formed," *News from the Republican National Committee*, April 1974, box 117, folder "Hispanic Americans, April–May 1974," Vice Presidential Papers, Office of Deputy Assistant for Nongovernmental Affairs, 1974, Ford Library.

146 His presence at: Benjamin Fernandez to "All State Chairmen" of the Republican National Hispanic Assembly, box 135, folder "Republican National Hispanic Assembly, Washington, DC, July 8, 1974," Vice Presidential Papers, Office of Deputy Assistant for Media Affairs, Speech File, Ford Library.

146 It was a first step: "Hispano Republicans Organize Nat'l Political Conference," *Atlanta Daily World*, July 30, 1974, A2.

146 In September 1974: "The President's Remarks Upon Signing the Proclamation in a Ceremony at the White House, September 4, 1974," *Weekly Compilation of Presidential Documents, Monday, July 9, 1974* (Washington, DC: General Services Administration, 1974), 1086.

147 Messages from Cuba's: "Cuba: The United States Stands Aside," *Washington Post*, November 13, 1974, A24.

147 US officials began: Robert Keatley, "Will Cuba Rejoin the Club?," *Wall Street Journal*, October 3, 1974, 16.

147 These all sounded: Mark Atwood Lawrence, "History from Below: The United States and Latin America in the Nixon Years," in Logevall and Preston, eds., *Nixon in the World*, chap. 13.

148 None of this was good: "Cuba: The United States Stands Aside."

148 The leaders of: Program for the "First Annual Banquet" of the Republican National Hispanic Assembly, July 29, 1976, box 53, folder 4, Rhodes Papers.

149 The RNHA also: Program for the "First Annual Banquet" of the Republican National Hispanic Assembly.

149 As a Ford campaign pamphlet: Gerald Ford campaign pamphlet, box F25, folder "Hispanic Americans," President Ford Committee Records, 1975–1976, Ford Library.

149 A radio ad asked: Spanish-language radio announcement written by

Thomas Aranda Jr., 1980 Campaign Papers, Series X-Citizen's Operations (Max Hugel), box 333, Reagan Library.

149 When "Hispanic leaders": Gerald Ford campaign pamphlet, box F25, folder "Hispanic Americans," President Ford Committee Records, 1975–1976, Ford Library.

150 Fernandez sent RNC head: Benjamin Fernandez to Mary Louise Smith, February 17, 1975, box 1011, Folder "Benjamin Fernandez," White House Central Files, Ford Library.

151 The RNHA saw: "Meeting with the Leadership of the Republican National Hispanic Assembly," Thursday, December 11, 1975, box 54, folder "Republican National Hispanic Assembly, 12/11/75," James M. Cannon Files, Ford Library.

151 Not only was: Benjamin Fernandez to Robert Hartmann et al., box 54, folder "Republican National Hispanic Assembly, 12/11/75," Ford Library.

151 All Hispanics were asking: Ibid.

152 The White House's task: "Meeting with the Leadership of the Republican National Hispanic Assembly," December 11, 1975, box 54, folder "Republican National Hispanic Assembly, 12/11/75," James M. Cannon Files, Ford Library.

153 Like the RNHA: Fernando De Baca, interview by Elizabeth Roach, "Ford Aide on Hispanic Status in US," *Washington Star*, October 13, 1975.

153 The RNHA had supported: Benjamin Fernandez to Robert Hartmann et al.

153 To show Hispanics: "Meeting with the Leadership of the Republican National Hispanic Assembly."

154 Casanova advised Ford: José Manolo Casanova to Gerald Ford, December 10, 1975, box 28, folder "Republican National Hispanic Assembly (1)," Robert T. Hartmann, Files, 1974–1977, Ford Library.

154 According to Casanova: José Manolo Casanova to Gerald Ford, December 10, 1975.

156 Shortly before the Florida primary: Francis-Fallon, *The Rise of the Latino Vote*, 284.

156 Because Ford seemed: Benjamin Fernandez to Douglas Bennett, December 19, 1975, box 1011, Folder "Benjamin Fernandez," White House Central Files, Ford Library.

157 Fernandez then sent: Benjamin Fernandez to the National Executive Committee and State Executive Committee of the Republican National Hispanic Assembly, January 27, 1976, box 21, folder "RNHA (2)," Gwen A. Anderson Files, 1974–1977, Ford Library.

157 On the road: Benjamin Fernandez in "Reception Due GOP Official," *Grand Rapids Press*, February 5, 1976.

158 Ford reflected on: William Safire, "Shucking the Tamale," *New York Times*, May 3, 1976, 31.

158 Newspapers big and small: Editorial, "The Hot Tamale," *Signal*, April 12, 1976, 7.

159 According to one historian: García, *Reagan's Comeback*, 100.

159 Another has written: Knaggs, *Two-Party Texas*, 194.

159 Such positive strides: Benjamin Fernandez to the National Executive Committee and State Executive Committee of the Republican National Hispanic Assembly, March 29, 1976, box 21, folder "RNHA (2)," Gwen A. Anderson Files, 1974–1977, Ford Library.

160 Chefs at the elite Mayflower: Klaus Tuschman to Roger Moure, June 17, 1976, box 2, folder "Republican National Hispanic Assembly," Thomas Aranda Files, 1976–1977, Ford Library.

160 Significantly, the RNHA: Col. Davidson to Elly Peterson, July 6, 1976, box 2, folder "Republican National Hispanic Assembly," Thomas Aranda Files, 1976–1977, Ford Library.

161 They pitched the event: Republican National Hispanic Assembly to Edward Debolt, June 17, 1976, box 2, folder "Republican National Hispanic Assembly," Thomas Aranda Files, 1976–1977, Ford Library.

162 He came from a district: Luján interview, October 23, 1975.

162 Luján supported: Ibid.

162 He spoke out: Ibid.

163 They "significantly contributed": Resolution by the Spanish-Speaking Advisory Committee on illegal immigration.

163 Ford's prepared remarks: Gerald Ford Remarks at the First Annual Banquet of the Republican National Hispanic Assembly, July 29, 1976, *Weekly Compilation of Presidential Documents, Monday, August 2, 1976*, vol. 12, no. 31 (Washington, DC: General Services Administration, 1976), 1225.

164 At the end: Program for First Annual Banquet, July 29, 1976, box 5,

folder "1976/07/29—Republican National Hispanic Assembly (1)," Thomas Aranda Files, 1976–1977, Ford Library.

164 As for his speech: Peggy Simpson, "Hispanics Demand Piece of GOP Action," *Kansas City Star*, August 26, 1976, 32.

164 Because of the RNHA: "Hispanics Vow Not to Be Exploited by Republicans as They Were by Nixon," *Valley Morning Star*, August 22, 1976.

165 Emboldened, Fernandez sent: Benjamin Fernandez to Stuart Spencer, August 25, 1976, box 5, folder "1976/07/29—Republican National Hispanic Assembly (1)," Thomas Aranda Files, 1976–1977, Ford Library.

165 If they had gotten: Confidential memorandum to Elly Peterson, n.d., box F24, folder "Cuban Americans (1)," President Ford Committee Records, 1975–1976, Ford Library.

166 One professor, Richard Santillan: Richard Santillan, "Chicanos and the American Bi-Centennial," *El Chicano*, June 5, 1975, 12.

166 As Americans planned: Ibid.

167 At the convention: Republican National Hispanic Assembly of Illinois to Stuart Spencer, September 2, 1976, box 2, folder "Republican National Hispanic Assembly," Thomas Aranda Files, 1976–1977, Ford Library.

168 On New Year's Eve: "President Advocates Statehood for Puerto Rico," *Charleston News and Courier*, January 1, 1977, 1–2.

168 The timing of: *NBC Nightly News*, December 31, 1976, clip #486878 from the Vanderbilt Television News Archive, accessed on December 30, 2018.

169 When Ford assumed office: Romero Barceló, *Estadidad es para los pobres*.

169 Ford claimed that: *NBC Nightly News*, December 31, 1976, clip #486878.

170 At the end of: Ibid.

170 When Ford said: "Upset Victor in Puerto Rico States His Case," *New York Daily News*, November 4, 1976, 22.

170 Ford knew that: *NBC Nightly News*, December 31, 1976, clip #486878.

170 In a letter Fernandez received: John Rhodes to Benjamin Fernandez, June 2, 1977, box 53, folder 4, Rhodes Papers.

CHAPTER 6: FERNANDEZ *IS* AMERICA

173 On the cover: *Time*, October 16, 1978, 48–61.

173 The article caricatured: Ibid., 61.

174 The differences among: Ibid., 48.

174 If Hispanics were: Ibid., 48, 52.

174 The story elsewhere: Ibid., 52, 55, 58.

175 On the Republican side: "Hispanics Vow Not to Be Exploited by Republicans as They Were by Nixon,'" *Valley Morning Star*, August 22, 1976.

176 Castillo and Carter's: James P. Sterba, "Administration Plans Amnesty for Aliens," *Atlanta Constitution*, April 18, 1977, 9A.

177 Most of the opposition: Frank del Olmo, "Latins Cool to Carter's Plans on Illegal Aliens," *Los Angeles Times*, February 23, 1977, A3.

177 As one Carter aide: Anthony Marro, "Carter Aides to Map New Policy on Aliens," *New York Times*, April 4, 1977, 1.

178 In December 1977: Joe Coudert, "Hispanic Presidential Run Viewed Possible," *Corpus Christi Caller*, March 11, 1978, 8A.

178 To document the tour: Widener interview, July 23, 2018.

179 In the New Year: Edgar Sanchez, "Latin in White House? Some See Possibility," *Palm Beach Post*, April 24, 1978, B2.

179 Fernandez interrupted his work: "Aliens, Energy Top Visit," *San Antonio Express*, August 31, 1978, 6C.

179 When he returned: Sanchez, "Latin in White House? Some See Possibility."

180 Tower had mastered: John Tower to Joseph A. Califano Jr., June 5, 1978, box 549, folder 14, Tower Papers; John Tower press release, July 15, 1978, box 549, folder 13, Tower Papers.

180 Hispanics in Texas recognized: Lilly Flores Vela, "Tower, Krueger Seeking Supporters at Convention," *Corpus Christi Caller*, June 18, 1978.

180 At a LULAC convention: Robert Montemayor, "Courtship of Mexican-American Vote Begins at LULAC Convention," *Dallas Times Herald*, May 21, 1978, 34A.

181 A young employee: "El Corrido de John Tower," box 547, folder 8, Tower Papers.

181 The firm "balked": Sosa, *The Americano Dream*, xiv.

181 When Fernandez announced: "'Boxcar Ben' Fernandez," https://www.youtube.com/watch?v=od8I21WT6sk, accessed on July 16, 2019.

182 A poll of: "Only 6 of 18 GOP Contenders Are Recognized by Half of Voters," *New York Times*, September 23, 1979.

182 Still, while some: Francis-Fallon, *The Rise of the Latino Vote*, 346; and "A Fernández for President?" *Agenda* (June 1978), 8.

183 But the article in *Time*: *Time*, October 16, 1978.

183 In his first months: "Carter to Name 12 Latins to Top Posts—Roybal," *Los Angeles Times*, March 7, 1977, A1.

184 Hispanics liked Kennedy: Don Sellar, "America's Hispanics Becoming a Force to Be Reckoned with," *Windsor Star*, September 24, 1979.

184 His premise was: "'Boxcar Ben' Fernandez," https://www.youtube.com/watch?v=od8I21WT6sk, accessed on July 16, 2019.

184 "Ben Fernandez is": Fernandez lecture, November 10, 1978.

184 If they weren't: Ibid.

185 He tried to rally: Francis-Fallon, *The Rise of the Latino Vote*, 350–51.

185 At a November: Ibid.

185 In his testimony: Fernandez testimony, November 8, 1973.

185 By the time: Casillas interview, February 19, 2019.

186 In fact, he: Carl Leubsdorf, "Fernandez Seeks GOP Nomination," *Baltimore Sun*, November 30, 1978, A9.

186 His yearbook said: Casillas interview, February 19, 2019.

187 Casillas's family had crossed: Ibid.

187 He said that his experience: Betty Renkor, "Born in a Boxcar, Seeks Presidency," *Southtown Star*, October 25, 1979, 7.

187 When a friend: George Will, "Benjamin Fernandez: A Natural Republican," *Washington Post*, August 23, 1979, A21.

187 In an interview: Frank del Olmo, "Fernandez' Last Quixotic Gesture," *Los Angeles Times*, June 3, 1980, 16.

188 He was convinced: Fernandez testimony, November 8, 1973.

188 Fernandez made a compelling case: "'Boxcar Ben' Fernandez," https://www.youtube.com/watch?v=od8I21WT6sk, accessed on July 16, 2019.

188 Yet New York: DiMartino interview, May 27, 2016.

189 The city had lost: Lee, *Building a Latino Civil Rights Movement*, chap. 6.

189 Fernandez thought he: Sanchez, "Latin in White House? Some See Possibility."

190 In the bicentennial year: "US, Cuba Mum on Issues," *Pensacola News Journal*, March 26, 1977, 4A.

190 In November 1978: García, *Havana USA*, 47; Torres, *In the Land of Mirrors*, 81–83.

190 Carter then commuted: García, *Havana USA*, chaps. 3 and 4; Torres, *In the Land of Mirrors*, chap. 4.

191 A greater number: García, *Havana USA*, 140; Torres, *In the Land of Mirrors*, chap. 4.

191 But as the: *Time*, October 16, 1978, 51–52.

192 Carter no longer: Westad, *Global Cold War*, 339–341.

193 He saw communism: Leubsdorf, "Fernandez Seeks GOP Nomination."

193 The United States needed: Renkor, "Born in a Boxcar."

193 Failing comprehensive reform: Jim Schutze, "6½-Mile-Long 'Tortilla Curtain' Planned to Stem Tide of Illegal Mexican Aliens," *Washington Post*, October 24, 1978, A6.

194 Three filed suits: Cadava, *Standing on Common Ground*, chap. 5; "Goldwater's Glass House," *Nuestro: The Magazine for Latinos*, April 1978, 10.

195 "Whatever we do . . .": Fernandez lecture, November 10, 1978.

195 What better way: Renkor, "Born in a Boxcar."

195 His longtime friend: Roger Langley, "GOP Tries Harder," *Jersey Journal*, October 16, 1979, 17.

196 His off-the-cuff: Casillas interview, February 19, 2019.

196 Casillas observed how: Ibid.

196 Sometimes Fernandez let: Ibid; Francis-Fallon, *The Rise of the Latino Vote*, 351.

197 We "didn't have": Ibid.

197 Along with "dominos": William Eaton, "GOP Vote in Puerto Rico Today," *Boston Globe*, February 17, 1980, 17.

197 Puerto Rico's *ley*: Baralt, *La razón del equilibrio*,

198 From that moment: Quiñones Calderón, *Historia Política de Puerto Rico, Tomo I*, 483–486.

198 Fernandez needed them: Renkor, "Born in a Boxcar."

198 Fernandez predicted that: "Can Fernandez Attract Hispanics to Cause?," *Arizona Daily Sun*, January 14, 1980, 4.

198 Then all of: Renkor, "Born in a Boxcar."

198 In many ways: "Can Fernandez Attract Hispanics to Cause?"

199 Fernandez said his: Will, "Benjamin Fernandez: A Natural Republican."

199 As one Puerto Rican Republican: Ramos interview, March 20, 2019.

199 Among the small: Will, "Benjamin Fernandez: A Natural Republican."

199 Bush, on the other hand: Guinot interview, February 8, 2019.

200 As it turned out: Ibid.

200 He called momentum: Ibid.

200 Bush and Guinot clicked: Ibid.

201 Bush traveled to: Ibid.

201 The result, according: Ward Sinclair, "Bush Wins Puerto Rico Race," *Washington Post*, February 18, 1980.

201 Jeb said that: Ibid.

202 Unfortunately for Fernandez: United Press International, "Carter Camp Predicts Tight New Hampshire Race," *Republic*, February 19, 1980.

202 After his stinging: "Fernandez Still in Race," *San Antonio Express*, March 21, 1980, 6A.

202 In his hometown: Del Olmo, "Fernandez' Last Quixotic Gesture."

203 Fernandez dropped out: Ibid.

203 It was "crushing": Mark Stein, "Fernandez Refuses to Drop Quest for Presidency," *Los Angeles Times*, March 6, 1980, 1.

CHAPTER 7: REAGAN'S *REVOLUCIÓN*

204 As a surrogate: Michael Daly, "Boxcar Ben Delivers the GOP's Minority Report," *New York Daily News*, July 15, 1980, 4.

204 Cuban Americans from Miami: Barry Bearak, "Cubanos en Detroit insisten en politica anticastrista," *El Nuevo Herald*, July 17, 1980, 1.

205 They were readying: Francis-Fallon, *The Rise of the Latino Vote*, 355.

205 Fernandez complained about: Daly, "Boxcar Ben Delivers the GOP's Minority Report."

205 After fielding enough: Fernando Oaxaca to Elizabeth Dole, August 1, 1980, box 333, folder "Voter Groups-Hispanics-Correspondence & Memos-Internal (1/7)," 1980 Campaign Papers, 1965–1980, Series X, Citizens' Operations (Max Hugel), Voter Group Files, Reagan Library.

205 They had assembled: Fernando Oaxaca to Bill Casey, July 27, 1980, box 222, folder "Media Campaign-[Hispanics] (Peter Dailey)," 1980 Campaign Papers, 1965–1980, Series V, Media, Advertising, Promotion (Dailey), Reagan Library.

205 Carter had the lowest: Gallup Historical Presidential Job Approval Statistics, Overall Averages, https://news.gallup.com/poll/116677/Presidential-Approval-Ratings-Gallup-Historical-Statistics-Trends.aspx, accessed on July 17, 2019.

206 The campaign finally: Summary of Reagan and Bush campaign visits in September 1980, box 333, folder "Voter Groups-Hispanics-Correspondence & Memos-Internal (2/7)," 1980 Campaign Papers, 1965–1980, Series X, Citizens' Operations (Max Hugel), Voter Group Files, Reagan Library.

206 Reagan again focused: "Fernandez Speaks Here—Hispanic Vote Critical, Says Former Candidate," *Odessa American*, September 29, 1980, D1.

206 Reagan spent Mexican Independence Day: "Fernandez Speaks Here."

207 Three weeks after: "Fernandez to Speak at Reagan Rally," *Seguin Gazette-Enterprise*, October 10, 1980, 1.

207 It was no less of: Francis-Fallon, *The Rise of the Latino Vote*, 365–78.

208 His close friend: Casillas interview, February 19, 2019.

208 They formulated the strategy: Fernando Oaxaca, "The Elements of Strategy: Attracting the Hispanic Vote—1980," May 1980, box 302, folder "National Citizens' Operation, Hispanics," 1980 Campaign Papers, 1965–1980, Series X, Citizens' Operations (Max Hugel), Subject Files, Reagan Library.

209 To those who thought: Fernando Oaxaca to Max Hugel, July 22, 1980, box 333, folder "National Citizens' Operation, Hispanics," 1980 Campaign Papers, 1965–1980, Series X, Citizens' Operations (Max Hugel), Subject Files, Reagan Library.

209 For the 1980 election: Santillan and Subervi-Veléz, "Latino Participation in Republican Party Politics in California," 302.

209 He also opposed quotas: "RR Prepared Remarks East Los Angeles," September 26, 1980, box 333, folder "Voter Groups-Hispanics-Correspondence & Memos-Internal (4/7)," 1980 Campaign Papers, 1965–1980, Series X, Citizens' Operations (Max Hugel), Voter Group Files, Reagan Library.

210 The campaign brought: Sosa interview, June 4, 2019.

210 Even though Armendariz: Fernando Oaxaca to Max Hugel, July 22, 1980, box 333, folder "National Citizens' Operation, Hispanics," 1980 Campaign Papers, 1965–1980, Series X, Citizens' Operations (Max Hugel), Subject Files, Reagan Library.

211 Oaxaca further visualized: Fernando Oaxaca to Max Hugel, July 22, 1980.

212 Oaxaca argued that: Fernando Oaxaca to William Casey, July 27, 1980,

box 222, folder "Media Campaign," Campaign Papers, 1965–1980, Series V, Media, Promotion, Advertising (Dailey), Reagan Library.

212 The goal of: Fernando Oaxaca to Max Hugel, July 22, 1980.

212 He amassed considerable: Brilliant, *The Color of American Has Changed*, chap. 8.

212 Reagan was also a westerner: Zapanta interview, August 7, 2015.

213 For fellow Hispanics: "Hispanic Strategy for Reagan," box 302, folder "National Citizens' Operation—Summaries, Hispanics," 1980 Campaign Papers, 1965–1980, Series X, Citizens' Operations (Max Hugel), Subject Files, Reagan Library.

213 Cuban Americans and Cuban exiles: Frank Gutierrez to Ronald Reagan, n.d., box 333, folder "Hispanics-Correspondence-Not Answered (Received Too Late) (1/3)," 1980 Campaign Papers, 1965–1980, Series X, Citizens' Operations (Max Hugel), Voter Groups, Reagan Library.

213 The new Refugee Act: Refugee Act of 1980, Public Law 96-212, March 17, 1980.

213 When they were still: Robert Pear, "US Policy on Cuban Refugees: Background and Prospects," *New York Times*, May 8, 1980, A12.

213 At first the refugees: Torres, *In the Land of Mirrors*, chap. 5; and Capó, "Queering Mariel."

214 Puerto Ricans, meanwhile: Oaxaca, "The Elements of Strategy: Attracting the Hispanic Vote—1980."

214 Finally, Mexican Americans: Ibid.

215 Puerto Ricans who called for: Ronald Reagan radio broadcast on Cuba, March 6, 1979, box 15, folder "R. Reagan Radio Broadcasts, 3/6/1979," 1980 Campaign Papers, 1965–1980, Series I—Hannaford California Papers, Reagan Library.

216 During a campaign stop: Dave McNeely, "Reagan Aims Texas Tour at Mexican-American Votes," *Longview Daily News*, September 17, 1980, 8D.

216 Farther down the campaign trail: Reagan quoted in Garry Wills, *Reagan's America*, preface.

216 Just before the convention: Alejandro Olais Olivas, "México, el País más Importante Para EU," *El Sonorense*, July 6, 1980, 1.

216 The Mexican lawyer: Agustín Navarro Vázquez to Steve Symms, October 18, 1980, with articles from *Impacto* attached, box 333, folder

"Voter Groups-Hispanics-Correspondence-Not Answered (Received Too Late) (1/3)," 1980 Campaign Papers, 1965–1980, Series X: Citizens' Operations (Max Hugel), Reagan Library.

217 Writing from his home: Mario Moya Roldán to Ronald Reagan, October 1980, box 333, folder "Voter Groups-Hispanics-Correspondence-Not Answered (Received Too Late) (3/3)," 1980 Campaign Papers, 1965–1980, Series X: Citizens' Operations (Max Hugel), Reagan Library.

217 A letter sent: Juan Thompson to Ronald Reagan, September 24, 1980, box 332, folder "Voter Groups (Hugel)-Hispanics-Correspondence-Answered 10/27/80 (2/2)," 1980 Campaign Papers, 1965–80, Series X: Citizens' Operations (Max Hugel), Voter Group Files, Reagan Library.

217 Her grandmother had just: Eolia P. Sibila to Ronald Reagan, October 10, 1980, box 333, folder "Voter Groups-Hispanics-Correspondence & Memos-External (4/5) (Hugel)," 1980 Campaign Papers, 1965–1980, Series X: Citizens' Operations (Max Hugel), Reagan Library.

217 She wanted to wish: Ondina Gonzalez to Ronald Reagan and George H. W. Bush, October 29, 1980, box 332, "Voter Groups-Hispanics-Correspondence-Misc. (Hugel)," 1980 Campaign Papers, 1965–1980, Series X: Citizens' Operations (Max Hugel), Reagan Library.

218 Reagan and Bush flipped: Torres, *In the Land of Mirrors*, chap. 5.

218 They did well: Frank del Olmo, "Politicos Point to Latino Vote as Proof—but of What?" *Los Angeles Times*, November 13, 1980, A21.

219 Just a week after: Stephen E. Nordlinger, "Reagan Victory Adds Urgency to Bilingual Fight," *Baltimore Sun*, November 12, 1980, A1.

219 In December 1980: Evan Maxwell, "Limited Amnesty for Illegal Aliens Proposed by Panel," *Los Angeles Times*, December 7, 1980, A1.

219 Causing them great consternation: "Reagan Likely to Tap Woman, Hispanic for Cabinet to Curb Controversy on Picks," *Wall Street Journal*, December 18, 1980, 3.

220 They convened at Blair House: "Recibe Reagan a dirigentes hispanos," *El Miami Herald*, December 11, 1980, 5.

221 As for the hard-liners: Torres, *In the Land of Mirrors*, chap. 5.

221 A former freedom fighter: Mirta Ojito, "Castro Foe's Legacy; Success, Not Victory," *New York Times*, November 30, 1997, WK5.

221 In response, Mas Canosa: Cuban American National Foundation Articles of Incorporation, July 6, 1981, www.canf.org, accessed on April 18, 2019.

222 As he liked to say: Torres, *In the Land of Mirrors*, 96.

222 Through the CIA: Westad, *The Global Cold War*, 341–45.

222 Growing numbers of: Cadava, *Standing on Common Ground*, chap. 5.

223 After Somoza's overthrow: "Somoza's Body Flown to Miami," *Atlanta Constitution*, September 20, 1980, 8A.

223 They began to train: "Nicaraguan Exiles Train in Florida for Guerrilla Assaults," *Atlanta Constitution*, March 17, 1981, 6A.

223 The recently formed: Carly Goodman, "How John Tanton's Network of Organizations Spread Anti-Immigrant Extremism," paper presented at the University of Chicago, April 12, 2019.

223 The rise in: Roger Conner to Members and Friends of FAIR, November 1981; and Roger Conner to Members and Friends of FAIR, December 1981, box 11, folder 3, Bonilla Collection.

224 Oaxaca, the cochair: Laurie Becklund, "2 GOP Latinos Support Foes of Measure on Immigration," *Los Angeles Times*, July 16, 1984, B3.

225 Tanton led US English: S. I. Hayakawa, quoted in Chavez, *An Unlikely Conservative*, 199.

225 Reagan had lunch: Ronald Reagan remarks at La Esquina de Tejas, May 20, 1983, https://www.youtube.com/watch?v=ZCmqceiup9I, accessed on April 19, 2019.

226 Mas Canosa introduced: "Discurso pronunciado por Jorge Mas Canosa, Presidente de la Fundación Nacional Cubano-Americana, en ocasión de la visita a Miami del Presidente Ronald W. Reagan, Dade County Auditorium, 20 de Mayo, 1983," www.canf.org, accessed on April 19, 2019.

226 Reagan walked to: "Text of President Reagan's Speech on Threat to Latin America," *New York Times*, May 21, 1983, 4.

227 The *Los Angeles Times* said: "Cheering Exiles Hear Reagan Blast Castro," *Los Angeles Times*, May 20, 1983, A1.

227 Reagan had pardoned: Alfonso Chardy, "Reagan Rips Castro Rule as 'Fascist,'" *Miami Herald*, May 21, 1983, 1A.

227 Another early stop: George Skelton, "Latinos Applaud Reagan Praise of Hard Work," *Los Angeles Times*, August 26, 1983, 1.

228 Nevertheless, more than 250: Santillan and Subervi-Veléz, "Latino Participation in Republican Party Politics in California," 302.

229 They organized a conference: Becklund, "2 GOP Latinos Support Foes of Measure on Immigration."

229 Because of Simpson-Mazzoli: Ibid.

229 The Cuban American appointees: Torres, *In the Land of Mirrors*, 99.

229 The Mexican American appointees: Chavez, *An Unlikely Conservative*, chap. 6.

230 The highest-ranking Hispanic: Santillan and Subervi-Veléz, "Latino Participation in Republican Party Politics in California."

230 Hispanics again charged: Lou Cannon, "Reagan Taps Hispanic for US Treasurer," *Washington Post*, September 13, 1983, A1.

231 Leaders of Mexico's: Zapanta interview, August 7, 2015.

231 On August 21, 1984: Susan Rasky, "'I Was Born a Republican': Katherine Davalos Ortega," *New York Times*, August 21, 1984, A20.

231 Just the evening before: Katherine Ortega 1984 Republican national convention keynote speech, https://www.c-span.org/video/?124532-101 /katherine-ortega-1984-republican-national-convention-keynote-speech, accessed on July 18, 2019.

231 Ortega had come a long way: Rasky, "'I Was Born a Republican.'"

232 When the applause died down: Katherine Ortega 1984 Republican national convention keynote speech.

232 As a keynote speaker: Ibid.

233 Where Carter and Mondale: Ibid.

233 These campaign themes: Ibid.

CHAPTER 8: IMMIGRATION DIVIDES

234 The maids, waiters, and waitresses: Robert Beier, "Reagan, Top Officials Praise Ortega Speech," *Albuquerque Journal*, August 22, 1984, A3.

234 "Apple pie was": Gary Deeb, "GOP Coverage Put Viewers to Sleep," *Idaho Statesman*, September 2, 1984, 28.

234 They were restless: Jack Colwell, "Hiler Resembles Seer After Keynote Speech," *South Bend Tribune*, August 22, 1984, 21.

234 David Brinkley wrote: Brinkley quoted in Mike Duffy, "Dallas Dullness Enlivened a Bit by Verbal Gems," *Detroit Free Press*, August 24, 1984, 4B.

234 Another writer put it: "With Conventions Over, Next Move's Up to Mondale," *Macon Telegraph News*, August 28, 1984, 9A.

235 In case Reagan: Mary Espinoza, "Not Humble Enough," *Fresno Bee*, August 30, 1984, B11.

235 In the words of: Juan Gonzalez, "Ortega Let Her People Down," *Adrian (MI) Daily Telegram*, September 17, 1984, 4.

235 He opened by saying: "Reagan's Speech to Hispanics," *Arizona Daily Star*, August 24, 1984, 10A.

235 He was just getting: Robert Beier, "Reagan Campaigns for Hispanic Vote at Dallas Luncheon," *Albuquerque Journal*, August 24, 1984, A12.

236 After the convention: Don Lewis, "US Treasurer speaks in NO to Republican Hispanic Group," *Advocate*, October 13, 1984, 6A.

236 The national chairman: Gonzalez, "Ortega Let Her People Down."

236 He wrote a column: Carlos Márquez-Sterling, "Un vicepresidente que puede ser presidente," *Diario las Américas*, December 14, 1984, 5A.

237 He liked Bush: Ibid.

237 Days after the column: Jeb Bush to Carlos Márquez-Sterling, December 19, 1984, box 2, folder 67, Márquez-Sterling Papers.

237 His victory, an article: Luis Feldstein Soto, "Reagan's Picnic at the Polls Had Some Hispanic Flavor," *Miami Herald*, November 11, 1984, 10A.

237 Hispanics had become: Charles Ericksen, "Bright Side for Hispanics in Politics," *Fresno Bee*, October 1, 1984, B13.

237 The Mexican American: Zapanta interview, August 7, 2015.

239 Two of the board's: Torres, *In the Land of Mirrors*, 97.

239 Named after José Martí: *Report by the Advisory Board for Radio Broadcasting to Cuba.*

240 A year in: "Second Thoughts on Radio Martí," *New York Times*, May 22, 1986, 26.

240 One listener in Cuba: *Report by the Advisory Board for Radio Broadcasting to Cuba.*

240 Radio Martí's director: Betancourt, quoted in Dick Capen, "All Share in Radio Martí's Goal," *Miami Herald*, October 4, 1987, 3C.

241 Democratic congressman from California: Paul Houston, "Latinos Oppose Employer Sanctions, Roybal Says," *Los Angeles Times*, April 25, 1985, B14.

241 Republicans claimed that: Wong, *Lobbying for Inclusion*, chap. 5.

241 By the Reagan years: Minian, *Undocumented Lives*, chap. 7.

242 In the two years: Charles E. Schumer, "Back to Immigration," *New York Times*, May 21, 1985, A27.

242 The goals for Democrats: Bob Secter, "New Alien Bill Delays Amnesty," *Los Angeles Times*, May 24, 1985, 1.

242 In the year: Resolution by the Northeast Region of the Republican National Hispanic Assembly, *Congressional Record—Senate*, May 6, 1985, 10536.

243 She moved over: Chavez, *An Unlikely Conservative*, 174, 181.

243 Even Mexican Americans: Lydia Chávez, "Fears Prompted Hispanic Votes in Bill's Support," *New York Times*, November 11, 1986, A12.

243 If employers feared: "Torpedo from the White House," *New York Times*, January 24, 1986, A26.

244 Members of the president's: Ibid.

244 But Luján's logic: Paul R. Wieck, "Luján Poll Shows Support for Alien Hiring Penalties," *Albuquerque Journal*, September 18, 1986, B11.

245 Hispanics were more divided: Christine Marie Sierra, "In Search of National Power: Chicanos Working the System on Immigration Reform, 1976–1986," in Montejano, ed., *Chicano Politics and Society in the Late Twentieth Century*, chap. 6.

245 The president acknowledged: "President Reagan's Remarks at Ceremony for Immigration Reform and Control Act in Roosevelt Room," November 6, 1986, https://www.youtube.com/watch?v=FvZ0QHpxmRs&t=24s, accessed on May 9, 2019.

245 As Reagan exited: Ibid.

246 Linda Chavez was on: Chavez, *An Unlikely Conservative*, 187–96.

246 Her Democratic opponent: W. Waldron, "Mikulski, Chavez Woo Ethnic Votes," *Baltimore Sun*, October 2, 1986.

246 They took the usual: Ileana Ros-Lehtinen, "Un nuevo reto para Dexter," *Diario las Américas*, June 19, 1988, 6B.

247 President Reagan himself: Ronald Reagan and Nancy Reagan address to the nation on the campaign against drug abuse, September 14, 1986, https://www.youtube.com/watch?v=pwpciZ7R8UU, accessed on July 20, 2019.

247 One was his call: Kevin Roderick, "On the Border: US Senate Candidates Focus on Policy to Combat Illegal Immigration," *Los Angeles Times*, May 30, 1986, OC3.

248 US English, the group that funded: John Trasvina, "Bilingual Americans," *San Francisco Examiner*, October 12, 1986, A13.

248 By 1986, US English: William Trombley, "Prop. 63 Roots Traced to Small Michigan City," *Los Angeles Times*, October 20, 1986, A3.

248 US English raised: Ibid.

249 Generally, US English: Ibid.

249 The opponents of Proposition 63: Ibid.

249 Hayakawa, though: Hayakawa quoted in HoSang, *Racial Propositions*, chap. 5.

249 The passage of Proposition 63: HoSang, *Racial Propositions*, chap. 5.

249 The "people behind": Trombley, "Prop. 63 Roots Traced to Small Michigan City."

250 Just days after: "GOP Panel: Make English 'Official,'" *Austin American-Statesman*, November 16, 1986.

250 He said it was: Gary Scharrer, "Hispanic Leader Says GOP Was No Help," *El Paso Times*, November 18, 1986, 4.

250 Finally, he warned: "GOP Panel: Make English 'Official.'"

250 She argued that bilingual education: Chavez, *An Unlikely Conservative*, 213–14.

250 The organization founded by: Ibid., 196–206.

251 After her failed Senate race: Ibid.

251 But accepting the position: Ibid.

251 Chavez's time as president: Ibid.

251 While Chavez traveled: Elaine Woo, "Entrepreneur Fernando Oaxaca, 76, Dies," *Los Angeles Times*, June 3, 2004.

252 In the days, weeks, and months: Chavez, "Fears Prompted Hispanic Votes in Bill's Support."

252 A month after: John Crewdson and Vincent Schodolski, "Tighter Border Could Strangle Mexico," *Chicago Tribune*, December 18, 1986, 1.

252 The skepticism about: Mick Rood, "Former Opponent Is Hired to Explain Immigration Law," *Arizona Daily Star*, April 10, 1987.

253 Because of how controversial: Ibid.

253 The main difficulty: Constanza Montana and Dianna Solis, "INS Starts Media Campaign to Promote New Amnesty Plan," *Wall Street Journal*, April 9, 1987, 19.

253 Herman Baca: Ibid.

253 Compounding his problem: Judy Wiessler, "Former Opponent of Alien Bill Will Direct Its Publicity Drive," *Houston Chronicle*, April 9, 1987, 4.

254 The US-backed president of El Salvador: Robert Pear, "Duarte Appeals to Reagan to Let Salvadorans Stay," *New York Times*, April 26, 1987, 1.

254 When she was still a Florida state senator: Ileana Ros-Lehtinen, "Humana demanda nicaragüense: derecho al trabajo," *Diario las Américas*, July 7, 1987, 4B.

254 The vast majority: García, *Seeking Refuge*, chap. 1.

255 Congressman Luján agreed: Manuel Luján Jr. to "Martha," March 11, 1985, box 49, folder 5, "Luján Clippings, 1985," Luján Congressional Papers.

255 Since 1982, Sanctuary Movement: Cadava, *Standing on Common Ground*, 205–6.

256 As sanctuary volunteers: David Siegel, "Contras' Dirty War Infects Innocents Far from the Fighting," *Los Angeles Times*, October 9, 1986, D5.

256 Meanwhile, Contra supporters: Ellen Hampton, "Embattled Contras Await Vote on Aid," *Miami News*, June 23, 1986, 9.

257 Iran-Contra, wrote: "House Wants Aid to Contras Halted," *El Paso Times*, March 12, 1987, 1.

257 Ros-Lehtinen proposed: Paul Anderson, "$13 Million in Aid to Nicaraguan Refugees Endorsed by HRS Committee," *Miami Herald*, May 8, 1987, 22A.

257 Dozens of Cuban Americans: Marshall Ingwerson, "Miami a Rebel Base Again," *Christian Science Monitor*, May 30, 1986, 3.

258 Some Cuban Americans: Frank del Olmo, "Miami: A Refuge for the Contras?," *Los Angeles Times*, March 23, 1987, C5.

258 Nevertheless, dozens of: Julia Preston and Joe Pichirallo, "Bay of Pigs Survivors Find Common Cause with Contras," *Washington Post*, October 26, 1986, A1.

258 Congressman Manel Luján Jr.: "Luján Supports Rebels in Nicaraguan Struggle," n.p., March 11, 1985, B7, box 49, folder 5, "Luján Clippings, 1985," Luján Congressional Papers.

258 Luján said he: "House Wants Aid to Contras Halted."

258 A month into: Manuel Luján Jr. to Gladys Winblad, June 29, 1987, box 49, folder 2, "Luján's Replies to Constituents' Letters, June–August 1987," Luján Congressional Papers.

259 In his correspondence: Colleen Heild, "Radio Ads Fail to Change Richardson," *Albuquerque Journal*, n.d. news clipping, box 49, folder 6, "Luján Clippings, 1986–1987," Luján Congressional Papers.

259 When constituents wrote: Manuel Luján Jr. to E. Gillon, January 26, 1987, box 49, folder 1, "Luján Replies to Constituents' Letters, January to May 1987," Luján Congressional Papers.

260 In a letter to: Manuel Luján Jr. to Ralph Bradley, August 10, 1987, box 49, folder 2, "Luján's Replies to Constituents' Letters, June–August 1987," Luján Congressional Papers.

260 Mas Canosa and: Luis Feldstein Soto, "VIPs Role in Detainees' Cause Diminished After Riots," *Miami Herald*, November 20, 1988, 27A.

261 What if one: Ibid.

261 The RNHA didn't know: Robert Joffee and Donna Gehrke, "Feds Vow to Press Some Deportations," *Miami News*, December 5, 1987, 1.

261 When Vice President Bush: Ileana Ros-Lehtinen, "George Bush entre nosotros," *Diario las Américas*, October 25, 1987, 6B.

261 More generally: Ileana Ros-Lehtinen, "Presencia hispana en la política," *Diario las Américas*, March 29, 1987, 4B.

261 She wrote another article: Ileana Ros-Lehtinen, "Comportamiento nacional del voto hispano," *Diario las Américas*, July 31, 1988, 9A.

262 Hispanics, Ros-Lehtinen wrote: Ileana Ros-Lehtinen, "Las próximas convenciones nacionales," *Diario las Américas*, May 15, 1988, 5B.

262 Some Cuban Americans: Dr. Luis Rodriguez Cepero and Dra. Dora G. Portela to Ileana Ros, March 18, 1988, box 3, "Varios Español, 1988," Ros-Lehtinen Collection.

262 But these sins: Ileana Ros-Lehtinen, "Lamentable coincidencia demócrata: manos fuera de Nicaragua," January 24, 1988, 3B.

262 They would turn: Ileana Ros-Lehtinen, "Posición de los candidatos presidenciales demócratas," *Diario las Américas*, March 13, 1988, 4B.

263 US English sought: Lozano, *An American Language*, 259–60.

263 Conservative but mainstream: Geraldo Cadava, "How Should Historians Remember the 1965 Immigration and Nationality Act?," *American Historian* (August 2015).

CHAPTER 9: LA FAMILIA BUSH

267 Ignoring the controversies: Ronald Reagan, quoted in 1988 RNC Report, 118.

267 The trick for Bush: Janet Cawley, "Bush to Take Risk with Tall Shadow," *Chicago Tribune*, August 12, 1988, 8.

268 "These are Jebby's kids . . .": "Bush Introduces 3 Grandchildren as 'Little Brown Ones,'" *Los Angeles Times*, August 16, 1988, 1.

268 Ruben Bonilla, the general counsel: "Bush Causes a Stir with Hispanics," *Philadelphia Inquirer*, August 17, 1988, A15.

268 Bonilla wanted to know: Maureen Santini, "'Little Brown Ones,'" *New York Daily News*, August 17, 1988, 22.

268 Bush countered their outcry: "Bush Defends 'Little Brown Ones' Term for Grandchildren, Tells 'Pride and Love,'" *Los Angeles Times*, August 17, 1988, 2.

268 They gave a series: "Hispanics Disagree on Bush Remark," *Philadelphia Daily News*, August 17, 1988, 4.

269 A Hispanic teen: Johnny Martinez and Rene Gracida, quoted in 1988 RNC Report, 53, 417, 500.

269 Leaders of the: Catalina Vasquez Villalpando, quoted in 1988 RNC Report, 44.

269 Florida state senator: Ileana Ros-Lehtinen in 1988 RNC Report, 31.

269 It wasn't every day: Ileana Ros-Lehtinen, "Termina la convención Republicana," *Diario las Américas*, August 21, 1988, 6B.

270 Meanwhile, the longtime Republican: Baralt, *La razón del equilibrio*, 263.

270 If there was any doubt: Columba Bush, quoted in 1988 RNC Report, 503.

271 When Jeb Bush was still: Michael Kruse, "Andover, Mexico, and the Making of Jeb Bush," *Politico Magazine*, https://www.politico.com/magazine/gallery/2015/05/andover-mexico-and-the-making-of-jeb-bush-000150?slide=0, accessed on July 24, 2019.

272 His employer after college: Maria Garcia, "The VP's Son Making His Own Mark in the GOP," *Chicago Tribune*, July 26, 1984, S1.

272 His father had taken: Paul Wieck, "Bush's Bid for Hispanic Votes Gets Boost from Son's Wife," *Albuquerque Journal*, October 18, 1988, A12.

273 As a student at Yale: David Hoffman, "Grappling with Image Problem, Bush Tries to Show 'What Makes Me Tick,'" *Washington Post*, July 18, 1988, A17.

273 It was equalizing rhetoric: Ibid.

273 The "Hispanics for Bush": Claudia Luther, "Bush Woos the Latino Vote at Breakfast and Makes Tortilla," *Los Angeles Times*, September 15, 1988, OC1.

274 Bush's Democratic opponent: Edward Walsh, "Dukakis Is Wearing Ethnicity on His Sleeve; First-Generation American Woos Growing Blocs of Asian and Hispanic Voters," *Washington Post*, May 22, 1987, A7.

274 It didn't matter that Dukakis: David Rogers, "Bush and Dukakis Are Waging 'a Battle for Texas' Soul,'" *Seattle Times*, September 28, 1988, B1.

274 In an interview with Univision: "Considera Ileana Ros que triunfó el anticomunismo en la Florida," *Diario las Américas*, November 10, 1988, 6B.

275 They met with Bush: Ileana Ros-Lehtinen, "Reunión con el presidente electo," *Diario las Américas*, December 25, 1988.

275 Bush was inaugurated: Jeannine Stein, "Inaugural Events Show Minority Support of GOP," *Los Angeles Times*, January 20, 1989, F1.

276 In early August: "1st Latino Named to Cabinet," *Los Angeles Times*, August 9, 1988, 1.

276 Cavazos's nomination hearing: Edward Kennedy in Lauro Cavazos nomination hearing, September 9, 1988.

276 Senators variously hailed: Lauro Cavazos nomination hearing, September 9, 1988.

277 *Los Angeles Times* reporter: "Bush Pledges to Include Minorities, Women, and Democrats in Cabinet," *Orange County Register*, November 30, 1988, A18.

277 Luján had been short-listed: Opening statement by J. Bennet Johnston in Manuel Luján nomination hearing, January 26, 1989, 1.

277 For most of his time: Pete Domenici and Bill Richardson in Manuel Luján nomination hearing, January 26, 1989, 18, 24.

277 After the Senate confirmed: "4 Bush Nominees Backed by Senate," *New York Times*, February 3, 1989, A11.

277 Bush appointed more: Randi Henderson, "To Her Health: Antonia Coello Novello Marks First Month as Surgeon General," *Baltimore Sun*, April 11, 1990, 1E.

278 About halfway through: George H. W. Bush, State of the Union address, February 9, 1989, https://www.youtube.com/watch?v=hV1O2FBXep0, accessed on May 16, 2019.

279 To shore up the sitting president's statement: Gerald Ford to Luis Ferré, February 14, 1989; Richard Nixon to Luis Ferré, March 24, 1989; and Ronald Reagan to Luis Ferré, June 22, 1989, all Partido Republican Nacional, box 5, folder 4, "Comunic. Pres. Bush," Ferré Archive.

279 The members of the PNP: Baralt, *La razón del equilibrio*, 265.

279 In relatively short order: Ibid., 268.

280 It didn't take long: "The End of Section 936," *Puerto Rico Report*, August 29, 2016, https://www.puertoricoreport.com/the-end-of-section-936/#.XOH3h6Z7lPM, accessed on May 19, 2019.

280 For about fifteen years: Baralt, *La razón del equilibrio*, 268.

280 Congress began to back away: Ibid., 269.

281 Pepper had died: "Bush Goes Out of His Way to Back Candidate," *Los Angeles Times*, August 17, 1989, 4.

281 The candidates talked: "Florida Elects Havana-Born State Senator as the First Cuban-American in Congress," *Los Angeles Times*, August 30, 1989, 4.

281 In the end: Adela Gooch, "Ethnic Divisions Dominate Florida's House Race," *Washington Post*, August 27, 1989, A3.

282 The day after: Adela Gooch, "Little Havana Exults Over Vote Outcome," *Washington Post*, September 1, 1989, A12.

282 During one of: Donnie Radcliffe, "Bush's Outstanding Hispanics," *Washington Post*, September 13, 1989, D4.

282 A Gallup poll: James P. Gannon, "First-Year Moves Shift Bush from Wimp to Winner," *Detroit News and Free Press*, January 14, 1990, 1A.

283 He was criticized: Jack Anderson and Dale Van Atta, "Education Secretary to Go?," *Daily Telegram*, January 2, 1990, 4.

283 When there were whispers: "Education Secretary Cavazos Defended by White House," *San Francisco Chronicle*, January 4, 1990, A10.

283 Bush's resolve: Baralt, *La razón del equilibrio*, 269.

283 Ferré complained bitterly: Ibid., 270.

284 Pat Buchanan, a former assistant: Patrick Buchanan, "Puerto Rico as Our 51st State?," *Washington Times*, February 26, 1990, D1.

285 The rest of what: Ibid.

285 Ferré responded immediately: Luis A. Ferré, "Puerto Rico Is Not 'Northern Ireland,'" *Washington Times*, March 16, 1990, F4.

285 He reminded Buchanan: Ibid.

286 Puerto Rican Republicans: Antonio M. Longo, "Buchanan Hit as 'Wrong' on Puerto Rico," *Washington Times*, March 23, 1990, F4.

286 Like a skilled debater: Patrick Buchanan, "Future State or Nation," *Washington Times*, May 16, 1990, F3.

286 Why, Ferré wondered: Baralt, *La razón del equilibrio*, 272.

286 Bush administration officials: Jaime B. Fuster, "Throwing Weight for Puerto Rican Statehood," *Washington Post*, November 4, 1990, C6.

287 Puerto Ricans continued: Baralt, *La razón del equilibrio*, 273.

287 To express their disappointment: Ibid.

288 As Bush and Russian leaders: Ramón Iglesias, "¿Estará al jugarse la carta de Cuba?," *Diario las Américas*, June 30, 1991.

288 When she launched: "Inician campaña de reelección de la congresista Ros-Lehtinen," *Diario las Américas*, October 24, 1991, 3B.

288 But Bush's "declarations on Cuba": José Sánchez-Boudy, "Las palabras de Bush sobre Cuba," *Diario las Américas*, June 11, 1991, 5A.

289 The growing skepticism: Rick Mendosa, "Bush Addresses Key Issues," *Hispanic Business* 13, no. 5 (May 1991).

289 Ferré had just written: Luis Ferré to George Bush, March 7, 1991, Partido Republicano Nacional, box 5, folder 4, "Communic. Pres. Bush," Ferré Archive.

290 Moreover, Hispanic high school dropout: Simon Tisdall, "Bush Fails His Own Education Presidency Test," *Guardian*, August 19, 1992, 9.

290 Bush's personal reputation: Mendosa, "Bush Addresses Key Issues."

291 Buchanan's repugnant views: Douglas Jehl, "Buchanan Raises Specter of Intolerance, Critics Say," *Los Angeles Times*, March 17, 1992, A1; and Manning Marable, "Buchanan: Racism on the Far Right," *Los Angeles Sentinel*, April 9, 1992, A6.

291 There was no love lost: Luis Ferré to David Carney, February 24, 1992, Partido Republicano Nacional, box 2, folder 1, Ferré Archive.

291 Republicans thought free trade: "Oposición al comercio con México mientras ese país siga comerciando con Cuba," *Diario las Américas*, December 11, 1991, 4A.

292 Hispanics met the Democratic nominee: Steven Komarow, "Candidates Court Possibly Crucial Hispanic Vote," *Seguin Gazette-Enterprise*, September 6, 1992, 10A.

292 On the eve of the: David Maraniss, "Bush Called on 'Sleaze' Vow After GOP Remarks," *Washington Post*, August 18, 1992, A17.

292 Villalpando said she was: Bruce Davidson, "Treasurer Sorry for 'Skirt-Chaser' Attack on Demos," *San Antonio Express-News*, August 18, 1992, 6A.

292 Over the next few months: Joe Davidson, "US Treasurer Takes Paid Leave After FBI Search," *Wall Street Journal*, November 2, 1992, B3.

293 Despite the efforts: Clarence Page, "Clinton's Wide Appeal Shatters Some Racial Myths," *Chicago Tribune*, March 25, 1992, 21.

293 Clinton also didn't cede: Jeffrey H. Birnbaum, "Cuban-American Contributors Open Checkbooks After Torricelli Exhibits an Anti-Castro Fervor," *Wall Street Journal*, August 3, 1992, A20.

CHAPTER 10: BUILDING THE WALL

294 The offending line: 1992 Republican Party platform, American Presidency Project, UC Santa Barbara, https://www.presidency.ucsb.edu /documents/republican-party-platform-1992, accessed on May 30, 2019.

294 That compelled him: Cathleen Decker, "Buchanan Seen but Not Heard in Texas Two-Step of Intentions," *Los Angeles Times*, August 17, 1992, A8.

294 Bush hoped Buchanan: Martin F. Nolan, "GOP's Right Wing Strengthens Hold on Party Platform," *Seattle Times*, August 13, 1992, A3; and Manford Rosenow, "Parties Take Opposite Tacks on Immigration," *Miami Herald*, September 14, 1992, 25.

295 The 1980 platform: 1980 Republican Party platform, American Presidency Project, UC Santa Barbara, https://www.presidency.ucsb.edu /documents/republican-party-platform-1980, accessed on July 27, 2019.

295 This shift was already evident: 1984 Republican Party platform, American Presidency Project, UC Santa Barbara, https://www.presidency .ucsb.edu/documents/republican-party-platform-1984, accessed on May 30, 2019; and 1988 Republican Party platform, American Presidency Project, UC Santa Barbara, https://www.presidency.ucsb.edu /documents/republican-party-platform-1988, accessed on May 30, 2019.

295 Bush's people insisted: Major Garrett, "Speakers to Present 'Program for Future,'" *Washington Times*, August 17, 1992, A1.

295 A campaign spokesman: Nancy Mathis, "Convention '92—GOP Backs Fence Along Mexico Border," *Houston Chronicle*, August 13, 1992, 26.

295 Hispanic Republicans agreed: Robert Dodge and Susan Feeney, "Party Platform OK'd Without Abortion Fight—Rights Advocates Sought Floor Debate on Language," *Dallas Morning News*, August 18, 1992, 1F.

295 Gaddi Vasquez: John Marelius, "Confusion Over Platform's Border Plank Drives GOP Delegates up the Wall," *San Diego Union-Tribune*, August 18, 1992, A7.

295 Secretary of the Interior: Richard Parker, "Border Plank Worries Hispanic Delegates," *Albuquerque Journal*, August 20, 1992, A1.

296 Gallegos dismissed as excuses: Ibid.

296 The divides between them: Larry Calloway, "Let's Have Four More Like Totally Changed Years! OK?" *Albuquerque Journal*, August 20, 1992, A6.

296 William Bennett, Bush's drug czar: Fred Knapp, "William Bennett Sings Praises of Conservatives," *Lincoln Journal*, September 18, 1992, 17.

296 Former Klansman David Duke: Patrick Buchanan, "Trendy GOP Routed in La.," *Chicago Sun-Times*, February 26, 1989; and Patrick Buchanan, "Duke's Victory Is GOP's Alarm Clock: Wake Up to Conservative Issues," *Press of Atlantic City*, October 19, 1991, A15.

297 Cardenas said the quest: Al Cardenas in 1992 RNC Report, 402–3.

297 Casanova said the Bushes: José Manolo Casanova in 1992 RNC Report, 402–3.

298 In the months after his inauguration: Richard Estrada, "A Border Program That Works," *Dallas Morning News*, November 26, 1993, 31A.

298 It didn't take long: NPR, "Proposed Border Fence Meets with Opposition and Support," *Weekend Edition Saturday*, February 5, 1994.

300 The new prostatehood governor: Baralt, *La razón del equilibrio*, 275.

300 One Cuban American: Ana Radelat, "Bienvenido, No More," *Washington Post*, November 6, 1994, C3.

301 For most of 1994: Beth Shuster, "Prop. 187 Backers Plan Recall Campaign," *Los Angeles Times*, November 15, 1994, B3.

301 Explaining the need: Beth Shuster, "Valley 187 Activist Shifts Passion to Implementation," *Los Angeles Times*, November 15, 1994, EVA1.

301 In 1991, she had walked: Roberto Suro, "California's SOS on Immigration," *Washington Post*, September 29, 1994, A1.

302 A former mayor: Wilson's Reelection Ads on Illegal Immigration, YouTube, https://www.youtube.com/watch?v=o0f1PE8Kzng, accessed on May 30, 2019.

302 Talk radio hosts: James O. Goldsborough, "Pete Won, His Party Lost," *San Diego Union-Tribune*, B7.

303 Hispanics in California: Newton, "Why Latinos Supported Proposition 187: Testing the Economic Threat and Cultural Identity Hypothesis."

303 Even more troubling: William Bennett and Jack Kemp, "The Fortress Party?," *Wall Street Journal*, October 21, 1994, A14.

303 When the head: Rosalva Hernandez, "Ezell Regrets 'Burrito' Remark: The Prop. 187 Leader Says He Shouldn't Have Made a Remark About a Talk-Show Host," *Orange County Register*, November 10, 1994, E1.

303 Because of such sentiments: Alex Nowrasteh, "Proposition 187 Turned California Blue," *Cato at Liberty*, a Cato Institute blog, https://www.cato.org/blog/proposition-187-turned-california-blue, accessed on July 26, 2019.

304 An editorial in: "An Alarm Bell from California," *Chicago Tribune*, October 4, 1994, D20.

304 They felt compelled to respond: Hugh Dellios and Karen Brandon, "Backlash Against Immigrants Beginning to Bite in California," *Chicago Tribune*, April 16, 1995, C4.

305 But Ros-Lehtinen wasn't confident: "Nicas luchan por obtener la residencia," *El Nuevo Herald*, October 21, 1995.

305 She and Díaz-Balart: Maria Morales, "Difícil el futuro migratorio de nicas," *El Nuevo Herald*, August 8, 1995.

305 It was a clear sign: "Nicas luchan por obtener la residencia."

305 Therefore, Díaz-Balart was ecstatic: Christopher Marquis, "'A Clear Message' to Cuba, House Vote Turns Screws Tighter," *Miami Herald*, September 22, 1995, 12A.

307 During the same month: "House Votes to Crack Down on Illegals," *Miami Herald*, March 22, 1996, 8A.

307 As a disincentive: Ibid.

307 The most controversial provision: Ibid.

307 As an editorial: "Bigots in the House," *Miami Herald*, March 22, 1996, 26A.

308 His critics said: Eric Schmitt, "House Votes to Let States Deny Public Education to Kids of Illegal Immigrants," *Miami Herald*, March 21, 1996, 1A.

308 It also became clear: Maria Morales, "Clinton desoye solicitud de Ileana Ros," *El Nuevo Herald*, June 29, 1996.

308 If they were denied: Maria A. Morales, "Misión de EU vigila inscripción de nicas," *El Nuevo Herald*, May 15, 1996, 3A.

308 Ros-Lehtinen and Díaz-Balart: "Bigots in the House."

308 The *Miami Herald* said: Schmitt, "House Votes to Let States Deny Public Education."

309 The platform approved: 1996 Democratic Party platform, American Presidency Project, UC Santa Barbara, https://www.presidency.ucsb .edu/documents/1996-democratic-party-platform, accessed on May 27, 2019.

310 The truth was that: 1996 Democratic Party platform.

310 He called the shift: B. Drummond Ayres Jr., "In New Role, Kemp Fights with His Past Over Ideology," *New York Times*, August 15, 1996, A19.

310 One poll showed: Jonathan Tilove, "Taking a Hard Line on Immigration," *Honolulu Star-Bulletin*, August 17, 1996, 9.

310 A Cuban American talk show host: Tim Collie, "Florida's Hispanics No Longer a Voting Bloc," *Tampa Tribune*, September 30, 1996, 1.

310 The former chairman: Jonathan Tilove, "GOP Risks Losing Hispanic, Asian Allies," *Cleveland Plain Dealer*, September 15, 1996, 1C.

311 They felt like: Divina Infusino, "No Room in the Tent," *Salon*, August 15, 1996, https://www.salon.com/1996/08/15/news_519/, accessed on July 26, 2019.

311 As the conservative firebrand: Tilove, "Taking a Hard Line on Immigration."

311 The anti-immigrant tide: Tilove, "GOP Risks Losing Hispanic, Asian Allies."

311 Applications for citizenship: Dellios and Brandon, "Backlash Against Immigrants Beginning to Bite in California."

312 The restrictive reforms: Minian, *Undocumented Lives*, chap. 8.

312 A campaign visit: "Capturen a asesinos, pide Dole," *El Nuevo Herald*, March 8, 1996, 11A.

312 Dole paid the requisite: Tom Fiedler, "New Look, Old Themes for Dole," *Miami Herald*, May 20, 1996, 5A.

313 The Republican Party's backing of statehood: "Puerto Rico Primary Means 14 Delegates," *Columbus (GA) Ledger-Enquirer*, March 3, 1996, A9.

313 One of the new leaders: Fonalledas interview, March 20, 2019.

314 Other signs of trouble: "Republicans Are Unjust to Cubans, Ignoring Ileana and Lincoln," n.d. clipping, box 4, folder "Varios Español 1996," Ros-Lehtinen Collection.

314 Members of the media: A. B. Stoddard, "Hispanics Wooed as Key Bloc in Hill Races," Hill, October 9, 1996, 1.

314 Even staunch allies: Magazine article, n.d., from *Ladies' Home Journal*, box 4, folder "Varios Inglés Jul–Dic 1996," Ros-Lehtinen Collection.

314 The Republican Party had "drifted": Stoddard, "Hispanics Wooed as Key Bloc."

315 "The America I know": Ros-Lehtinen, 1996 RNC Report, 582.

315 The new governor: George W. Bush, 1996 RNC Report, 449–450.

316 In early 1997: Mabell Dieppa, "Inmigrantes legales protestan contra ley," *El Nuevo Herald*, April 6, 1997.

316 Hearing their complaints: "Legisladores buscan atenuar impacto de leyes de welfare," *El Nuevo Herald*, March 31, 1997, 7A.

316 One leader of: Robert Salladay, "GOP Wary of Losing More Hispanics," *Honolulu Star-Bulletin*, October 21, 1997, 10.

316 Nobody denied that: Louis Aguilar, "Immigration Law Gets an Overhaul," *Milwaukee Journal Sentinel*, November 27, 1997, 28.

316 The RNHA itself: Santana interview, January 30, 2019; Garza interview, May 17, 2016; and Fuentes interview, March 28, 2019.

317 The head of the liberal: Aguilar, "Immigration Law Gets an Overhaul."

317 Both Giuliani and Riordan: Ibid.

317 They "dismantled" much: Jodi Wilgoren, "Republicans Struggle to Find Balance on Immigration Issues," *Tampa Tribune*, November 27, 1997, 34.

317 Díaz-Balart said: Ibid.

318 To Mark Krikorian: Rosanna Ruiz, "Undocumented Immigrants Get Room to Breathe," *Fort Worth Star-Telegram*, November 19, 1997, 1.

318 Liberal Hispanics believed: Wilgoren, "Republicans Struggle to Find Balance."

318 Two years after: Scott McConnell, "Barrio Blues: Republicans Will Never Win Hispanic Votes Pandering," *National Review*, August 17, 1998, 31–33.

319 But the majority: Mike Madrid, "The Latin Vote," *National Review*, September 28, 1998, 2.

319 The Republican Party went back: McConnell, "Barrio Blues."

319 They also heeded: Fuentes interview, March 28, 2019.

319 He had done well: Mark Barabak, "GOP Can't Afford to Write Off California," *Los Angeles Times*, August 31, 2005, B1.

320 When Bush launched: Fuentes interview, March 28, 2019.

320 He also asked: Luis A. Ferré to George W. Bush, November 10, 1999, Partido Republicano Nacional, box 3, folder 4, "Bush, George W. (Hijo)/ 11–97," Ferré Archive.

321 *Time* magazine printed: *Time*, May 1, 2000.

321 The González case: Lizette Alvarez, "Hundreds Protest Plan to Return Boy to Cuba," *New Orleans Times-Picayune*, January 7, 2000, 1.

322 Bush sided: Armando Villafranca, Steve Lash, Bennett Roth, R. G. Ratcliffe, "Elian's Attorneys Get More Time to Try to Derail Boy's Deportation," *Houston Chronicle*, March 31, 2000, A1; and Bennett Roth, "Fight Over Elian Turns into Battle for Florida," *Houston Chronicle*, April 2, 2000, A1.

322 In November, they punished Gore: William Schneider, "Elián González Defeated Al Gore," *Atlantic*, May 1, 2001, https://www.theatlantic.com /politics/archive/2001/05/elian-gonzalez-defeated-al-gore/377714/, accessed on July 27, 2019.

CONCLUSION: THE FUTURE FOR HISPANIC REPUBLICANS

324 They all remembered: Barreto, interview by author, January 31, 2019; Aguilar, interview by author, January 29, 2019; and Garza, interview by author, May 17, 2016.

324 A statement by: "Secretary of State Colin Powell Holds News Conference with Mexico's Foreign Minister," CNN, January 30, 2001, http://transcripts.cnn.com/TRANSCRIPTS/0101/30/se.08.html, accessed on July 30, 2019.

326 McCain supporter Fernando de Baca: "C de Baca Steps Down After All," *Albuquerque Journal*, September 26, 2008, https://www.abqjournal .com/21465/updated-at-705am-c-de-baca-steps-down-after-all.html, accessed on July 30, 2019.

326 Four years later: Lucy Madison, "Romney on Immigration: I'm for 'Self-Deportation,'" CBS News, January 24, 2012, https://www.cbs news.com/news/romney-on-immigration-im-for-self-deportation/, accessed on July 30, 2019.

326 Just a couple of months: Sarah Huisenga, "Herman Cain Acknowledges His Electric Border Fence Idea Isn't a Joke After All," CBS News, October 17, 2011, https://www.cbsnews.com/news/herman-cain-ack nowledges-his-electric-border-fence-idea-isnt-a-joke-after-all/, accessed on July 30, 2019.

328 When he lost: Jan Crawford, "Adviser: Romney 'shellshocked' by loss," CBS News, November 8, 2012, https://www.cbsnews.com/news /adviser-romney-shellshocked-by-loss/, accessed on July 30, 2019.

328 In its analysis: Jennifer Rubin, "GOP Autopsy Report Goes Bold," *Washington Post*, March 18, 2013, https://www.washingtonpost.com /blogs/right-turn/wp/2013/03/18/gop-autopsy-report-goes-bold/?utm _term=.0995fde24a60, accessed on July 30, 2019.

329 After the *New York Times* reported: Alan Rappeport, "Jeb Bush Listed Himself as 'Hispanic' on Voter Form," *New York Times*, April 6, 2015, https://www.nytimes.com/politics/first-draft/2015/04/06/jeb-bush -listed-himself-as-hispanic-on-voter-form/, accessed on July 30, 2019.

330 An older generation: Evan Osnos, "The Opportunist," *New Yorker*, November 30, 2015, https://www.newyorker.com/magazine/2015/11/30 /the-opportunist, accessed on July 30, 2019.

331 The profile of the segment of: "The Beginning of Now," *This American Life*, NPR, April 28, 2017, https://www.thisamericanlife.org/615/the -beginning-of-now, accessed on July 30, 2019.

332 It took almost five: Carlos Conde, "A Memorable Presidential Candidate," *El Paso Times*, September 15, 2000, 15A.

332 An obituary in the: Elaine Woo, "Fernando Oaxaca, 76, Founder of Republican Latino Group," *Los Angeles Times*, June 3, 2004, https:// www.latimes.com/archives/la-xpm-2004-jun-03-me-oaxaca3-story.html, accessed on July 30, 2019.

333 The Los Angeles paper: Alejandra Reyes-Velarde, "Romana Acosta Bañuelos, First Latina US Treasurer and Mexican American Pioneer, Dies at 92," *Los Angeles Times*, January 22, 2018, https://www.latimes .com/local/obituaries/la-me-romana-acosta-banuelos-20180119-story .html, accessed on July 30, 2019.

333 The Democratic governor: Zach Montague, "Manuel Luján Jr., Ex-Congressman and Interior Secretary, Dies at 90," *New York Times*, April 26, 2019, https://www.nytimes.com/2019/04/26/us/manuel-lujan -jr-dead.html, accessed on July 30, 2019.

333 Luján had already: Dillon Mullan, "Former US Rep. Luján Also Served as Interior Secretary," *Santa Fe New Mexican*, April 25, 2019, https:// www.santafenewmexican.com/news/local_news/former-u-s-rep-lujan -also-served-as-interior-secretary/article_a360cb40-5698-5b47-983c -d2315094ccb0.html, accessed on July 30, 2019.

334 Before the 2016 Republican national convention: Lionel Sosa, "Farewell, My Grand Old Party," *San Antonio Express-News*, June 22, 2016, https://www.mysanantonio.com/opinion/commentary/article/Farewell -my-Grand-Old-Party-8307172.php, accessed on July 30, 2019.

334 During the general: Alan Rappeport, "'He Used Us as Props': Conservative Hispanics Deplore Donald Trump's Speech," *New York Times*, September 1, 2016, https://www.nytimes.com/2016/09/02/us/politics /gop-hispanic-reaction-trump.html, accessed on November 9, 2019.

335 In the Puerto Rico primary: Regis interview, March 23, 2019.

335 As Daniel Garza put it: Garza interview, May 17, 2016.

BIBLIOGRAPHY

ARCHIVES

Andow Collection: Paul Andow Collection, 1963–1964, LULAC Archives, Nettie Lee Benson Latin American Collection, University of Texas Libraries, University of Texas at Austin.

Bonilla Collection: Ruben Bonilla Collection, 1973–1984, LULAC Presidential Papers Project, Nettie Lee Benson Latin American Collection, University of Texas at Austin.

Eisenhower Library: Dwight D. Eisenhower Presidential Library, Abilene, KS.

Ferré Archive: Archivo Histórico Luis A. Ferré Aguayo, Museo de Arte de Ponce, Ponce, PR.

Ford Library: Gerald Ford Presidential Library, Ann Arbor, MI.

Luján Congressional Papers: Manuel Luján Congressional Papers, 1968–1992, MSS 142 BC, University of New Mexico Center for Southwest Research, Albuquerque, NM.

Márquez-Sterling Papers: Carlos Márquez-Sterling Papers, CHC 5133, Cuban Heritage Collection, University of Miami, Coral Gables, FL.

Nixon Library: Richard Nixon Presidential Library, Yorba Linda, CA.

Peña Papers: Abe M. Peña Papers, 1894–2004 (MS 0434), New Mexico State University Library, Las Cruces, NM.

Reagan Library: Ronald Reagan Presidential Library and Museum, Simi Valley, CA.

Reveles Papers: Roberto Reveles Papers (MSS 351), Archives & Special Collections, Arizona State University Libraries, Tempe, AZ.

Rhodes Papers: John J. Rhodes Papers, MSS3, Arizona State University Libraries, Archives & Special Collections, Tempe, AZ.

Robles Papers: Robert Benitez Robles Papers (MS 563), Special Collections, University of Arizona, Tucson, AZ.

Ros-Lehtinen Collection: Ileana Ros-Lehtinen Collection, CHC 5174, Cuban Heritage Collection, University of Miami, Coral Gables, FL.

Ruiz Papers: Manuel Ruiz Papers, 1931–1995, M0295, Stanford University, Special Collections & University Archives, Palo Alto, CA.

Tower Papers: John G. Tower Papers, Southwestern University, A. Frank Smith, Jr. Library Center, Special Collections and Archives, Georgetown, TX.

INTERVIEWS AND ORAL HISTORIES

Aguilar, Alfonso, interviewed by author, January 29, 2019 (Aguilar interview, January 29, 2019).

Aponte-Parsi, Ricardo, interviewed by author, March 20, 2019 (Aponte-Parsi interview, March 20, 2019).

Barreto, Héctor, interviewed by author, January 31, 2019 (Barreto interview, January 31, 2019).

Casillas, Frank, interviewed by author, February 19, 2019 (Casillas interview, February 19, 2019).

Díaz, Guarioné, interviewed by author, March 27, 2019 (Díaz interview, March 27, 2019).

DiMartino, Rita, interviewed by author, May 27, 2016 (DiMartino interview, May 27, 2016).

Fernandez, Benjamin, "Interview no. 358," November 10, 1978, Institute of Oral History, University of Texas at El Paso (Fernandez lecture, November 10, 1978).

Fonalledas, Zoraida, interviewed by author, March 20, 2019 (Fonalledas interview, March 20, 2019).

Fuentes, José, interviewed by author, March 28, 2019 (Fuentes interview, March 28, 2019).

Garza, Daniel, interviewed by author, May 17, 2016 (Garza interview, May 17, 2016).

Guinot, Luis, Jr., interviewed by author, February 8, 2019 (Guinot interview, February 8, 2019).

Luján, Manuel, interviewed by Oscar J. Martinez, October 12, 1975, Institute of Oral History, University of Texas at El Paso (Luján interview, October 23, 1975).

Oaxaca, Fernando, interviewed by Oscar J. Martinez, October 23, 1975, Institute of Oral History, University of Texas at El Paso (Oaxaca interview, October 23, 1975).

Ramos, Héctor, interviewed by author, March 20, 2019 (Ramos interview, March 20, 2019).

Regis, John, interviewed by author, March 23, 2019 (Regis interview, March 23, 2019).

Santana, Hannie, interviewed by author, January 30, 2018 (Santana interview, January 30, 2019).

Sosa, Lionel, interviewed by author, June 4, 2019 (Sosa interview, June 4, 2019).

Widener, Jeff, interviewed by author, July 23, 2018 (Widener interview, July 23, 2018).

Zapanta, Al, interviewed by author, August 7, 2015 (Zapanta interview, August 7, 2015).

NEWSPAPERS AND NEWS MEDIA

Advocate (Baton Rouge, LA)

AGENDA

Albuquerque Journal

Arizona Daily Star

Arizona Daily Sun (Flagstaff, AZ)

Austin American-Statesman

Baltimore Sun

Boston Globe

Charleston News and Courier

Chicago Sun-Times

Chicago Tribune

El Chicano

Christian Science Monitor

Columbus Ledger (Columbus, GA)

Corpus Christi Caller

Daily Telegram (Adrian, MI)

Dallas Morning News

Del Rio Herald News

Denver Post

Detroit Free Press

Diario las Américas

El Paso Times

Evansville Courier (Evansville, IN)

Fort Worth Star-Telegram

Fresno Bee

Guardian (London, UK)

Honolulu Star-Bulletin

Houston Chronicle

Idaho Statesman

Jersey Journal (Jersey City, NJ)

Lincoln Journal (Lincoln, NE)

Longview Daily News (Longview, TX)

Los Angeles Sentinel

Los Angeles Times

Macon Telegraph (Macon, GA)

Miami Herald

Miami News

National Review

New York Daily News

New York Herald Tribune

New York Times

Odessa American

Orange County Register

Oregonian (Portland, OR)

Palm Beach Post (Palm Beach, FL)

PBS Video, *Summer of Judgment: The Watergate Hearings* (1983)

Pensacola News Journal

Philadelphia Inquirer

Plain Dealer (Cleveland, OH)

Press of Atlantic City

Puerto Rico Report

Record (Hackensack, NJ)

Republic (Columbus, OH)

San Antonio Express-News and *San Antonio Express*

San Diego Union-Tribune

San Francisco Chronicle

Santa Fe New Mexican

Seattle Times
Seguin Gazette-Enterprise (Seguin, TX)
Signal (Santa Clarita, CA)
El Sonorense
South Bend Tribune
Southtown Star (Tinley Park, IL)
Tampa Tribune
Times-Picayune (New Orleans, LA)
Trabajadores (Cuba)
Valley Morning Star (Harlingen, TX)
Wall Street Journal
Washington Post
Washington Times
Windsor Star (Windsor, ON, Canada)

GOVERNMENT HEARINGS AND DOCUMENTS

Advisory Board for Broadcasting to Cuba, *Report by the Advisory Board for Radio Broadcasting to Cuba* (Washington, DC: US Government Printing Office, 1986).

Federal Register 48, no. 169 (1983): 39205–39446 (Washington, DC: US Government Printing Office, August 30, 1983).

Fernandez, Benjamin, testimony before the Senate Select Committee on Presidential Campaign Activities, originally televised on WETA, November 8, 1973, available at the American Archive of Public Broadcasting: http://americanarchive.org/exhibits/watergate/the-watergate-coverage (Fernandez testimony).

Manuel Luján Jr. nomination, hearing before the Committee on Energy and Natural Resources, US Senate, 101st Cong., 1st sess., on the nomination of Manuel Luján Jr. to be secretary of the interior, January 26, 1989 (Washington, DC: US Government Printing Office, 1989) (Manuel Luján nomination hearing, January 26, 1989).

Marumoto, William, testimony before the Senate Select Committee on Presidential Campaign Activities, originally televised on WETA, November 7, 1973, available at the American Archive of Public Broadcasting: http://americanarchive.org/exhibits/watergate/the-watergate-coverage (Marumoto testimony).

Nomination of Donald E. Kirkendall and Catalina Vasquez Villalpando, hearing before the Committee on Finance, US Senate, 101st Cong., 1st sess., on the nomination of Donald E. Kirkendall to be inspector general of the US Department of the Treasury and Catalina Vasquez Villalpando to be treasurer of the United States, November 15, 1989 (Washington, DC: US Government Printing Office, 1990) (Catalina Villalpando nomination hearing, November 15, 1989).

Nomination of Lauro Cavazos, hearing before the Committee on Labor and Human Resources, US Senate, 100th Cong., 2nd sess., on Lauro Cavazos to be secretary, Department of Education, September 9, 1988 (Washington, DC: US Government Printing Office, 1988) (Lauro Cavazos nomination hearing, September 9, 1988).

Official report of the proceedings of the 34th Republican National Convention held in New Orleans, Louisiana, August 15, 16, 17, 18, 1988, resulting in the nomination of George H. W. Bush, of Texas, for president, and the nomination of J. Danforth Quayle, of Indiana, for vice president, reported by Asher & Associates, Inc., official reporter, published by the Republican National Committee (1988 RNC Report).

Official report of the proceedings of the 35th Republican National Convention held in Houston, Texas, August 17, 18, 19, 20, 1992, resulting in the nomination of George H. W. Bush, of Texas, for president, and the nomination of J. Danforth Quayle, of Indiana, for vice president, reported by Court Reporting Services, Inc., official reporter, published by the Republican National Committee (1992 RNC Report).

Official report of the proceedings of the 36th Republican National Convention held in San Diego, California, August 12, 13, 14, 15, 1996, resulting in the nomination of Robert Dole, of Kansas, for president, and the nomination of Jack Kemp, of California, for vice president, reported by Alderson Reporting Company, Inc., official reporter, published by the Republican National Committee (1996 RNC Report).

Presidential Campaign Activities of 1972, hearings before the Select Committee on Presidential Campaign Activities of the US Senate, 93rd Cong., 1st sess., Watergate and Related Activities, Book 13 (Washington, DC: US Government Printing Office, 1973) (Presidential Campaign Activities of 1972).

Public Law 89-732, An act to adjust the status of Cuban refugees to that of lawful permanent residents of the United States, and for other purposes, November 2, 1966.

Report of the Select Commission on Western Hemisphere Immigration, January 1968 (Washington, DC: US Government Printing Office, 1968).

Weekly Compilation of Presidential Documents 25, no. 37 (1989): 1337–85 (Washington, DC: US Government Printing Office, September 18, 1989).

BOOKS AND ARTICLES

Arnaz, Desi. *A Book* (New York: Morrow, 1976).

Ayala, César, and Rafael Bernabe. *Puerto Rico in the American Century: A History Since 1898* (Chapel Hill: University of North Carolina Press, 2007).

Baralt, Guillermo A. *Desde el mirador de Próspero: La vida de Luis A. Ferré, Tomo I: 1904–1968* (San Juan, PR: Fundación El Nuevo Día, 1996).

———. *La razón del equilibrio: La vida de Luis A. Ferré, Tomo II: 1968–1998* (San Juan, PR: Fundación El Nuevo Día, 1998).

Barreto, Matt, and Gary Segura. *Latino America: How America's Most Dynamic Population Is Poised to Transform the Politics of the Nation* (New York: PublicAffairs, 2014).

Bean, Frank, and Marta Tienda. *The Hispanic Population of the United States* (New York: Russell Sage Foundation, 1987).

Behnken, Brian. *Fighting Their Own Battles: Mexican Americans, African Americans, and the Struggle for Civil Rights in Texas* (Chapel Hill: University of North Carolina Press, 2014).

Bernstein, Carl, and Bob Woodward. *All the President's Men* (New York: Simon & Schuster, 1974).

Brilliant, Mark. *The Color of America Has Changed: How Racial Diversity Shaped Civil Rights Reform in California, 1941–1978* (New York: Oxford University Press, 2010).

Britto, Lina. *Marijuana Boom: The Rise and Fall of Colombia's First Drug Paradise* (Berkeley: University of California Press, 2020).

Burt, Kenneth. *The Search for a Civic Voice: California Latino Politics* (Claremont, CA: Regina Books, 2007).

Cadava, Geraldo. *Standing on Common Ground: The Making of a Sunbelt Borderland* (Cambridge, MA: Harvard University Press, 2013).

Capó, Julio. "Queering Mariel: Mediating Cold War Foreign Policy and US Citizenship among Cuba's Homosexual Exile Community, 1978–1994," *Journal of American Ethnic History* 29, no. 4 (Summer 2010): 78–106.

Cavazos, Lauro F. *A Kineño Remembers: From the King Ranch to the White House* (College Station: Texas A&M University Press, 2008).

Chasteen, John Charles. *Born in Blood & Fire: A Concise History of Latin America*, 4th ed. (New York: Norton, 2016).

Chavez, Linda. *An Unlikely Conservative: The Transformation of an Ex-Liberal (or How I Became the Most Hated Hispanic in America)* (New York: Basic Books, 2002).

Dávila, Arlene. *Latinos Inc.: The Marketing and Making of a People* (Berkeley: University of California Press, 2012).

De Zavala, Lorenzo. *Journey to the United States of North America*, trans. Wallace Woolsey (Austin, TX: Shoal Creek, 1980).

Dean, John W. *The Nixon Defense: What He Knew and When He Knew It* (New York: Viking, 2014).

Fitz, Caitlin. *Our Sister Republics: The United States in an Age of American Revolutions* (New York: Liveright, 2017).

Foley, Neil. *Mexicans in the Making of America* (Cambridge, MA: Belknap Press of Harvard University Press, 2014).

Francis-Fallon, Benjamin. *The Rise of the Latino Vote: A History* (Cambridge, MA: Harvard University Press, 2019).

García, Gilbert. *Reagan's Comeback: Four Weeks in Texas That Changed American Politics Forever* (San Antonio: Trinity University Press, 2012).

Garcia, Ignacio. *Viva Kennedy: Mexican Americans in Search of Camelot* (College Station: Texas A&M University Press, 2000).

García, María Cristina. *Havana USA: Cuban Exiles and Cuban Americans in South Florida, 1959–1994* (Berkeley: University of California Press, 1996).

———. *Seeking Refuge: Central American Migration to Mexico, the United States, and Canada* (Berkeley: University of California Press, 2006).

Gillingham, Paul, and Benjamin T. Smith. *Dictablanda: Politics, Work, and Culture in Mexico, 1938–1968* (Durham, NC: Duke University Press, 2014).

Goldwater, Barry. *The Conscience of a Conservative* (Washington, DC: Regnery Gateway, 1990).

Gonzales, Felipe. *Política: Nuevomexicanos and American Political Incorporation, 1821–1910* (Lincoln: University of Nebraska Press, 2016).

Grandin, Greg. *The Blood of Guatemala: A History of Race and Nation* (Durham, NC: Duke University Press, 2000).

———. *The End of the Myth: From the Frontier to the Border Wall in the Mind of America* (New York: Metropolitan Books, 2019).

Gutierrez, David. *Walls and Mirrors: Mexican Americans, Mexican Immigrants, and the Politics of Ethnicity* (Berkeley: University of California Press, 1995).

Haldeman, H. R. *The Haldeman Diaries: Inside the Nixon White House* (New York: G. P. Putnam's Sons, 1994).

Hernández, Kelly Lytle. *Migra! A History of the US Border Patrol* (Berkeley: University of California Press, 2010).

Hero, Rodney. *Latinos and the U.S. Political System: Two-Tiered Pluralism* (Philadelphia, PA: Temple University Press, 1999).

HoSang, Daniel Martinez. *Racial Propositions: Ballot Initiatives and the Making of Postwar California* (Berkeley: University of California Press, 2010).

Hunt, E. Howard, with Greg Aunapu. *American Spy: My Secret History in the CIA, Watergate, and Beyond* (Hoboken, NJ: Wiley, 2007).

Igo, Sarah. *The Known Citizen: A History of Privacy in Modern America* (Cambridge, MA: Harvard University Press, 2018).

Immerwahr, Daniel. *How to Hide an Empire: A History of the Greater United States* (New York: Farrar, Straus & Giroux, 2019).

Joseph, Gilbert, and Daniela Spenser. *In from the Cold: Latin America's New Encounter with the Cold War* (Durham, NC: Duke University Press, 2008).

Katznelson, Ira. *Fear Itself: The New Deal and the Origins of Our Time* (New York: Liveright, 2014).

Keller, Renata. *Mexico's Cold War: Cuba, the United States, and the Legacy of the Mexican Revolution* (New York: Cambridge University Press, 2015).

Knaggs, John. *Two-Party Texas: The John Tower Era, 1961–1984* (Fort Worth, TX: Eakin Press, 1986).

Kruse, Kevin, and Julian Zelizer. *Fault Lines: A History of the United States Since 1974* (New York: Norton, 2019).

Lee, Sonia Song-Ha. *Building a Latino Civil Rights Movement: Puerto Ricans, African Americans, and the Pursuit of Racial Justice in New York City* (Chapel Hill: University of North Carolina Press, 2014).

Lepore, Jill. *These Truths: A History of the United States* (New York: Norton, 2018).

Loaeza, Soledad. *Acción Nacional: El apetito y las responsabilidades del triunfo* (México, DF: El Colegio de México, 2010).

Logevall, Fredrik, and Andrew Preston, eds. *Nixon in the World: American Foreign Relations, 1969–1977* (New York: Oxford University Press, 2008).

Lozano, Rosina. *An American Language: The History of Spanish in the United States* (Berkeley: University of California Press, 2018).

Machado, Manuel A. *Listen Chicano! An Informal History of the Mexican-American* (Chicago: Nelson-Hall, 1978).

Manela, Erez. *The Wilsonian Moment: Self-Determination and the International Origins of Anticolonial Nationalism* (New York: Oxford University Press, 2007).

McPherson, Alan. *Yankee No! Anti-Americanism in US–Latin American Relations* (Cambridge, MA: Harvard University Press, 2006).

Minian, Ana. *Undocumented Lives: The Untold Story of Mexican Migration* (Cambridge, MA: Harvard University Press, 2018).

Montejano, David, ed. *Chicano Politics and Society in the Late Twentieth Century* (Austin: University of Texas Press, 1999).

Mora, G. Cristina. *Making Hispanics: How Activists, Bureaucrats, and Media Constructed a New American* (Berkeley: University of California Press, 2014).

Newton, Lina Yvette. "Why Latinos Supported Proposition 187: Testing the Economic Threat and Cultural Identity Hypothesis" (Irvine: University of California Center for the Study of Democracy, 1988).

Ngai, Mae. *Impossible Subjects: Illegal Aliens and the Making of Modern America* (Princeton, NJ: Princeton University Press, 2004).

Nugent, Walter. *Color Coded: Party Politics in the American West, 1950–2016* (Norman: University of Oklahoma Press, 2018).

Office of the Historian and Office of the Clerk, US House of
    Representatives. *Hispanic Americans in Congress, 1822–2012*
    (Washington, DC: US Government Printing Office, 2013).

Oropeza, Lorena. *The King of Adobe: Reies López Tijerina, Lost Prophet of
    the Chicano Movement* (Chapel Hill: University of North Carolina
    Press, 2019).

———. *¡Raza Sí! ¡Guerra No! Chicano Protest and Patriotism During the
    Viet Nam War Era* (Berkeley: University of California Press, 2005).

Pérez, Louis A., Jr. *On Becoming Cuban: Identity, Nationality, and Culture*
    (Chapel Hill: University of North Carolina Press, 1999).

Perlstein, Rick. *Before the Storm: Barry Goldwater and the Unmaking of the
    American Consensus* (New York: Hill and Wang, 2001).

———. *The Invisible Bridge: The Fall of Nixon and the Rise of Reagan*
    (New York: Simon & Schuster, 2014).

Pycior, Julia. *LBJ & Mexican Americans: The Paradox of Power* (Austin:
    University of Texas Press, 1997).

Quiñones Calderón, Antonio. *Historia Política de Puerto Rico, Tomo I &
    Tomo II* (San Juan, PR: Credibility Group, 2002).

Rabe, Stephen. *Eisenhower and Latin America: The Foreign Policy of
    Anticommunism* (Chapel Hill: University of North Carolina Press,
    1988).

———. *The Most Dangerous Area in the World: John F. Kennedy Confronts
    Communist Revolution in Latin America* (Chapel Hill: University of
    North Carolina Press, 1999).

Ramirez, Henry M. *A Chicano in the White House: The Nixon No One
    Knew* (Self-pub., 2014).

———. *Nixon and the Mexicans: How a Young Man Encountered the
    Diaspora of 1913–1930 and Made a Difference* (Laytonsville, MD:
    Depomo Press, 2018).

Richardson, Heather Cox. *To Make Men Free: A History of the Republican
    Party* (New York: Basic Books, 2014).

Rigueur, Leah Wright. *The Loneliness of the Black Republican: Pragmatic
    Politics and the Pursuit of Power* (Princeton, NJ: Princeton University
    Press, 2015).

Rodgers, Daniel. *Age of Fracture* (Cambridge, MA: Belknap Press of
    Harvard University Press, 2012).

Romero Barceló, Carlos. *Estadidad es para los pobres* [Statehood is for the poor] (San Juan, PR: 1978).

Santillan, Richard, and Federico A. Subervi-Vélez. "Latino Participation in Republican Party Politics in California," Byran O. Jackson, ed., *Racial and Ethnic Politics in California* (IGS Press, 1991), 285–319.

Segura, Gary M. "Latino Public Opinion and Realigning the American Electorate," *Daedalus* 141, no. 4 (Fall 2012): 98–113.

Shermer, Elizabeth Tandy. *Barry Goldwater and the Remaking of the American Political Landscape* (Tucson: University of Arizona Press, 2013).

Sosa, Lionel. *The Americano Dream: How Latinos Can Achieve Success in Business and in Life* (New York: Dutton, 1998).

Torres, María de los Angeles. *In the Land of Mirrors: Cuban Exile Politics in the United States* (Ann Arbor: University of Michigan Press, 1999).

Westad, Odd Arne. *The Global Cold War* (Cambridge, UK: Cambridge University Press, 2007).

Wong, Carolyn. *Lobbying for Inclusion: Rights Politics and the Making of Immigration Policy* (Stanford, CA: Stanford University Press, 2006).

Wills, Garry. *Reagan's America: Innocents at Home* (New York: Penguin Books, 2000).

INDEX